ENGINEERING LIBR

AF572650

Thermodynamics for Engineers

CRC MECHANICAL ENGINEERING SERIES

Series Editor Frank Kreith

Published

Entropy Generation Minimization
Adrian Bejan

Finite Element Method Using MATLAB
Young W. Kwon and Hyochoong Bang

Fundamentals of Environmental Discharge Modeling
Lorin R. Davis

Intelligent Transportation Systems: New Principles and Architectures
Sumit Ghosh and Tony Lee

Mathematical and Practical Modeling of Materials Processing Operations
Olusegun Johnson Ileghus, Manabu Iguchi, and Walter E. Wahnsiedler

Mechanics of Composite Materials
Autar K. Kaw

Mechanics of Fatigue
Vladimir V. Bolotin

Nonlinear Analysis of Structures
M. Sathyamoorthy

Practical Inverse Analysis in Engineering
David M. Trujillo and Henry R. Busby

Viscoelastic Solids
Roderic S. Lakes

Thermodynamics for Engineers
Kau-Fui Wong

To be Published

Distributed Generation: The Power Paradigm for the New Millenium
Marie Borbely and Jan F. Kreider

Engineering Experimentation
Euan Somerscales

Energy Audit of Building Systems: An Engineering Approach
Moncef Krarti

Finite Element Method Using MATLAB, Second Edition
Young W. Kwon and Hyochoong Bang

Introduction Finite Element Method
Chandrakant S. Desai and Tribikram Kundu

Mechanics of Solids and Shells
Gerald Wempner and Demosthenes Talaslidis

Principles of Solid Mechanics
Rowland Richards, Jr.

Thermodynamics for Engineers

Kau-Fui Vincent Wong, Ph.D., P.E.
University of Miami

CRC Press
Boca Raton London New York Washington, D.C.

Library of Congress Cataloging-in-Publication Data

Wong, Kau-Fui Vincent.
Thermodynamics for engineers / Kau-Fui Vincent Wong.
p. cm. -- (Mechanical engineering series)
Includes bibliographical references and index.
ISBN 0-8493-0232-3 (alk. paper)
1. Thermodynamics. I. Title. II. Advanced topics in mechanical engineering series

TJ265 .W56 2000
621.402′1--dc21 00-029781
CIP

Direct all inquiries to CRC Press LLC, 2000 N.W. Corporate Blvd., Boca Raton, Florida 33431.

International Standard Book Number 0-8493-0232-3
Library of Congress Card Number 00-029781
Printed in the United States of America 1 2 3 4 5 6 7 8 9 0
Printed on acid-free paper

Preface

In this first edition, the basic objective is to present a comprehensive treatment of engineering thermodynamics from the classical aspect, so as to provide a sound basis for all engineering students to prepare them to use thermodynamics in professional practice. The book is written for a one-semester first course in thermodynamics for all undergraduates in engineering.

A major difficulty encountered by engineering undergraduates in studying thermodynamics is the use of thermodynamic tables to find the correct properties of substances. This difficulty is removed by emphasizing the use of computer-aided thermodynamic tables from the onset.

Another complexity that need not exist is understanding the different concepts and trying to memorize many special cases where special rules apply. This difficulty is removed by unifying the treatment of the first law and the second law entropy as well as the second law availability. Using balance equations in each of the above laws emphasizes the commonality between the laws and allows for easier comprehension and use.

In the spirit of commonality, the first law for control volumes is introduced first, as the most general system used in the book. The first law for control masses is derived as a special case of the general system. This technique eliminates the step of going from the control mass system to the control volume system, often covered in at least two chapters in most books. The present book covers the same material in one chapter. This approach bridges the chasm in the students' minds; it allows the fundamental concepts to be presented and understood at one time.

The special features of the book are as follows:

- Clutter and unnecessary details have been eliminated. Only essential facts and methods have been provided to make the work of the student easier.
- Limit the manual interpolation of thermodynamic tables. Support the use of computer-aided thermodynamic tables.
- Allow expansion of the property tables by the reader.
- Minimize the differences between the ideal gas treatment and the vapor treatment, by referring to computer-aided thermodynamic tables for all properties.
- Unify the treatment of the conservation of energy, the creation of entropy, and the destruction of availability by using a balance equation for each of them.

- Introduce the control volume first as the more general system, and the control mass as a special case of the control volume. This gives the students a fast start, especially in comprehending the first law of thermodynamics.
- Introduce the second law ratio to measure thermal environmental impact. A parameter that directly measures the impact of the environment has hardly ever been mentioned in other textbooks. Include a discussion on the air preheater and the economiser in the vapor cycle chapter, to make the overall discussion of vapor cycles more practical and complete. This second law ratio is illustrated with the air preheater.
- Logic diagrams have been introduced to help the students understand the logical steps used in solving thermodynamic problems.
- A chapter on heat transfer is included. This topic is important to electrical engineers, for instance, and unlike mechanical engineers, they will not take another course that will offer them this material. This provides the option of teaching nonmechanical engineers a topic that is often included in the syllabus of Part I of the professional engineer's examination.

In Chapter 1, the concepts, definitions, and the laws of thermodynamics are presented in a concise manner. A systematic problem-solving approach is also discussed.

The properties of pure substances are discussed in Chapter 2. The emphasis here, by way of solved examples and problems, is on the students' understanding of the phase diagrams, and their differences from the ideal gas phase as characterized by the ideal gas equation.

In Chapter 3, mass conservation and the first law of thermodynamics are presented. The control volume is introduced first as the general system, and the control mass is shown to be a special case.

The second law of thermodynamics and entropy is treated in Chapter 4. The one control volume worked example, used in Chapter 3, is investigated in Chapter 4 using the second law in entropy equation.

The availability (exergy) analysis is the topic of Chapter 5. The control volume worked example, used in Chapters 3 and 4, is also studied here using the second law in availability equation.

Chapter 6 discusses vapor power systems, and Chapter 7 the principles of energy (heat) transfer. The approach in Chapter 7 is basic, but the material is made useable to students who do not need to go into depth in that topic.

The special features of the book are geared towards making thermodynamics a less difficult field for undergraduates in engineering. One unique feature is that students can add to the thermodynamic tables, either by using smaller intervals in the existing databases or increasing the ranges in the databases. Smaller intervals provide better accuracy, and larger ranges allow wider applicability. This participatory nature of the textbook can be accomplished by any reader with access to Excel. The laborious and sometimes confusing task of manual tables is completely eliminated for students, and adding to the databases is not only a good learning process but also provides a sense of accomplishment to students on their completion.

In streamlining the presentation of the thermodynamic laws, students of thermodynamics are exposed to the big picture regarding the usefulness of thermodynamics in the real world, rather than being bogged down with unnecessary details. The interesting aspects of thermodynamics and their importance to all engineers are emphasized. It is to this end that this book has been written.

Author

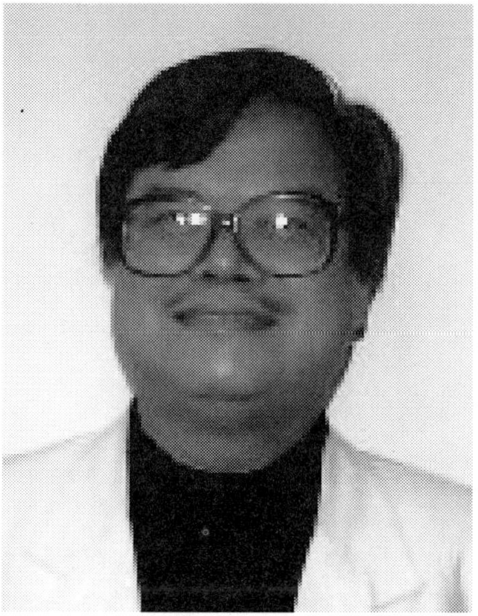

Kau-Fui Vincent Wong grew up in a small town, Segamat, Malaysia. As a young boy, he always wanted to be an engineer because of his love for mathematics. He fulfilled the first stage of his wish when he completed the baccalaureate degree in mechanical engineering at the University of Malaya in Kuala Lumpur, Malaysia, in 1973. The second stage of his wish was to go to the U.S.A. to do graduate work, because of the leadership position of the U.S.A. in scientific and engineering research. This second stage was realized when he completed the requirements for the Ph.D. at Case Western Reserve University, Cleveland, Ohio, U.S.A., in 1976, the bicentennial year. It was a personally fulfilling year for the author, enhanced that much more by the bicentennial celebrations taking place all around him. The third stage of his wish was to teach young men and women of the world to become the engineers of tomorrow. The realization of this stage began when he accepted a position at the University of Miami, Coral Gables, Florida, U.S.A., in 1979. The fourth stage of his wish was to write a textbook for the young men and women to whom he had chosen to dedicate his professional life. The current book represents the first written by the author for students studying to be engineers. The year 2000 is a joyous year as the world ushers in a new millennium, and so it is for the author as he launches his first book.

Conversion Table

Dimensions	SI/English	English/SI
Mass and Density	1 kg = 2.2046 lb_m	1 lb_m = 0.4536 kg
	1 g/cm^3 = 62.428 lb/ft^3	1 lb_m/ft^3 = 0.016018 g/cm^3
		1 lb_m/ft^3 = 1.6018 kg/m^3
Length	1 cm = 0.3937 in.	1 in. = 2.54 cm
	1 m = 3.2808 ft	1 ft = 0.3048 m
Volume	1 cm^3 = 0.061024 in.^{3}	1 in.^{3} = 16.387 cm^3
	1 m^3 = 35.315 ft^3	1 ft^3 = 0.028317 m^3
	1 l = 10^{-3} m^3	1 gal = 0.13368 ft^3
	1 l = 0.0353 ft^3	1 gal = 0.0037854 m^3
Velocity	1 m/s = 3.2808 ft/s	1 ft/s = 0.3048 m/s
	1 km/h = 0.62137 mi/h	1 mi/h = 1.6093 km/h
Force	1 N = 1 kg.m/s^2	1 lb_f = 32.174 lb_m.ft/s^2
	1 N = 0.22481 lb_f	1 lb_f = 4.4482 N
Pressure	1 Pa = 1 N/m^2	1 lb_f/in.^{2} = 6894.8 Pa
	= 1.4504 × 10^{-4} lb_f/in.^{2}	1 atm = 14.696 lb_f/in.^{2}
	1 bar = 100 kPa = 10^5 N/m^2	1 lb_f/in.^{2} = 144 lb_f/ft^2
	1 atm = 1.01325 bar	
Energy and Specific Energy	1 J = 1 N.m = 0.73756 ft.lb_f	1 ft.lb_f = 1.35582 J
	1 kJ = 737.56 ft.lb_f	1 Btu = 778.17 ft.lb_f
	1 kJ = 0.9478 Btu	1 Btu = 1.0551 kJ
	1 kJ/kg = 0.42992 Btu/lb_m	1 Btu/lb_m = 2.326 kJ/kg
	1 kcal = 4.1868 kJ	
Rate of Energy Transfer	1 W = 1 J/s = 3.413 Btu/h	1 Btu/h = 0.293 W
	1 kW = 1.341 hp	1 hp = 0.7457 kW
		1 hp = 2545 Btu/h
		1 hp = 550 ft.lb_f/s
Specific Heat	1 kJ/(kg.K) = 0.238846 Btu/(lb_m.°R)	1 Btu/(lb_m.°R) = 4.1868 kJ/(kg.K)
	1 kcal/(kg.K) = 1 Btu/(lb_m.°R)	
Other Equivalents		
Standard Atmospheric Pressure	1 atm = 1.01325 bar	
	= 14.696 lb_f/in.^{2}	
Temperature Conversions	T (°R) = 1.8T (K)	
	T (K) = T (°C) + 273.15	
	T (°R) = T (°F) + 459.67	
Standard Acceleration of Gravity	g = 32.174 ft/s^2	
	= 9.80665 m/s^2	
Universal Gas Constant	$\overline{R}$ = 8.314 kJ/(kmol.K)	
	= 1.986 Btu/(lbmol.°R)	
	= 1545 (ft.lb_f)/(lbmol.°R)	

This book is dedicated to my parents,

Kee-Lim Charles Wong

and

Min-Moy Mary (Leong) Wong

1

Concepts, Definitions, and the Laws of Thermodynamics

CONTENTS

1.1 Introduction

Thermodynamics is the science of energy. The topics covered will be about energy and the relationships among the properties of matter.

Thermodynamics is an energy science that is a key to the design of important and interesting energy systems. These systems include automotive engines, heat pumps, airplanes, rockets, space stations, power stations, gas turbines, fuel cells, air conditioners, artificial kidneys, firefighting equipment, chemical refineries, lasers, refrigerators, cryogenic systems, solar heating systems, computers, and energy-efficient buildings. Thermodynamics is the foundation of the design of engineering systems.

The existence of energy can be deduced from the physical effects of energy transfer. For instance, the hoisting of a weight increases the energy of the weight. This could be accomplished by the flow of electricity through a motor, or the flow of steam through a turbine. Quantitative relations do exist between the amount of electricity or steam used and the amount of weight raised through a given elevation. The study of these relationships is part of thermodynamics.

A system may be investigated either from a microscopic or a macroscopic point of view. In the microscopic viewpoint, the position and velocity of the molecules comprising the system are specified in detail. The behavior of the system is the sum of the behavior of each molecule. Such a study is known as statistical thermodynamics. In the macroscopic viewpoint, we are concerned with the gross or average effects of many molecules. We can feel these effects and measurements can be made by instruments. We can measure the system or parts of the system, the dimensions of which are large compared to the distance between molecules. In this book, we use this macroscopic point of view. There is no need to specify molecular detail in *classical thermodynamics*. In *statistical thermodynamics*, the basic building blocks of matter are postulated. For instance, free molecules are postulated for gas, and strongly interacting and vibrating molecules are postulated for solids. After that, statistical thermodynamics is developed by generating relationships for macroscopic properties.

1.2 Definitions

A *system* is an artificially defined body or bodies of matter that has been demarcated in space by *boundaries*. The purpose of defining a system is to define the subject of a thermodynamic analysis. An *open system* is one where there is flow of matter or mass in or out of the boundaries. A *closed system* is one where there is no flow of mass in or out of the boundaries. If, in addition, there is no transfer of energy between a system and its surroundings, the system is called an *isolated system*. Sometimes, a closed system is called a constant mass system or a *control mass* system; an open system may be referred to as a *control volume* system.

If we consider a quantity of water, we know that this water may exist in various forms. If we start off with a liquid, the water can become a vapor when it is heated, or a solid when it is cooled. These are known as different phases of water. A *phase* is a quantity of matter that is homogeneous throughout. In each phase the substance may exist at various temperatures and pressures, or in various states. The *state* may be identified by certain observable, macroscopic properties. The state of a substance may be viewed as its condition of aggregation.

The *property* of a system is a macroscopic characteristic of a system to which numerical values can be assigned at any time without knowledge of the history of the system. Some familiar ones are temperature, pressure, density, mass, and volume. Other properties will be introduced later. An *intensive property,* at a given state, is one that is independent of the mass of the system. An example is the temperature. An *extensive property* is one that depends on the mass of the system; mass and volume are extensive properties.

An *equilibrium state* is a condition of a system in which there are no spontaneous changes within the system. A *steady state* is a condition of a system in which none of its properties change with time. A *process* is the transformation from one state to another. It is the series of conditions taken by a substance to go from one condition of aggregation to another. A *thermodynamic cycle* is one in which a process, or processes, begins and ends at the same state. In general, a cycle describes the processes of a substance that starts at an initial condition of aggregation, goes to at least one other condition of aggregation, and then returns to the original condition of aggregation via a different series of conditions.

Work and *heat* are quantities of energy transfer that occur at a system boundary. Work and heat are not properties of a system, and they cannot be possessed by a system. Work and heat are forms of energy that occur at a system boundary.

Work can always be reduced to a raising of a weight in a gravitational field, with no other effects. Heat cannot be so reduced. Heat transfer is normally accompanied by a change in temperature field. In other words, if any system is analyzed in detail, work may be simplified to energy transfer that increases the potential energy of a weight only. We cannot perform the same simplification with heat. Work done by a system is considered positive, while work done on a system is negative. Heat supplied to a system is considered positive, while heat leaving a system is negative.

A system can undergo a process where both heat and work take place at the system boundary. When no heat transfer is involved at the system boundary, the process is called an *adiabatic* process.

Figure 1.1 is a schematic of a steam power plant system with a boiler, a turbine, a condenser, and a pump. Heat is added to the boiler to convert the water into steam. The turbine is where the steam is expanded to turn a shaft, thus doing work. (This shaft may be coupled to an electric generator to generate electricity.) The condenser is where the expanded steam from the turbine is condensed into liquid water, by giving up heat. The water changes from the vapor phase to the liquid phase in the condenser by cooling. (See Chapter 6 for more details about the condenser.) The pump increases the pressure of the water from the condenser so that it may mix with the water in the boiler which is at higher pressure; work has to be done on the pump by supplying electricity.

In this example, the dotted line around the whole system is a *boundary* which defines a *closed system*, because no mass enters or leaves this system. The dot-dash line around the boiler, however, defines an *open system* because mass enters the boiler and leaves the boiler. Similarly, a boundary could be drawn around the turbine to define an open system, and also around the condenser, and also around the pump. In other words, the four open systems can be combined together to form a closed system. The working substance in the system, water, undergoes a *cycle* because it begins and ends at the same state.

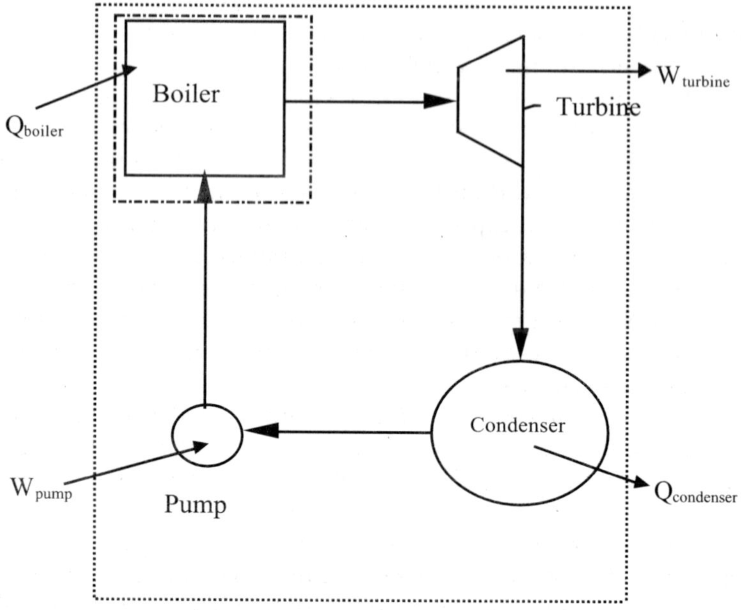

FIGURE 1.1
Schematic of a steam power plant.

Note that both heat and work can cross a system boundary even though the system is a closed one. It is mass that determines whether a system is closed or open. The heat that enters the boiler is considered positive, whereas the heat that leaves the condenser is considered negative. The work done by the turbine is considered positive, whereas the work (electricity supplied) done on the pump is negative.

1.2.1 A Note about Units

Two sets of units are commonly used today: the metric SI (from Le **S**ysteme International d'Unites), which is also called the International System, and the English system. In this book, both units are used, though the dominant one is the SI units.

In SI, the units of mass, length, and time are the kilogram (kg), meter (m), and second (s), respectively. In the English system, the corresponding units are the pound-mass (lb_m), foot (ft), and second (s). The relationship between the principal units are

$$1\ lb_m = 0.45359\ kg$$
$$1\ ft = 0.3048\ m.$$

The volume unit is m^3 in SI, and ft^3 in the English system. The *specific volume* of a substance is defined as the volume per unit mass of the substance, and has the units of m^3/kg (SI) or ft^3/lb_m (English).

In SI, the force unit is the newton (N), and it is the value of the force required to accelerate a mass of 1 kg at a rate of 1 m/s^2. In the English system, the force unit is the pound-force (lb_f) and it is the value of the force required to accelerate a mass of 32.174 lb_m (1 slug) at a rate of 1 ft/s^2. In other words,

$$\begin{aligned} 1\ N &= 1\ kg.m/s^2 \\ 1\ lb_f &= 32.174\ lb_m.ft/s^2. \end{aligned}$$

Unlike mass, weight (W) is a force. It is the gravitational force on a body,

$$W = mg$$

where m is the mass of the body and g is the local gravitational acceleration (g is 9.807 m/s^2 or 32.174 ft/s^2 at sea level at mid-latitude). At this location, a mass of 1 kg will weigh 9.807 N.

Work, a form of energy, may be defined as force times distance. It has the unit newton-meter or joule, 1 J = 1 N.m. The kilojoule (kJ) is 1000 joules. In the English system, the energy unit is the Btu (British thermal unit). This is the energy required to raise the temperature of 1 lb_m of water at 68°F by 1°F. The conversion is 1 Btu = 1.055 kJ.

The English system is no longer used in England or the rest of the U.K. The English system of units is principally used in the U.S. For this reason, the symbol US is used as an indicator of English units for all the tables in the Appendices and the computer software that accompanies this book.

1.3 Pressure

Pressure is defined as

$$P = \lim_{\delta A \to \delta A'} \frac{\delta F_n}{\delta A} \tag{1.1}$$

where $\delta A'$ is the smallest area over which we can consider the fluid as a continuum. Note that δF_n is normal to δA. The pressure relative to the atmosphere is called the gage pressure. The pressure relative to a perfect vacuum is called the absolute pressure.

When the pressure in a system is greater than the local atmospheric pressure (Figure 1.2), P_{atm}, the term gage pressure is used

$$P(gage) = P(absolute) - P_{atm}(absolute).$$

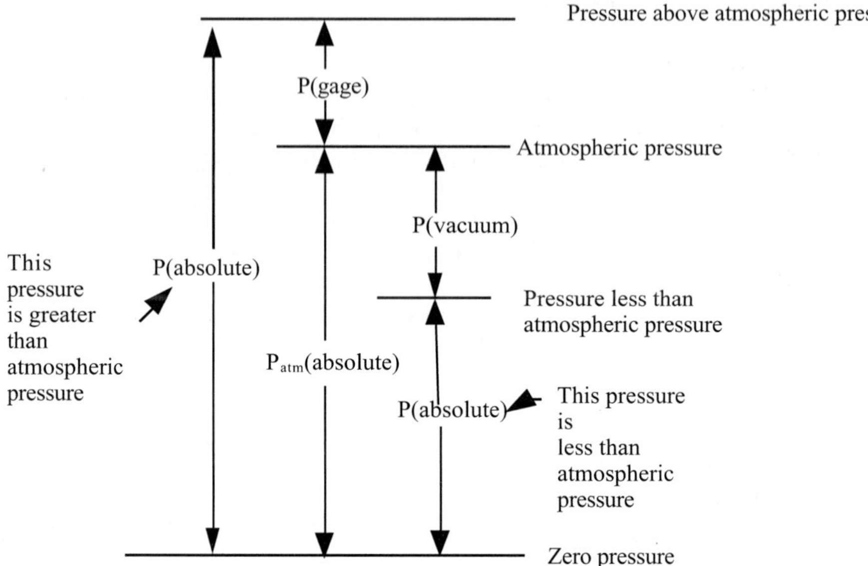

FIGURE 1.2
Absolute, atmospheric, gage, and vacuum pressures.

When the pressure in a system is less than the local atmospheric pressure, the term vacuum pressure is used

$$P(\text{vacuum}) = P_{\text{atm}}(\text{absolute}) - P(\text{absolute}).$$

In SI, the unit of stress and pressure is the pascal

$$1\ \text{Pa} = 1\ \text{N/m}^2.$$

Multiples of the pascal are often used, such as the kilopascal, the bar, and the megapascal:

$$1\ \text{kPa} = 10^3\ \text{N/m}^2$$
$$1\ \text{bar} = 10^5\ \text{N/m}^2$$
$$1\ \text{MPa} = 10^6\ \text{N/m}^2.$$

The English units for stress and pressure are pounds force per square inch, $\text{lb}_f/\text{in.}^2$, and pounds force per square foot, lb_f/ft^2. For absolute pressures, the following abbreviations are frequently used: psia for pounds force per square inch absolute, and psfa for pounds force per square foot absolute. The atmospheric pressure depends on the location. A standard reference value may be defined as 1 standard atmosphere (atm) = 1.01325×10^5 N/m^2 or 14.696 lb$_f$/in.^{2}

Example 1.1

Problem: Express the following absolute pressures in terms of gage pressure: (a) 125 kPa, (b) 25 kPa. The barometric reading is 100 kPa.

Solution

(a) Gage pressure = 125 kPa – 100 kPa = 25 kPa.

(b) Gage pressure = 25 kPa – 100 kPa = – 75 kPa or 75 kPa vacuum.

Example 1.2

Problem: A pressure gage connected to a pipeline reads 2.0 bars at a site where the barometric reading is 75.8 cmHg. Calculate the absolute pressure in the pipeline. Take the density of mercury ρ_{Hg} to be equal to 13,600 kg/m^3.

Solution: Consider a fluid column of height h, and cross-sectional area A. The force on the area A at the bottom of the column is due to the weight of the column. The weight of the column is given by its volume multiplied by its specific weight,

$$F = W = \gamma A h$$

where

$$\gamma = \text{specific weight of the fluid.}$$

The pressure is the force divided by the area, so the pressure at the bottom of the fluid column is given by

$$P = F/A = W/A = \gamma h.$$

Thus, the barometric pressure is

$$\begin{aligned} P_{atmos} &= (13{,}600 \text{ kg/m}^3)(0.758 \text{ m})(9.81 \text{ m/s}^2) \\ &= 101.1 \times 10^3 \text{ N/m}^2 = 101.1 \text{ kPa.} \end{aligned}$$

The absolute pressure in the pipeline is

$$\begin{aligned} P_{abs} &= 2 \text{ bars} + 101.1 \text{ kPa} \\ &= 200 \text{ kPa} + 101.1 \text{ kPa} = 301.1 \text{ kPa.} \end{aligned}$$

1.4 Forms of Work

1.4.1 Mechanical Forms of Work

The work done by a constant force F on a body which is moved a distance s in the direction of the force is given by

$$W = Fs. \tag{1.2}$$

In general if the force is not constant, the work is obtained by integration,

$$W = \int_1^2 F ds. \tag{1.3}$$

There are two requisites for a work interaction between a system and its surroundings. There must be (1) a force acting on the boundary, and (2) the boundary must move. If there are forces on the boundary, but there is no displacement of the boundary, there is no work interaction. Similarly, if there is a displacement of the boundary without a force to oppose or drive this movement, there is no work interaction.

Consider the piston-cylinder device as shown in Figure 1.3. Select the system to be the gas, with the system boundary as shown. It is often convenient to use physical boundaries as system boundaries. Assume that the gravity force is negligible. The gas is assumed to undergo a series of equilibrium states, also known as a quasiequilibrium process or quasistatic process. The gas moves the piston upward a small distance ds. The total force on the piston is the pressure multiplied by the area of the piston,

$$F = PA. \tag{1.4}$$

The differential work done by the gas on the piston is

$$\delta W = PAds. \tag{1.5}$$

Since the volume element is $dV = Ads$,

$$\delta W = PdV. \tag{1.6}$$

If the piston moves from position s_1 to position s_2, the total work done may be found by integration,

$$W_{1-2} = \int_{s_1}^{s_2} PAds = \int_{V_1}^{V_2} PdV. \tag{1.7}$$

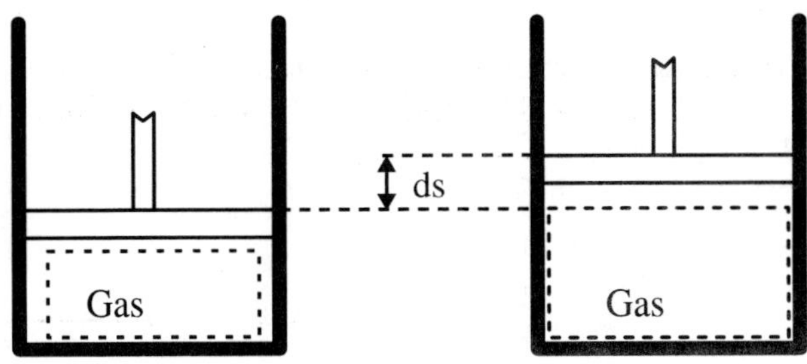

FIGURE 1.3
Work as energy transferred by a moving boundary.

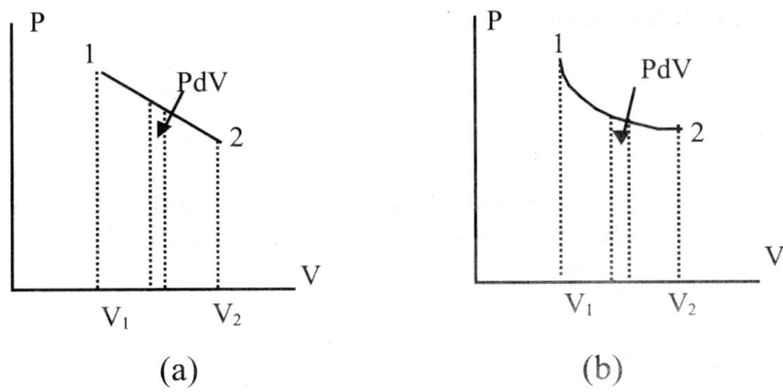

FIGURE 1.4
The amount of work is dependent on the path between two states.

Consider Figure 1.4, which shows states 1 and 2 in (a) and (b), but the paths connecting the two states are different. Consequently, the area under the curve in (a) is different from that in (b). In other words, the amount of work is dependent on the *path* between the two end points. A *path* refers to the specification of a series of states through which the system passes. Therefore, work is a path function, as contrasted to a *point* function, which depends only on the end states. For instance, pressure is a point function. The integral of an exact differential, dP (as compared to an inexact differential δW that depends on the path), is given by

$$\int_{P_1}^{P_2} dP = P_2 - P_1 \tag{1.8}$$

where P_1 is the pressure at state 1 and P_2 is the pressure at state 2.

Example 1.3

Problem: A gas in a piston-cylinder equipment undergoes a constant pressure process. The initial volume is 0.025 m^3, and the final volume is 0.05 m^3. Calculate the work for the process, in kilojoules.

Solution

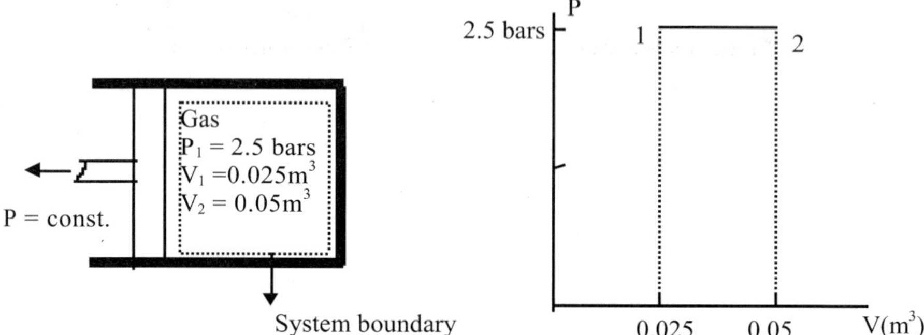

Schematic for Example 1.3.

The system is selected as shown by the system boundary. The work is indicated by the area under the horizontal line joining the initial and final states in the P-V diagram. Thus,

$$\text{Work} = \int_{V_1}^{V_2} PdV = P(V_2 - V_1)$$

$$\text{Work} = (2.5\ \text{bars})(0.05 - 0.025)\text{m}^3\left(\frac{10^5\,\text{N/m}^2}{1\ \text{bar}}\right)\left(\frac{1\ \text{kJ}}{10^3\,\text{N.m}}\right) = +6.25\ \text{kJ}.$$

1.4.2 Spring or Elastic Work

The force on an elastic substance is often related to the displacement of the substance by Hooke's law, which is

$$F = k_s x = k_s(L_f - L_i) \tag{1.9}$$

where k_s is the spring constant, x is the displacement, L_i is the initial unstressed length of the substance, and L_f is the final length of the substance. The force F is taken to be negative when the final length is less than the unstressed length. The work done in stretching or compressing the substance (e.g., spring, wire, bar) is

$$W_{spring} = \int F dX = \int_1^2 k_s (L_f - L_i) d(L_f - L_i)$$
$$= \frac{k_s}{2}\left[(L_2 - L_i)^2 - (L_1 - L_i)^2\right]. \tag{1.10}$$

Example 1.4

Problem: A linear spring has a spring constant of 100 kN/m. Its unstressed length is 0.1 m. The spring is then stretched to 0.15 m. Determine the work done on the spring.

Solution

$$\text{Work} = \frac{k_s}{2}\left[(L_2 - L_i)^2 - (L_1 - L_i)^2\right].$$

Since the spring is initially unstressed, $L_1 = L_i$.

$$\text{Work} = \frac{k_s}{2}(L_2 - L_i)^2 = \frac{100}{2}(0.15 - 0.1)^2 \frac{\text{kN}}{\text{m}}.\text{m}^2 = +0.125 \text{ kJ}.$$

1.4.3 Electrical Work

For electrons moving in a wire under electromotive forces, work is done. When N coulombs of electrons move through a potential difference V, the electrical work done is

$$W_e = VN. \tag{1.11}$$

The rate form of the equation is

$$\dot{W}_e = VI \tag{1.12}$$

where $\dot{W}_e$ is the electrical power (rate of work done) and I is the current or the number of coulombs per unit time. If V and I are functions of time, the electrical work done in an interval Δt is given by

$$W_e = \int_1^2 VI \, dt. \tag{1.13}$$

Example 1.5

Problem: A charger for batteries supplies energy to a battery for 3 h at 12 V and 2.5 A. What is the amount of work done by the charger?

Solution

Electrical work is done by the charger on the battery:

$$W_e = \int_1^2 VI\,dt.$$

Voltage and current are constant in this case. So,

$$W_e = VI\int_1^2 dt = VI(t_2 - t_1)$$

$$W_e = (12\text{ V})(2.5\text{ A})(10,800\text{ s}) = 324\text{ kJ}.$$

Note that if the voltage and current change with time, then their functional dependence on time is required and the integration is more complex.

1.4.4 Work of Polarization and Magnetization

From the theory of electromagnetism, work is done on a substance contained within an electric or magnetic field when the field is changed. In the case of a dielectric material which lies in an electric field, the work done on the system to increase the polarization of the dielectric is

$$\delta W_{polar} = V\vec{E}.d\vec{P} \tag{1.14}$$

where V is the volume, $\vec{E}$ is the electric field strength, and $\vec{P}$ is the polarization of the dielectric. A similar equation is applicable to work done to increase the magnetization of a substance owing to a change in the magnetic field,

$$\delta W_{magnet} = V\mu_o\vec{H}.d\vec{M} \tag{1.15}$$

where V is the volume, $\vec{H}$ is the magnetic field strength, $\vec{M}$ is the magnetization per unit volume, and μ_o is the permeability of free space.

Example 1.6

Problem: For a paramagnetic substance, Curie's law provides that $M = CH/T$, where C is a proportionality constant. For a quasistatic, constant temperature process, show that the work done per unit volume is given by

$$W = \mu_o \frac{C}{2T}\left(H_f^2 - H_i^2\right).$$

Solution

Since M = CH/T, dM = C/T dH, for constant T

$$\delta W_{magnet} = V\mu_o \vec{H}.d\vec{M}$$

$$W = V\mu_o \int_i^f H.C/T\, dH = V\mu_o C/T \int_i^f HdH.$$

Considering per unit volume, the work done is

$$W = \mu_o \frac{C}{T}\left[\frac{H^2}{2}\right]_i^f = \mu_o \frac{C}{2T}\left(H_f^2 - H_i^2\right).$$

1.4.5 Torsion Work

The internal energy of a solid rod (or bar) may be increased by fixing one end and applying a force at some radial distance r and at right angle to its axis. The product of the force and the radial distance r is the torque τ applied to the system. The material will go through an angular displacement θ. The torsion work is given by

$$\delta W_{torsion} = \tau d\theta. \tag{1.16}$$

Example 1.7

Problem: A motor applies a constant torque of τ = 20 N.m to an external load. The shaft of the motor rotates at a constant angular speed of 1000 rpm ($\dot{\theta}$ = 1000 × 2π/60 = 104.7 rad/s). Calculate the rate of work done by the motor.

Solution

Since the angular speed of the motor shaft is given, the rate of work done or power rather than the work is computed. From $\delta W_{torsion} = \tau d\theta$,

$$\delta \dot{W}_{torsion} = \tau d\dot{\theta}$$

$$\dot{W} = \int \tau d\dot{\theta} = \tau\dot{\theta} = (20\text{N.m})(104.7 \text{ rad/s}) = 2094 \text{ W} = 2.094 \text{ kW}.$$

1.4.6 Work in Changing a Surface Area

Surface tension is the property of a fluid which describes the force per unit length required to maintain a surface at a specified area. The change in surface energy for a differential area dA change is a measure of the surface work done, and it is calculated by

$$\delta W_{surface} = \sigma dA \tag{1.17}$$

where σ is the surface tension.

Example 1.8

Problem: The value of the surface tension of soapy water may be taken to be $\sigma = 0.07$ N/m, approximately. Calculate the work done in blowing a bubble of soapy water 2 cm in radius.

Solution

$$\delta W_{surface} = \sigma dA$$
$$W = \int \sigma dA = \sigma \int dA = \sigma 8\pi r^2.$$

Note that the surface area created is approximately twice the area of the spherical bubble. There is a surface area inside the bubble and another on the outside; the thickness of the soapy film is considered small and negligible. Thus,

$$W = 0.07 \text{ N/m } (8\pi 0.02^2) \text{ m}^2 = 7.04 \times 10^{-4} \text{ J}.$$

1.5 Temperature

The temperature of a substance is a measure of the level of energy in the substance. The zeroth law of thermodynamics forms the basis of the temperature scale. This law states that if two bodies are in thermal equilibrium with a third body, they are in thermal equilibrium with each other. In other words, no net heat energy flows between the two bodies in thermal equilibrium. This gives definite meaning to the concept of temperature, since two bodies at the same temperature at steady state are considered to be in thermal equilibrium. For instance, a body at a particular temperature in Washington, D.C., will be at thermal equilibrium with another body at the same temperature in Zurich, if both of them are brought into contact with each other.

There are two commonly used scales for measuring temperature. These are the Fahrenheit scale named after Gabriel Fahrenheit, and the Celsius scale named after Anders Celsius.

Until 1954, the scales were based on two fixed points, the ice point and the steam point. At the Tenth Conference on Weights and Measures in 1954, the Celsius scale was redefined in terms of a single fixed point and a definition of the magnitude of the degree. The triple point of water (solid, liquid, and vapor phase of water exist together in equilibrium) is assigned the value 0.01°C, and the steam point is experimentally found to be 100°C.

The absolute zero temperature is a practically unachievable temperature where a perfect crystalline substance has no energy. A temperature scale based on this absolute zero temperature scale as datum is called an absolute temperature scale. The absolute scale related to the Celsius scale is referred to as the Kelvin scale (after Lord Kelvin, 1824–1907), and is designated K. The relation between these scales is

$$\mathrm{K} = {}^{\circ}\mathrm{C} + 273.15. \tag{1.18}$$

Note that by convention degree Kelvin or °K is not used in this absolute temperature scale. The absolute scale related to the Fahrenheit scale is referred to as the Rankine scale, and is designated R. The relationship between these scales is

$${}^{\circ}\mathrm{R} = {}^{\circ}\mathrm{F} + 459.67. \tag{1.19}$$

1.6 Heat

Heat transfer is energy that moves across the boundary of a system due to a difference in temperature between the system and its surroundings. Heat is transferred from a higher temperature to a lower temperature. A system does not possess heat, it possesses energy, and heat is the transfer form of energy that crosses a system boundary. As such, heat is not a property of the system.

Consider a block of gold and an equal mass of silver. Let the gold be hot and the silver be cold. When the two blocks are placed in contact with each other, energy moves from the hot block to the cold block. This energy that is transferred from the hot body to the cold body is termed heat. Heat will be transferred until the two blocks become of the same temperature, or reach thermal equilibrium as it is called. The hot gold has lost energy and the cold silver has gained energy.

Like work, heat is a path function, not a point function. In other words, heat is a function of the process. The amount of heat transferred is dependent on the path taken to go from state 1 to state 2. Consider the gold block in the previous discussion to be at 100°C, and the silver block to be at 30°C. When they are brought together, they will reach thermal equilibrium at some temperature between 100°C and 30°C. They are then allowed to cool in an environment of 20°C so that they are eventually at a temperature of 20°C. Contrast this with the situation of the silver block being cooled from 30°C to 20°C in an environment of 20°C. Heat is gained by the silver block from the gold, then heat is lost to the surroundings in the first scenario. The first scenario is clearly different from the second scenario in which the silver block lost heat to the surroundings; note that the silver did not gain heat at all. However, the end states of the silver block are identical in both scenarios. Hence, this practical example illustrates that heat is a path function.

1.7 Laws of Thermodynamics and Mass Conservation

There are four laws of thermodynamics; the zeroth law, the first law, the second law, and the third law. The zeroth law of thermodynamics states that if two bodies are in thermal equilibrium with a third body, they are in thermal equilibrium with each other and thus at the same temperature. This law is the basis of the temperature scale. The third body can be a thermometer. As a result of the zeroth law, the thermometer may be used for comparing the temperatures of two bodies without the bodies being in contact with each other.

The first law of thermodynamics is the statement of the conservation of energy, which states that energy can neither be created nor destroyed.

The second law of thermodynamics is the statement of the creation of entropy, which states that entropy can be created, but not destroyed.

The third law of thermodynamics states that the entropy of a pure crystalline substance is zero at the absolute zero temperature. This law is the basis of the absolute temperature scale.

These laws of thermodynamics cannot be proved. The validity of the laws rests upon the fact that neither they, nor any deductions made from them, have ever been disproved by experience.

The conservation of mass principle is used with the laws of thermodynamics. The statement of the conservation of mass is that mass is neither created nor destroyed.

1.8 The Systematic Problem-Solving Approach

It is important to have a systematic approach to solving thermodynamic problems. This methodology is a step-by-step guideline to aid your thinking. The major thermodynamic principles to be used in thermodynamic problems are as follows:

Mass conservation

First law of thermodynamics or energy conservation

Second law of thermodynamics

The rest of this section is a discussion of the steps used in the systematic approach.

Step 1: identify the system — The physical system to be analyzed should be identified. Factors that aid in identifying the system are the quantities that need to be determined. The system selected should involve these quantities, in general. A sketch should then be drawn of the system, unless the system is very simple and easily visualized without the sketch. It does not have to be a detailed drawing; a schematic of the system is usually adequate. A boundary should be drawn around it so that the system to which the thermodynamic principles are applied is clearly defined.

Step 2: write the given information — Write the given information with the proper thermodynamic symbols on the sketch. At this point, also identify the information requested with the proper thermodynamic symbols. Put a question mark next to these symbols as indication that these are data that need to be determined. With this listing, one can see the entire picture depicting the problem. Be sure to include all boundary interactions like work and heat, using the proper arrows to indicate their correct directions.

Step 3: list any assumptions — Any assumptions made to solve the problem should be stated. Only reasonable assumptions should be made. One should not bother with insignificant or unimportant assumptions. Some of the more common assumptions in thermodynamics include the following:

- A process is a quasiequilibrium process
- Change in kinetic energy can be neglected
- Change in potential energy can be neglected
- An insulated system allows no heat transfer
- A gas behaves like an ideal gas

These simplifying assumptions make the solution of thermodynamic problems less complicated.

Step 4: apply thermodynamic principles — Mass conservation is normally required in the simplest of thermodynamic problems. Apply this principle first. The first law of thermodynamics should be applied next. The second law of thermodynamics may or may not be required, depending on the nature of the problem. The zeroth law of thermodynamics is the basis of the temperature scale, and is not used explicitly in problem solving. Similarly, the third law of thermodynamics is the basis of the absolute temperature scale, and is not used explicitly in problem solving.

Step 5: determine the required unknowns — All the properties at any state can be determined with the use of thermodynamic tables or relations,(see Chapter 2). Since the state of a substance is defined by any two independent intensive properties (Chapter 2, Section 2.1) all the properties of a substance are known once the state is defined. In a process, usually either the initial state is completely known or the final state is completely known. All the properties at the known state are also known. One of the two defining properties of the unknown state should normally be known. The unknown property that defines the unknown state is related to the process. The process will usually provide the necessary information to determine this unknown property. Once this property has been determined, the unknown state is completely defined; all the properties at this state can then be found from the thermodynamic tables or relations.

Substitute property data and other numerical values into the equations, and obtain the required answers. Pay particular attention to the units of the different quantities when substituting numerical values. Inconsistent units are the cause of many errors in problem solving. Any unreasonable results should be an indication of possible errors. The analysis should be checked, and corrections made.

Problems

Concepts and Definitions

1.1 Explain the meanings of the following terms: system, boundary, open system, closed system, isolated system, phase, property, intensive property, extensive property.

1.2 Explain the meanings of the following terms: steady state, thermodynamic cycle, process, work, heat.

1.3 Sketch a possible system boundary for studying each of the following:

a. Automotive engine in motion.
b. A pot of soup heating.
c. A room air conditioner in operation.
d. A personal computer in use.
e. A satellite in orbit.
f. A wood fire in a residential fireplace.
g. A motorboat in motion.
h. A power plant in operation.

1.4 Consider a kettle of water that is being heated. Let the water in the kettle be a system. Identify locations on the system boundary where the system interacts with its surroundings. Describe changes that occur within the system with time. Enlarge the system to include the kettle and the fire that is heating the kettle. Repeat the exercise of identifying the interactions between the system and its surroundings, and describe changes that occur over time.

Pressure

1.5 Explain the difference between absolute pressure and gage pressure.

1.6 A pressure gage attached to a tank reads 600 kPa at a site where the barometric reading is 98 kPa. Compute the absolute pressure in the tank.

1.7 A pressure gage connected to a pipeline reads 2.3 bars at a site where the barometric reading is 76 cmHg. Determine the absolute pressure in the pipeline. Take $\rho_{Hg} = 13{,}600\ kg/m^3$.

1.8 A pressure gage attached to a vessel reads 100 psi at a site where the barometric reading is 29.2 inHg. Find the absolute pressure in the vessel. Take $\rho_{Hg} = 848\ lb_m/ft^3$.

1.9 A vacuum gage connected to an equipment indicates 25 kPa at a site where the barometric reading is 75 cmHg. Calculate the absolute pressure in the equipment. Take $\rho_{Hg} = 13{,}600\ kg/m^3$.

1.10 A vacuum gage attached to a piece of equipment reads 38 kPa at a site where the atmospheric pressure is 101 kPa. Compute the absolute pressure in the equipment.

1.11 A vacuum gage connected to a pipeline indicates 0.5 bar at a site where the barometric reading is 75 cmHg. Find the absolute pressure in the pipeline. Take $\rho_{Hg} = 13{,}600\ kg/m^3$.

1.12 A vacuum gage connected to a vessel reads 4 psi at a site where the barometric reading is 29.2 inHg. Calculate the absolute pressure in the vessel. Take $\rho_{Hg} = 848\ lb_m/ft^3$.

1.13 The barometer may be used to measure the height of tall buildings. If the barometric readings at the top and at the bottom of a building are 740 and 760 mmHg, respectively, find the height of the building. Assume an average air density of $1.17\ kg/m^3$.

1.14 Find the pressure exerted on the surface of an underwater research vessel moving 200 m below the free surface of the sea. Assume that the barometric pressure is 101 kPa and the specific gravity of seawater is 1.025.

1.15 An underwater reconaissance vessel moves 600 ft below the free surface of the sea. The barometric pressure is 14.7 psia and the specific gravity of seawater is 1.025. Find the pressure on the surface of the vessel.

1.16 Calculate the pressure exerted on a diver at 10 m below the free surface of the pool. Assume a barometric pressure of 101 kPa.

1.17 A manometer attached to an oxygen tank uses oil ($\rho = 800\ kg/m^3$) as the manometric fluid. If the oil-level difference between the two columns is 30 cm and the atmospheric pressure is 97 kPa, compute the absolute pressure of the oxygen.

1.18 Oil ($\rho = 49\ lb_m/ft^3$) is used as the manometric fluid in a manometer connected to an oxygen tank. If the oil-level difference between the two columns is 15 in. and the atmospheric pressure is 14.7 psia, calculate the absolute pressure of the oxygen.

1.19 Both a manometer and a gage are attached to a gas tank. The pressure gage reads 90 kPa. Find the distance between the two fluid levels of the manometer if the fluid is (a) water ($\rho = 1000\ kg/m^3$) or is (b) mercury ($\rho = 13{,}600\ kg/m^3$).

1.20 A vacuum gage shows that the pressure in a closed gas chamber is 0.1 bar (vacuum). The pressure of the atmosphere is measured as 760 mm by a column of mercury. The acceleration of gravity is $9.8\ m/s^2$ and the density of mercury is $13.6\ g/cm^3$. Find in bars, the absolute pressure within the chamber.

1.21 A manometer attached to a gas tank indicates that the pressure within is greater than the surrounding atmosphere. The manometric fluid is mercury, which has a density of $13.6\ g/cm^3$. The manometer shows a reading of 3 cm. The atmospheric pressure is 95 kPa, and the acceleration due to gravity is $9.8\ m/s^2$. Determine in kPa (a) the absolute pressure of the gas, and (b) the gage pressure of the gas.

1.22 Evaluate, in bars, the change in atmospheric pressure between sea level and the top of a mountain which is 10,000 ft above sea level. Assume that the acceleration due to gravity is $32.2\ ft/s^2$, and the specific volume of air is $14.05\ ft^3/lb_m$.

1.23 Refrigerant-134a enters the compressor of a refrigeration system at an absolute pressure of 80 kPa. A pressure gage shows that the pressure at the compressor exit is 1800 kPa (gage). The atmospheric pressure is 101 kPa. Find the change in absolute pressure from inlet to exit.

Forms of Work

1.24 A working fluid at 500 kPa, 400°C (673.15 K) is in a piston-cylinder equipment with an initial volume of $0.9\ m^3$. The fluid undergoes an expansion process which may be described as PV = constant. The final pressure is 200 kPa. Calculate the work for this process.

1.25 A gas in a piston-cylinder equipment undergoes a constant pressure process. The initial pressure is 3 bars, the initial volume is $100\ cm^3$, and the final volume is $300\ cm^3$. Determine the work for the process.

1.26 Air in a piston-cylinder assembly is compressed from $0.02\ m^3$ to $0.01\ m^3$, at a pressure of 150 kPa. Compute the work done on the air.

1.27 Carbon dioxide in a piston-cylinder equipment is compressed from $200\ cm^3$ to $100\ cm^3$, at a pressure of 200 kPa. Find the work done on the carbon dioxide.

1.28 Air is compressed from 9 ft^3 to 5 ft^3 at a pressure of 1 atm in a piston-cylinder device. Calculate the work done on the air.

1.29 A fluid undergoes a two-step process. First, it is expanded from 0.8 to 1 m^3 at a constant pressure of 190 kPa. Then it is expanded from 1 to 1.7 m^3 with a linearly increasing pressure from 190 kPa to 400 kPa. Find the work done at the boundary.

1.30 Gas at 150 kPa is contained in a piston-cylinder device. At this initial state, a linear spring (spring constant = 100 kN/m) is touching the piston but not exerting any force on it. Heat is then transferred to the gas, causing the gas to expand, the piston to rise and compress the spring until the volume inside is three times the intial volume of 0.01 m^3. The cross-area of the piston is 0.3 m^2. Find (a) the final pressure of the gas, (b) the work done against the spring, and (c) the total work done by the gas.

1.31 A linear spring has a spring constant of 50 kN/m. Its unstressed length is 0.05 m. The spring is then stretched to 0.1 m. Find the work done on the spring.

1.32 A charger for batteries supplies energy to a battery for 1 h at 10.5 V and 3 A. What is the amount of work done by the charger?

1.33 A battery is being charged by 13.4 V at a current of 5 A. Calculate the instantaneous rate of work and the total work done over 3 h 15 min.

1.34 A motor supplies a constant torque of $\tau = 10$ N.m to an external load. The shaft of the motor rotates at a constant angular speed of 1200 rpm ($\theta = 125.7$ rad/s). Compute the rate of work done by the motor.

1.35 The surface tension of soapy water may be valued at $\sigma = 0.07$ N/m, approximately. Determine the work done in blowing a bubble of soapy water 5 cm in diameter.

1.36 Consider a film of ethanol maintained on a wire frame, as the control mass. The film has an initial area of 10 mm by 26 mm, and a surface tension of 22.3 mN/m at 20°C (293.15 K). Calculate the work done in increasing the area of the ethanol to 10 mm by 32 mm.

Temperature

1.37 Explain in your own words, the zeroth law of thermodynamics.

1.38 (a) Write the relationship between the Celsius scale and the Kelvin scale. Do the same for (b) the Fahrenheit scale and the Rankine scale; (c) the Fahrenheit scale and the Celsius scale; (d) the Rankine scale and the Kelvin scale.

1.39 Write the relationship between the Fahrenheit scale and the Kelvin scale. Do the same for the Celsius scale and the Rankine scale.

1.40 The average temperature of a human body is 98.6°F. What is this in the Celsius scale?

1.41 An engineering system is at a temperature of 20°C. Express this in °R and °F.

1.42 A vessel is at a temperature of 22°C. Express this in K.

1.43 A body rises in temperature by 12°C. Express this rise in K.

1.44 A tank rises in temperature by 15°F. Express this in °C and K.

1.45 An engine cools by 10°C in a process. Express this drop in temperature in °R and °F.

1.46 An instrument cools by 22°F in a process. Express this drop in temperature in °C and K.

1.47 Two temperatures are given in terms of the Celsius scale. Convert these temperatures to the Kelvin scale. Show that the difference between the two temperatures in the Celsius scale is the same as that in the Kelvin scale.

1.48 The Fahrenheit scale is used to express two temperatures. Convert these temperatures to the Rankine scale. Show that the difference between the two temperatures in the Fahrenheit scale is the same as that in the Rankine scale.

1.49 A new absolute temperature scale is being considered. On this scale, 0°C is 100°A and 100°C is 300°A. What are the temperatures in °C corresponding to 200°A and 250°A? What is the ratio of the size of the kelvin to the °A?

1.50 Two liquid-in-glass thermometers are used to measure the temperature of the same object. If one is in the Celsius scale and the other in the Fahrenheit scale, and both have the same numerical reading, what is the temperature, in degree Rankine?

Computer, Design, and General Problems

1.51 Write an essay regarding the objectives and principles of statistical thermodynamics. How does it differ from the macroscopic approach to thermodynamics?

1.52 Write an essay on one of the following technologies:

a. Waste-to-energy power plants.

b. Nuclear power plants.

c. Fuel cells.

d. Hydrogen as a fuel.

e. Solar energy used for power generation.

Investigate the history and current status of each technology. Discuss the main technical and societal issues related to the increased use of the technology.

1.53 Discuss the present uses of the following forms of energy: (a) solar energy, (b) hydropower, and (c) wind energy. Discuss the impact of using more of these renewable forms of energy.

1.54 Write an essay on the use of different devices used by mankind throughout history to measure mass.

1.55 Select one of the mass measurement devices of the previous question. Modify it with modern-day tools and devices, and discuss how you have improved the design.

1.56 Temperatures are measured using many different devices. List these devices, and write an essay on the principles of these devices.

1.57 Pressures are measured using many different devices. List these devices, and write an essay on the principles of these devices.

1.58 A manometer is often used to measure the pressure of a system relative to atmospheric pressure. An inclined manometer allows for a greater length of liquid between the two arms for any pressure difference, thus allowing for better precision. Design an inclined manometer that wraps around like a screw, which increases the length of liquid even more. Discuss the advantages and disadvantages of such a device.

1.59 Write a computer program to list corresponding pressures in $kPa_{absolute}$ and $lb_f/in.^2_{gage}$ from 0 $kPa_{absolute}$ to 1000 $kPa_{absolute}$ in increments of 10 kPa.

1.60 Write a computer program to list corresponding temperatures in °C, K, °F, and °R from 100°C to 200°C in increments of 5 degrees.

1.61 Design an experiment using two solid blocks of equal mass. The environment should be approximately constant in temperature during the duration of the experiment. Heat one block to 100°C, and the other to about 10°C above the environment, say 30°C. Select combinations of materials for the two blocks so that their equilibrium temperature is halfway in between, that is, at 65°C.

2

Properties of Pure Substances

CONTENTS

A *pure substance* is one that is uniform and invariable in chemical composition. An example is water, or a mixture of solid ice and liquid water. Air is not a pure substance. Air is composed of various amounts of nitrogen, oxygen, carbon dioxide, water vapor, and a small percentage of other gases. In many thermodynamic problems, air is treated as a pure substance for simplicity.

2.1 The State Principle

The state of a system is defined by the value of its properties. Based on considerable practical experience, it can be concluded that when a sufficient number of properties are specified, the rest of the properties automatically assume certain values. The conclusion is embodied in the *state principle*:

> *The state of a compressible system is defined by two independent, intensive properties.*

This principle is applicable when there are no electrical, magnetic, or surface tension effects. These effects are the result of external force fields, and they have to be characterized when present.

The state principle states that the two properties specified be independent to define the state. Two properties are independent if one can be changed over a range of values while the other is kept at a constant value. For instance, specific volume and temperature are independent properties for a compressible system and together they may be used to define the state. Pressure and temperature are independent properties for single-phase systems, but are not independent for multiphase systems. For example, water boils at 100°C at atmospheric pressure, but at a higher temperature when the pressure is increased beyond atmospheric. This fact shows that pressure and temperature are not sufficient to define the state of a two-phase system. For instance, pressure and specific volume may be used to define the state of a two-phase system.

2.2 The P-v-T Surface

A substance can exist in three different phases: solid, liquid, and gas. We all know that water can exist as solid ice, liquid water, or water vapor. Simple experiments can be done to show the existence of these phases of water. The data obtained in experiments relating the pressure to the temperature and the specific volume can be presented as a three-dimensional surface with the pressure being a function of the specific volume and temperature. This surface would be a P-v-T surface, and an example is shown in Figure 2.1 for a substance that contracts on freezing. For a substance that expands on freezing (like water), the solid-liquid surface is at a smaller specific volume than for the solid surface. The regions where there is only one phase are labeled liquid, solid, and vapor. In regions where two phases exist, the labels are liquid-vapor (L-V), solid-liquid (S-L), and solid-vapor (S-V). The triple line is a line of constant temperature and pressure, where all three phases can exist.

Saturated vapor are states bordering the vapor region where phase change from vapor to liquid begins. *Saturated liquid* are states bordering the liquid region where phase change from liquid to vapor begins. *Saturated solid* are states bordering the solid region where phase change from solid to vapor or solid to liquid begins.

The P-v, T-v, and P-T diagrams, as shown in Figure 2.2, are the projections of the P-v-T surface into the P-v plane, the T-v plane, and the P-T plane, respectively. Distortions have been made so that more of the regions are displayed. The *triple line* is the line which represents the states where saturated solid, liquid, and vapor phases can coexist at equilibrium. The triple line in Figure 2.1 appears to be a point when viewed parallel to the v axis, so it is called the *triple point*. The *critical point* is the point at the top of the two-phase

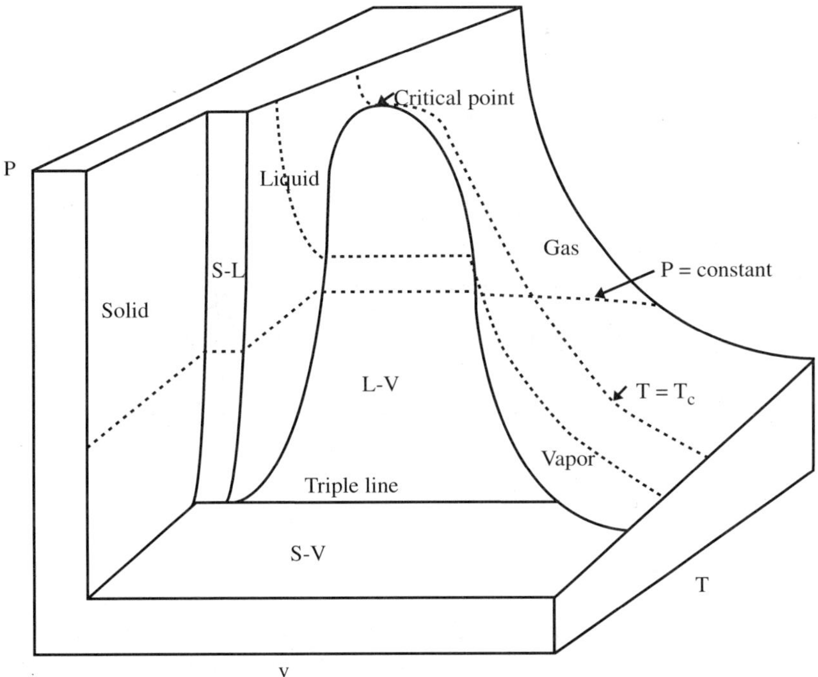

FIGURE 2.1
The P-v-T surface of a substance that contracts on freezing.

dome where the saturated liquid and vapor lines meet. At pressures greater than that of the critical point, the liquid changes to a vapor without a constant-temperature vaporization process.

Superheated vapor is a vapor whose temperature is greater than the saturation temperature at the given pressure. *Subcooled liquid* is a liquid whose temperature is less than the saturation temperature at the given pressure. This is sometimes also called a *compressed liquid.*

The *quality* is the ratio of the mass of vapor to the total mass of a two-phase liquid-vapor mixture. When applied to steam during the steam age, it is easy to see the origin of the term. A 100% quality is saturated steam, and 0% quality is saturated liquid; the saturated steam possesses greater internal energy (Section 2.4.2) than the saturated liquid water and hence is of greater quality. The quality is also called the dryness fraction. Saturated steam has a dryness fraction of one; in other words, there is no liquid or "wetness" in saturated steam. Along that trend of reasoning, saturated liquid water has a dryness fraction of zero. There is no vapor or "dryness" in saturated liquid.

Note that in Figure 2.2, Part (b), the constant-temperature line in the liquid-vapor region is a horizontal line on the P-v diagram. This means that the temperature and pressure are not independent in the liquid-vapor region. In other words, at any given pressure, there exists only one temperature at

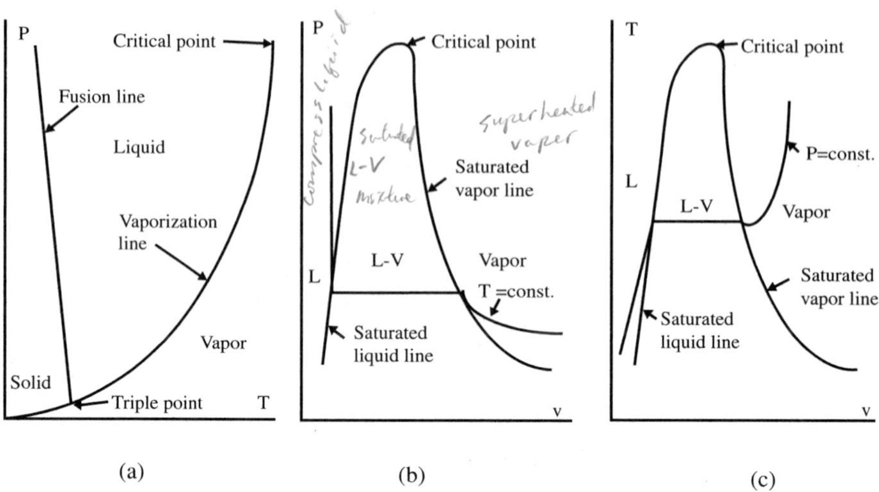

FIGURE 2.2
The P-v, T-v, and P-T diagrams.

which the liquid phase can exist in equilibrium with the vapor phase. Note also in Figure 2.2, Part (c) the constant pressure line in the liquid-vapor region is a horizontal line on the T-v diagram. This feature also indicates that the temperature and pressure are not independent in the liquid-vapor region.

At any given temperature, if the pressure is higher than the saturation pressure at that temperature, then the state has to be in the compressed liquid region. This fact is shown by the isotherm rising in the compressed liquid region in Figure 2.2, Part (b). Of the three regions shown, states with higher pressures than the corresponding saturation pressure can only exist in the compressed liquid region. At any given temperature, if the pressure is lower than the corresponding saturation pressure, then the state has to be in the superheated vapor region. This behavior is indicated by the isotherm descending in the superheated vapor region in Figure 2.2, Part (b).

At any given pressure, if the temperature is lower than the saturation temperature at that pressure, then the state has to be in the subcooled (compressed) liquid region. This characteristic is shown by the isobar descending in the compressed liquid region in Figure 2.2, Part (c). Of the three regions shown, states with lower temperatures than the corresponding saturation temperature can only exist in the compressed liquid region. At any given temperature, if the temperature is higher than the corresponding saturation temperature, then the state has to be in the superheated vapor region. This fact is indicated by the isobar rising in the superheated vapor region in Figure 2.2, Part (c).

Most substances contract on cooling; however, water does not. As water cools below 4°C, it expands. It continues to expand as it freezes at 0°C. Since it is less dense than the liquid water around it, ice rises to the surface. When ponds and lakes freeze, the ice forms a layer over the liquid water below. Sheltered from the wind and more heat loss, that water is able to maintain a steady 4°C temperature, so more freezing is impeded. The life forms in the water can survive for another season.

Imagine what would have happened if ice were denser than liquid water. As the temperatures of Earth's early ponds and lakes fell below freezing, the top layer of ice would have sunk, allowing the next layer to also freeze and sink. As the body of water gradually froze solid, life would have had a very difficult time making it past the first winter.

Figure 2.3 is a P-T diagram for water, which is a substance that expands on freezing. The line A-B represents sublimation, that is, the conversion of solid ice to water vapor, without passing through the liquid state. The line C-D passes through the triple point. The initial solid state, when heated, turns to liquid and vapor at the triple point. On further heating, the liquid portion converts to vapor. The line E-F passes from the solid phase through the liquid phase to the vapor phase.

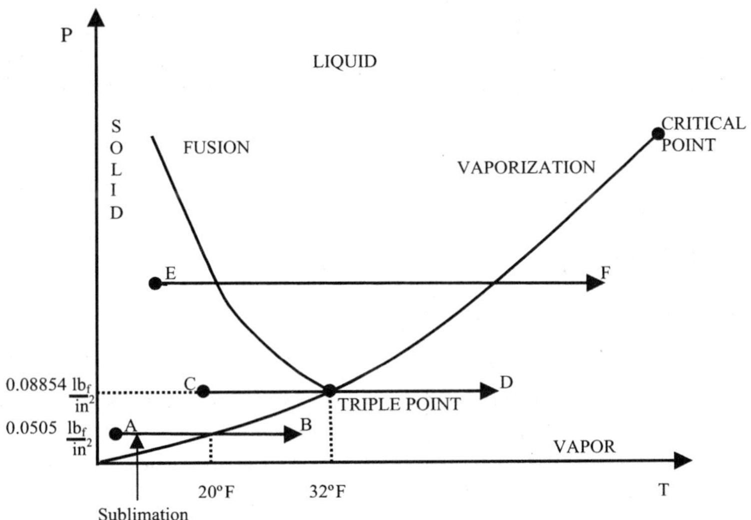

FIGURE 2.3
P-T diagram for water.

2.3 Phase Change

It is interesting to observe what happens when a substance like water undergoes phase change. Consider a mass of liquid water maintained at a pressure of 0.1 MPa, initially at a temperature of 20°C. Heat is being added to the system. In the first stage, the temperature increases and the specific volume

(volume per unit mass) increases. When the temperature reaches 99.63°C, a change of phase occurs. Some of the liquid starts changing into vapor. The temperature remains constant whereas the specific volume increases considerably. This phenomenon occurs because the *saturation temperature* of the water at 0.1 MPa is 99.63°C. When the last drop of liquid has vaporized, the temperature again rises with additional heat added, and the specific volume also increases.

FIGURE 2.4
Phase change.

2.4 Thermodynamic Property Data

Thermodynamic property data occur in several forms, including tables, graphs, equations, and personal computer programs. This book encourages and emphasizes the use of computer programs.

There are tables dedicated to the properties of water; these are often called *steam tables*. There are tables for refrigerants and for gases. There are tables in Standards International (SI) units and in English units.

2.4.1 Pressure, Specific Volume, and Temperature

There are tables for properties of water vapor and for that of liquid water. They are also called superheated vapor tables and compressed liquid tables, respectively. The saturation tables list property values for the saturated liquid and vapor states.

Since the values tabulated are discrete, the states encountered in the problems are often not on the grid of values provided. The values of the properties then have to be interpolated between adjacent table entries. Linear interpolation is most often used. This is one good reason why the use of computerized programs is encouraged and emphasized in this book. The computerized programs do not require interpolation.

In the saturation tables, the saturated liquid and vapor states are marked by subscripts f and g, respectively. There is one set of tables where the temperature is the independent variable, that is, the temperatures are listed in the first column in convenient increments. The second column lists the corresponding saturation pressure. The other columns give the other properties. There is another set of tables where the pressure is the independent variable, that is, the pressures are listed in the first column at convenient increments. The second column lists the corresponding saturation temperature.

The specific volume of a mixture of liquid and vapor may be found as discussed below. The total volume of the mixture is the sum of the volumes of the liquid and vapor

$$V = V_{liq} + V_{vap}.$$

If the mass of the mixture is m, then an average specific volume is

$$v = \frac{V}{m} = \frac{V_{liq}}{m} + \frac{V_{vap}}{m}.$$

As the liquid is in the saturated liquid phase and the vapor is in the saturated vapor phase, $V_{liq} = m_{liq}v_f$ and $V_{vap} = m_{vap}v_g$, thus

$$v = \left(\frac{m_{liq}}{m}\right)v_f + \left(\frac{m_{vap}}{m}\right)v_g.$$

The quality of the mixture is defined as $x = m_{vap}/m$, so $m_{liq}/m = 1 - x$. The expression becomes

$$v = (1-x)v_f + xv_g = v_f + x(v_g - v_f) = v_g - (1-x)(v_g - v_f). \tag{2.1}$$

The quantity $(v_g - v_f)$ is often written as v_{fg}. Note that Equation (2.1) is a mass-averaged equation. In other words, the property of the mixture is a combination of the property of the saturated liquid and the corresponding property of the saturated vapor, weighted by their relative masses. The range of values for the quality is from zero to one. The property, quality, does not exist in the single phase regions, like the superheated vapor region or the compressed liquid region.

Example 2.1

Problem: What is the quality of water at 20°C with a specific volume of 50.0 m^3/kg?

Solution

From the computerized tables or tables in the Appendix,

$$v_g = 57.79 \text{ m}^3/\text{kg}$$
$$v_f = 1.002 \times 10^{-3} \text{ m}^3/\text{kg}.$$

The quality of the water is $x = \dfrac{v - v_f}{v_g - v_f} = 0.87.$

Example 2.2

Problem: Refrigerant-134a is in a closed vessel at 2 bars and is of 80% quality. Find the volume of the refrigerant, if its mass is 10 kg.

Solution

From the computerized tables,

$$v_g = 0.100 \text{ m}^3/\text{kg}$$
$$v_f = 0.755 \times 10^{-3} \text{ m}^3/\text{kg}.$$

The specific volume of the refrigerant mixture of liquid and gas is

$$v = (1 - x)\, v_f + x v_g = 0.0802 \text{ m}^3/\text{kg}.$$

Notice that the specific volume of the mixture lies between that of the saturated liquid v_f, and that of the saturated vapor, v_g.

Thus, the volume of 10 kg of the refrigerant is

$$10 \text{ kg} \times 0.0802 \text{ m}^3/\text{kg} = 0.802 \text{ m}^3.$$

Example 2.3

Problem: A closed system comprises a two-phase liquid-vapor mixture of ammonia in equilibrium at 20°C. The ratio of the mass of vapor to that of liquid ammonia is 1:4. Determine the quality of the ammonia and the total volume of the system, in cubic meters if the total mass is 10 kg.

Solution

The quality of the mixture,

$$x = \frac{\text{mass of vapor}}{\text{total mass}} = \frac{1}{1+4} = 0.2.$$

From the computerized tables,

$$v_g = 0.1492 \text{ m}^3/\text{kg}$$
$$v_f = 1.638 \times 10^{-3} \text{ m}^3/\text{kg}.$$

The specific volume of the ammonia mixture of liquid and gas is

$$v = (1 - x)v_f + xv_g = 0.03115 \text{ m}^3/\text{kg}.$$

Notice that the specific volume of the mixture lies between that of the saturated liquid v_f, and that of the saturated vapor, v_g.

The volume of the system = mass × specific volume = 0.3115 m³.

Example 2.4

Problem: The saturation temperature of water T_s corresponding to a pressure of 1 MPa is 179.91°C. The specific volumes of the saturated liquid and the saturated vapor, respectively, are $v_f = 0.0011273$ m³/kg and $v_g = 0.19444$ m³/kg. Deduce the state of the water for the following cases and explain:

(i) $P = 1$ MPa, $v = 0.2$ m³/kg
(ii) $P = 1$ MPa, $v = 0.0011$ m³/kg
(iii) $T = 179.91$°C, $P = 1.01$ MPa
(iv) $P = 1$ MPa, $T = 180$°C
(v) $P = 1$ MPa, $T = 170$°C
(vi) $P = 1$ MPa, $v = 0.1$ m³/kg.

Solution

(i) Since $v > v_g$, the state is superheated vapor.
(ii) Since $v < v_f$, the state is compressed liquid.
(iii) Since $P > P_s$, the state is compressed liquid.
(iv) Since $T > T_s$, the state is superheated vapor.
(v) Since $T < T_s$, the state is subcooled (compressed) liquid.
(vi) Since $v_f < v < v_g$, the state is a mixture of liquid and vapor.

$$\text{Quality } x = \frac{v - v_f}{v_g - v_f} = \frac{v - v_f}{v_{fg}} = 0.51.$$

2.4.2 Specific Internal Energy and Enthalpy

The energy of the molecules of a physical system is represented by the internal energy. This internal energy has a kinetic part which is due to the velocity of the molecules, and a potential part which is due to the attractive forces existing between the molecules.

As an example, consider a mug of water at 90°C and atmospheric pressure and an equal mass of water at 10°C and atmospheric pressure. We have the physical intuition that the water at 90°C has more energy than that at 10°C.

This form of energy that is manifested by the difference in temperature, in this case, is called internal energy.

Internal energy is represented by the symbol U, and the change in internal energy as ΔU. The specific internal energy is denoted as u on a unit mass basis, or as $\overline{u}$ on a unit mole basis.

Often, the sum of the internal energy U, and the product of pressure P, and volume V occurs in thermodynamics. It is given the name *enthalpy,* H. It is defined as

$$H = U + PV. \tag{2.2}$$

The specific enthalpy is expressed as

$$h = \frac{H}{m} = u + Pv. \tag{2.3}$$

In the thermodynamic property tables presented in the Appendix, the specific internal energy can always be calculated from the enthalpy using $u = h - Pv$. The specific internal energy and the specific enthalpy are extensive properties of a substance, like the specific volume. Hence, in the two-phase region, these extensive properties are calculated like the specific volume. The specific internal energy of a two-phase liquid-vapor mixture is calculated in terms of the quality by

$$u = (1-x)u_f + xu_g = u_f + x(u_g - u_f) = u_g - (1-x)(u_g - u_f). \tag{2.4}$$

The specific enthalpy of a two-phase liquid-vapor mixture is calculated in terms of the quality by

$$h = (1-x)h_f + xh_g = h_f + x(h_g - h_f) = h_g - (1-x)(h_g - h_f). \tag{2.5}$$

Note that Equations (2.4) and (2.5), like Equation (2.1), are mass-averaged equations. In fact, the property of any liquid-vapor mixture is obtained from its corresponding saturated liquid and saturated vapor properties using this mass-averaging formula. In addition, the property of any solid-vapor mixture is obtained from its corresponding saturated solid and saturated vapor properties using this mass-averaging formula. Finally, the property of any solid-liquid mixture is obtained from its corresponding saturated solid and saturated liquid properties using this mass-averaging formula.

Example 2.5

Problem: Ammonia at 0°C (273.15 K) has a quality of 0.15. Determine its specific internal energy.

Solution

From the tables in the Appendix,

$$h_g = 1442.2 \text{ kJ/kg}$$
$$h_f = 180.36 \text{ kJ.kg}$$
$$u_g = h_g - Pv_g = 1318 \text{ kJ/kg}$$
$$u_f = h_f - Pv_f = 179.7 \text{ kJ/kg.}$$

The specific internal energy of the mixture of liquid and gas is

$$u = (1 - x)u_f + xu_g = 350.4 \text{ kJ/kg.}$$

From the computerized tables, u is also given as 350.4 kJ/kg for ammonia at 0°C and quality of 0.15. Notice that the value of the specific internal energy of the mixture lies between that of the saturated liquid u_f, and that of the saturated vapor, u_g.

Example 2.6

Problem: Water at 1.0 bar has a quality of 90%. Determine its specific enthalpy.

Solution

From the tables,

$$h_g = 2675.5 \text{ kJ/kg}$$
$$h_f = 417.44 \text{ kJ/kg.}$$

The specific enthalpy of the mixture of liquid and gas is

$$h = (1 - x)h_f + xh_g = 2450 \text{ kJ/kg.}$$

Notice that the value of the specific enthalpy of the mixture lies between that of the saturated liquid h_f, and that of the saturated vapor, h_g.

Example 2.7

Problem: The saturation temperature of ammonia T_s corresponding to a pressure of 14.696 $lb_f/in.^2$ is –28°F. The specific internal energies of the saturated liquid and the saturated vapor, respectively, are $u_f = 12.59$ Btu/lb_m, $u_g = 552.52$ Btu/lb_m. Deduce the state of the ammonia for the following cases and explain:

(i) $P = 14.696\ lb_f/in.^2$, $u = 600\ Btu/lb_m$
(ii) $P = 14.696\ lb_f/in.^2$, $u = 10\ Btu/lb_m$
(iii) $T = -28°F$, $P = 15\ lb_f/in.^2$

(iv) $P = 14.696\ lb_f/in.^2$, $T = -10°F$
(v) $P = 14.696\ lb_f/in.^2$, $T = -30°F$
(vi) $P = 14.696\ lb_f/in.^2$, $u = 100\ Btu/lb_m$.

Solution

(i) Since $u > u_g$, the state is superheated vapor.
(ii) Since $u < u_f$, the state is compressed liquid.
(iii) Since $P > P_s$, the state is compressed liquid.
(iv) Since $T > T_s$, the state is superheated vapor.
(v) Since $T < T_s$, the state is subcooled (compressed) liquid.
(vi) Since $u_f < u < u_g$, the state is a mixture of liquid and vapor.

$$\text{Quality } x = \frac{u - u_f}{u_g - u_f} = \frac{u - u_f}{u_{fg}} = 0.16.$$

Example 2.8

Problem: The saturation temperature of water T_s corresponding to a pressure of 0.101 MPa is 99.9°C. The specific enthalpies of the saturated liquid and the saturated vapor, respectively, are $h_f = 418.6$ kJ/kg, $h_g = 2676$ kJ/kg. Deduce the state of the water for the following cases and explain:

(i) $P = 0.101$ MPa, $h = 2700$ kJ/kg
(ii) $P = 0.101$ MPa, $h = 400$ kJ/kg
(iii) $T = 99.9°C$, $P = 0.15$ MPa
(iv) $P = 0.101$ MPa, $T = 101°C$
(v) $P = 0.101$ MPa, $T = 90°C$
(vi) $P = 0.101$ MPa, $h = 1000$ kJ/kg.

Solution

(i) Since $h > h_g$, the state is superheated vapor.
(ii) Since $h < h_f$, the state is compressed liquid.
(iii) Since $P > P_s$, the state is compressed liquid.
(iv) Since $T > T_s$, the state is superheated vapor.
(v) Since $T < T_s$, the state is subcooled (compressed) liquid.
(vi) Since $h_f < h < h_g$, the state is a mixture of liquid and vapor.

$$\text{Quality } x = \frac{h - h_f}{h_g - h_f} = \frac{h - h_f}{h_{fg}} = 0.258.$$

2.4.3 Reference Values

The values of u, h, and entropy s listed in the tables are not determined by direct experimentation but are calculated from other data that can be more easily found experimentally. The discussion of this computation requires the second law of thermodynamics, and is not discussed in this introductory book on thermodynamics. When analyzing a system using the conservation of energy, it is the differences in internal, kinetic, and potential energy between two states that are significant, and not the values of these energy quantities at these two states.

Consider the situation of a kid playing with a ball outside his house which is located on a hill. At one instant, the ball is on the ground and at another instant, it is on the roof of the house. The numerical value of the potential energy of the ball measured from the surface of the earth is different from the value relative to the sea level. However, the difference in potential energy between the two elevations of the ball is the same regardless of the datum chosen, because the datum cancels in the calculation.

In like manner, values can be given to specific internal energy and enthalpy relative to arbitrary reference values at arbitrary reference states. There is no problem as long as the calculations performed involve only differences in properties, since the reference value cancels. When chemical reactions occur among the substances under study, care must be taken regarding the reference states and values.

A discussion of how property values are assigned when analyzing reactive systems is not in the scope of this book.

2.5 Specific Heats and Their Relationships

The specific heat is the energy needed to increase the temperature of a unit mass of a substance by one degree. In SI, it is one degree Celsius. In the English system, it is one degree Fahrenheit. The physical interpretation of the specific heat at constant volume C_v is the energy needed to increase the temperature of a unit mass of a substance by one degree as the volume is kept constant. Similarly, the specific heat and constant pressure C_p is the energy needed to increase the temperature of a unit mass of a substance by one degree as the pressure is kept constant.

For pure, simple compressible substances, the intensive properties C_v and C_p are defined as

$$C_v = \left.\frac{\partial u}{\partial T}\right)_v \tag{2.6}$$

$$C_P = \left.\frac{\partial h}{\partial T}\right)_P \quad (2.7)$$

where u = u(T,v), h = h(T,P), and the differentiation is held at constant volume for C_v and at constant pressure for C_p. Both properties can be obtained by direct measurements, or by measurements via statistical mechanics. Take, for example, the heating of water. If water is heated in a rigid container, the specific heat at constant volume applies since the volume of the rigid container is constant. If water is heated in a container capped by a piston, the specific heat at constant pressure applies since the pressure exerted by the piston on the water is constant.

The change in internal energy and enthalpy may be calculated as

$$\Delta u = u_2 - u_1 = \int_1^2 C_v(T)dT \approx C_{v,av}(T_2 - T_1) \quad (2.8)$$

$$\Delta h = h_2 - h_1 = \int_1^2 C_P(T)dT \approx C_{P,av}(T_2 - T_1). \quad (2.9)$$

For incompressible substances, i.e., solids and most liquids,

$$C_v = C_p = C, \text{ a constant.} \quad (2.10)$$

The change in internal energy and enthalpy, in this case, may be calculated as

$$\Delta u = \int_1^2 C(T)dT \approx C_{av}(T_2 - T_1) \quad (2.11)$$

$$\Delta h = \Delta u + v\Delta P. \quad (2.12)$$

For an ideal gas, the specific internal energy depends only on the temperature. Hence, the specific heat C_v as defined in Equation (2.6), is a function only of the temperature.

$$C_v = \frac{du}{dT} \quad \text{(ideal gas)} \quad (2.13)$$

For an ideal gas, the specific enthalpy depends only on the temperature. Hence, the specific heat C_p as defined in Equation (2.7), is a function only of the temperature.

$$C_p = \frac{dh}{dT} \quad \text{(ideal gas)} \quad (2.14)$$

For ideal gases (see Section 2.7)

$$Pv = RT.$$

Since by definition,

$$h = u + Pv,$$

then

$$h = u + RT \qquad \text{(ideal gas)} \qquad (2.15)$$

$$dh = du + RdT \qquad \text{(ideal gas).} \qquad (2.16)$$

Substituting $C_p dT$ for dh and $C_v dT$ for du,

$$C_p = C_v + R \qquad \text{(ideal gas).} \qquad (2.17)$$

When the specific heats are given on a molar basis, the universal gas constant R_u should be used.

$$\overline{C_p} = \overline{C_v} + R_u \qquad \text{(ideal gas)} \qquad (2.18)$$

The specific heat ratio k is defined as

$$k = \frac{C_p}{C_v} \qquad \text{(ideal gas).} \qquad (2.19)$$

The specific heat ratio is dependent on temperature, though this dependence may be small. Air has a value of 1.4 at room temperature. Combining Equations (2.18) and (2.19), it may be shown that

$$C_p = \frac{kR}{k-1} \qquad \text{(ideal gas)} \qquad (2.20)$$

$$C_v = \frac{R}{k-1} \qquad \text{(ideal gas).} \qquad (2.21)$$

2.6 Processes

A *process* is a change in state that can be stated as any change in the properties of a system (see Chapter 1, Section 1.2). There are names for special processes.

An *isobaric* process is one at constant pressure. An *isothermal* process is one at constant temperature. A constant volume process is *isometric* or *isochoric*. An *isentropic* process is one at constant entropy. A *polytropic* process is one that follows the equation PV^n = constant. An *adiabatic* process is one without heat transfer.

In Figure 2.5 a P-v diagram is shown for various processes undergone by an ideal gas. An isobaric process appears as a horizontal line, while an isometric process is a vertical line. The isothermal process is a hyperbolic curve. The isentropic process is shown as a curve with a steeper gradient than the isothermal-process curve at any given volume.

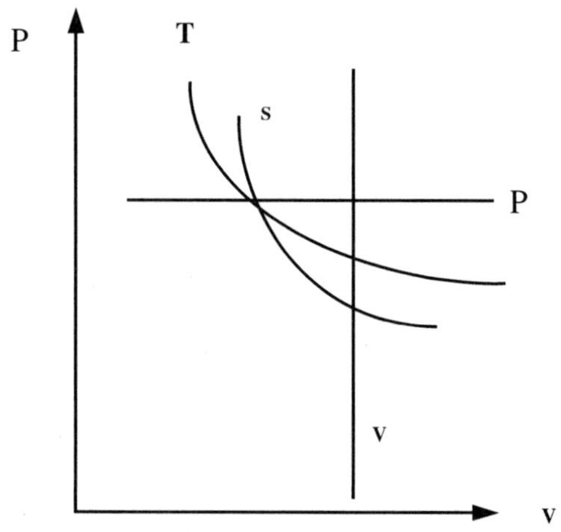

FIGURE 2.5
P-v diagram indicating various processes.

In Figure 2.6 a P-T diagram is shown for various processes undergone by an ideal gas. An isobaric process appears as a horizontal line, while an isothermal process is a vertical line. The isometric process for an ideal gas is a straight line with a positive gradient. The isentropic process is shown as a curve.

In Figure 2.7 a T-s diagram is shown for various processes undergone by an ideal gas. An isothermal process appears as a horizontal line, while an isentropic process is a vertical line. The isobaric process is shown as a curve. The isometric process is also shown as a curve with a steeper gradient than the isobaric-process curve at any given entropy.

2.6.1 Polytropic Processes

The path of polytropic processes is described by the equation

$$Pv^n = \text{constant}. \tag{2.22}$$

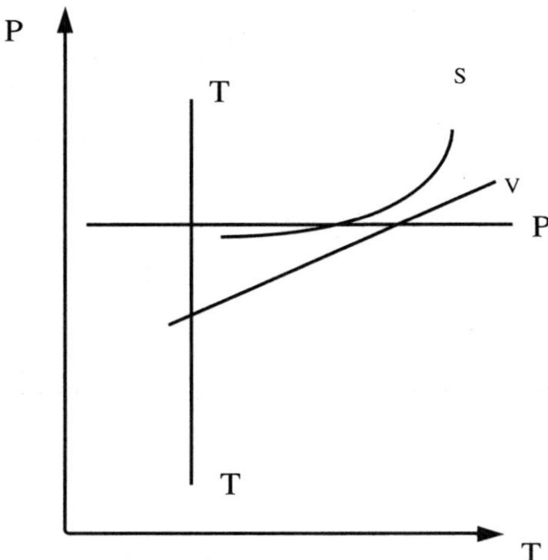

FIGURE 2.6
P-T diagram indicating various processes.

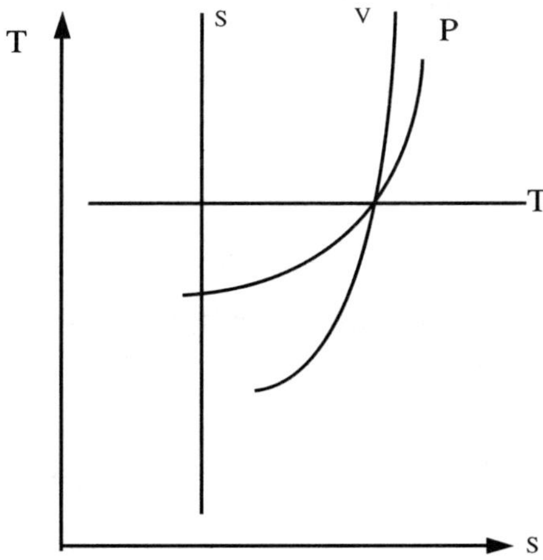

FIGURE 2.7
T-s diagram indicating various processes.

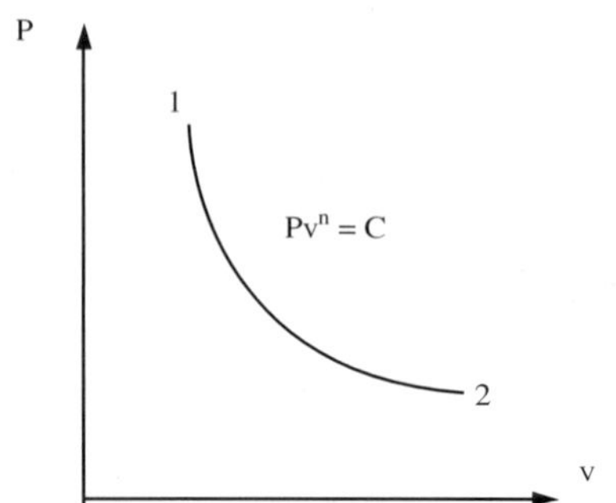

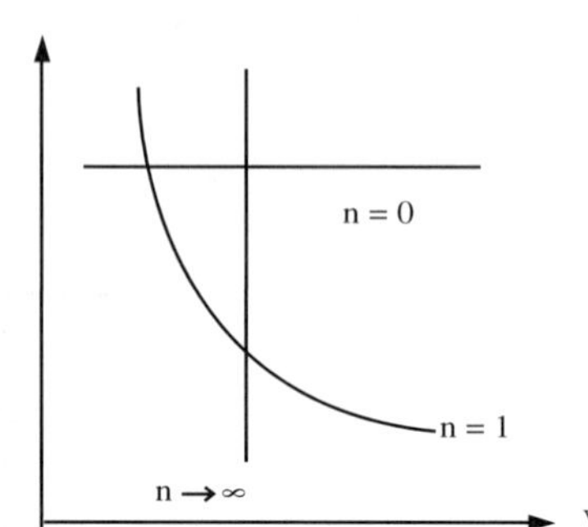

FIGURE 2.8
P-v diagram.

When $n = 0$, $P =$ constant for an isobaric process.

When $n \to \infty, \dfrac{dP}{dv} = -n\dfrac{P}{v} \Rightarrow dv \to 0$ for an isometric or isochoric process.

When $n = 1$, $Pv =$ constant for an isothermal process involving an ideal gas.

When $n = k$, $Pv^k =$ constant for an isentropic process involving an ideal gas, where k is the ratio of specific heats.

For a polytropic process between two states,

$$P_1 v_1^n = P_2 v_2^n \tag{2.23}$$

$$\frac{P_2}{P_1} = \left(\frac{v_1}{v_2}\right)^n. \tag{2.24}$$

As indicated above, n may take any value from $-\infty$ to $+\infty$.

For a polytropic process

$$\int_1^2 P dv = \frac{P_2 v_2 - P_1 v_1}{1 - n} \qquad (n \neq 1) \tag{2.25}$$

for any value of n except $n = 1$. When $n = 1$,

$$\int_1^2 P dv = P_1 v_1 \ln \frac{v_2}{v_1} \qquad (n = 1). \tag{2.26}$$

Equations (2.23) through (2.26) are applicable to any fluid undergoing a polytropic process. If an ideal gas assumption is used in addition, more relations apply. When the ideal gas equation (Pv = RT, Section 2.7) is added into Equations (2.24) through (2.26), the following relations are obtained, respectively:

$$\frac{T_2}{T_1} = \left(\frac{P_2}{P_1}\right)^{(n-1)/n} = \left(\frac{v_1}{v_2}\right)^{n-1} \quad \text{(ideal gas)}, \tag{2.27}$$

$$\int_1^2 Pdv = \frac{mR(T_2 - T_1)}{1-n} \quad \text{(ideal gas, } n \neq 1\text{)}, \tag{2.28}$$

$$\int_1^2 Pdv = mRT \ln \frac{v_2}{v_1} \quad \text{(ideal gas, } n = 1\text{)}. \tag{2.29}$$

For an ideal gas, an isothermal process corresponds to the case of n = 1, and an isentropic process corresponds to one where n = k, the ratio of specific heats.

Example 2.9

Problem: A gas in a piston-cylinder equipment undergoes a polytropic process such that PV^n = constant. The initial pressure is 10 bars, the initial volume is 0.01 m^3, and the final volume is 0.03 m^3. Calculate the work for the process, in kilojoules, if (a) n = 2.0, (b) n = 0.0, and (c) n = 1.0.

Solution

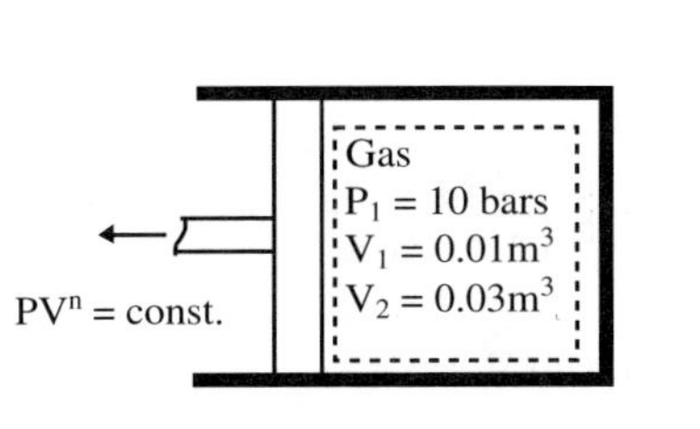

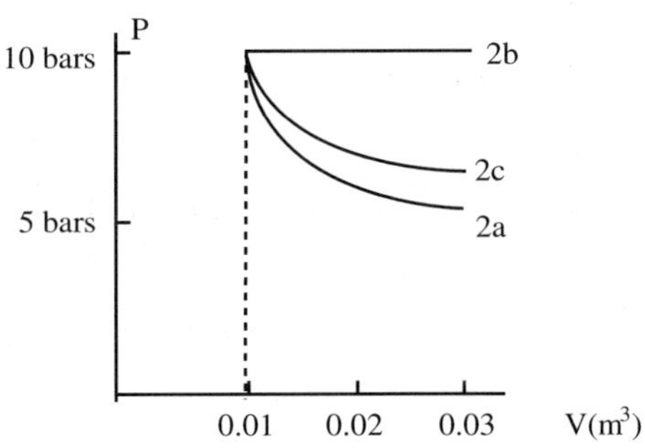

Schematics for Example 2.9.

$$\text{Work} = \int_{V_1}^{V_2} PdV = \int_{V_1}^{V_2} \frac{\text{constant}}{V^n} dV = \text{constant}\left(\frac{V_2^{1-n} - V_1^{1-n}}{1-n}\right).$$

The constant in the relation can be determined at either state,

$$\text{constant} = P_1V_1^n = P_2V_2^n.$$

The work is then

$$W = \frac{(P_2V_2^n)V_2^{1-n} - (P_1V_1^n)V_1^{1-n}}{1-n} = \frac{P_2V_2 - P_1V_1}{1-n}.$$

This relation is valid for all n except n = 1.0, which is considered in Part (c).

(a) To calculate W, the pressure at state 2 is required.

$$P_2 = P_1\left(\frac{V_1}{V_2}\right)^n = (10 \text{ bars})\left(\frac{0.01}{0.03}\right)^2 = 1.11 \text{ bars}.$$

Thus,

$$W = \left(\frac{(1.11 \text{ bars})(0.03 \text{ m}^3) - (10)(0.01)}{1-2}\right)\left(\frac{10^5 \text{N/m}^2}{1 \text{ bar}}\right)\left(\frac{1 \text{ kJ}}{10^3 \text{ N.m}}\right)$$
$$= +6.67 \text{ kJ}.$$

(b) For n = 0, P = constant,

$$\text{Work} = \int_{V_1}^{V_2} PdV = P(V_2 - V_1)$$

$$W = (10 \text{ bars})(0.03 - 0.01)\left(\frac{10^5 \text{N/m}^2}{1 \text{ bar}}\right)\left(\frac{1 \text{ kJ}}{10^3 \text{N.m}}\right) = +20 \text{ kJ}.$$

(c) For n = 1.0, PV = constant, and P = constant/V,

$$\text{Work} = \int_{V_1}^{V_2} \frac{\text{constant}}{V} dV = \text{constant}\left(\ln\frac{V_2}{V_1}\right) = (P_1V_1)\ln\frac{V_2}{V_1}$$

$$W = (10 \text{ bars})(0.01 \text{ m}^3)\left(\frac{10^5 \text{N/m}^3}{1 \text{ bar}}\right)\left(\frac{1 \text{ kJ}}{10^3 \text{ N.m}}\right)\ln\left(\frac{0.03}{0.01}\right)$$
$$= +10.99 \text{ kJ}.$$

Notice that the area under the curves in the P-V diagram represents the work for the process. The relative areas for the three cases are in agreement with the computed results.

2.6.2 Reversible Processes

A process is *reversible* if its direction can be reversed at any stage by an infinitesimal change in external conditions. All naturally occurring processes are irreversible, for example, a clock running down and not rewinding itself. All practical processes can be made to approach closely a reversible process by a suitable choice of conditions.

The purpose of the concept of a reversible process is to establish a standard for the comparison of actual processes. The reversible process provides the maximum accomplishment. For instance, the reversible process yields the greatest amount of work or requires the least amount of work to bring about a given change. Reversible heat transfer is viewed as heat transfer that takes place with an infinitesimal temperature difference; all real heat transfer processes take place with finite temperature differences.

Since the reversible process represents a succession of equilibrium states, each only a differential step from its neighbor, it can be represented as a continuous line or curve on a phase diagram. The irreversible process cannot be so represented. We can note the terminal states and indicate the general direction of change, but the complete path of the irreversible process is indeterminate. Irreversibilities always decrease the efficiency of processes.

The reversible process represents a standard of perfection that cannot be exceeded. An irreversible process is one in which dissipative effects occur or one which is not executed quasistatically. A quasistatic process is an idealized process in which the departure from equilibrium is negligible. Examples of factors that make processes irreversible are as follows:

1. Friction
2. Free expansion (unrestrained expansion)
3. Inelastic deformation
4. Heat transferred across a finite temperature difference
5. Mixing of substances
6. All chemical reactions
7. Sudden change of phase
8. Losses in electrical resistors
9. Hysteresis effects

When a real process can be represented by at least a number of equilibrium states that can be plotted on a phase diagram, the process is called quasistatic. Under these circumstances, the mechanical work performed may be approximated rather accurately by the integral of PdV. If the process cannot be represented by many equilibrium states, then the integral of PdV is a poor approximation of the actual work done. For instance, the free expansion of the helium in a balloon after the balloon is popped cannot be represented by a series of many equilibrium states. The work in this case should not be calculated by using the integral of PdV.

2.7 The Ideal-Gas Equation of State

In 1662, Robert Boyle, an English scientist, observed during his experiments with a vacuum chamber that the pressure of gases is inversely proportional to their volume. J. Charles and J. Gay-Lussac, French scientists, in 1802 experimentally determined that at low pressures the volume of a gas is proportional to its temperature. In other words,

$$P = R\left(\frac{T}{v}\right), \quad \text{or}$$

$$Pv = RT \tag{2.30}$$

where the constant of proportionality R is the gas constant. Equation (2.30) is the ideal-gas equation of state or the ideal-gas relation, and a gas that behaves like the relation is called an ideal gas.

The gas constant R is different for each gas and is defined as

$$R = \frac{R_u}{M} \tag{2.31}$$

where

R_u = universal gas constant

M = molar mass or molecular weight of the gas.

R_u is the same for all gases, and its value is

$$R_u = \begin{cases} 8.314 \text{ kJ/(kmol.K)} \\ 8.314 \text{ (kPa.m}^3\text{)/(kmol.K)} \\ 0.08314 \text{ (bar.m}^3\text{)/(kmol.K)} \\ 1.986 \text{ Btu/(lbmol.}^\circ\text{R)} \\ 10.73 \text{ (psia.ft}^3\text{)/(lbmol.}^\circ\text{R)} \\ 1545 \text{ (ft.lb}_f\text{)/(lbmol.}^\circ\text{R).} \end{cases}$$

The ideal-gas equation is used when the relation between P, v, and T is simple and may be represented by one equation. The P-v-T surface shown in Figure 2.1 is obviously too complicated to be represented by one simple equation. So near the saturation region, thermodynamic property tables have to be used for any simple substance. Thermodynamic property tables for common gases and gaseous mixtures, like nitrogen and air, are easily available and may be used instead of the ideal gas relation. For less common gases that exhibit ideal behavior, the ideal-gas P-v-T equation is really useful for determining their properties.

At low pressures and high temperatures (relative to their critical points), real gases behave like an ideal gas. Common gases such as air, nitrogen, oxygen, hydrogen, the rare gases, and carbon dioxide may be treated as ideal gases for most practical temperature and pressure ranges of interest with little error. Water vapor and refrigerant vapors, as a rule of thumb, should not be treated as ideal gases. Property tables should be used for them.

2.8 Compressibility Factor

Gases deviate from ideal-gas behavior significantly at states near the saturation region and the critical point. This deviation can be accurately accounted for by the use of a correction factor called the compressibility factor Z. It is defined as

$$Z = \frac{Pv}{RT} \tag{2.32}$$

or

$$Pv = ZRT. \tag{2.33}$$

It can also be calculated as

$$Z = \frac{v_{actual}}{v_{ideal}} \tag{2.34}$$

where $v_{ideal} = RT/P$. For ideal gases, $Z = 1$. For real gases, Z can take on any positive value.

Different gases behave differently at a given temperature and pressure, but they behave very similarly at temperatures and pressures normalized with respect to their critical temperatures and pressures. The normalization is performed as

$$\text{the reduced pressure} \qquad P_R = \frac{P}{P_{cr}}$$
$$\text{and the reduced temperature} \qquad T_R = \frac{T}{T_{cr}}. \tag{2.35}$$

The Z factor for all gases is approximately the same at the same reduced pressure and temperature. This is called the principle of corresponding states.

The compressibility factors for reduced temperatures and pressures, Figures A–1 to A–3 in the Appendices, were developed by Nelson and Obert (1954). Nelson and Obert's plots are called the generalized compressibility chart. The horizontal and vertical coordinates of the chart are the reduced pressure and the compressibility factor, respectively. The chart may be used to find the properties of a gas; note that the critical pressure, temperature, and specific volume are unique at the critical point for each gas.

When P and v, or T and v, are given instead of P and T, the generalized compressibility chart can still be used to determine the third property, but it would take trial and error. Hence, we define another reduced property called the pseudoreduced specific volume v_R as

$$v_R = \frac{v_{actual}}{RT_{cr}/P_{cr}}. \tag{2.36}$$

Note that v_R is related to T_{cr} and P_{cr} instead of v_{cr}.

Example 2.10

Problem: Determine the specific volume of water vapor at 1 MPa and 250°C, using (a) the ideal-gas equation of state, and (b) the generalized compressibility chart. Evaluate the values obtained by comparing them to the experimental volume of 0.2327 m³/kg.

Solution

From the table on critical constants, the relevant properties of water are obtained:

$$R = 0.4615 \text{ kPa.m}^3/(\text{kg.K})$$
$$P_{cr} = 22.09 \text{ MPa}$$
$$T_{cr} = 647.3 \text{ K}.$$

(a) From the ideal-gas equation of state, the specific volume is

$$v = \frac{RT}{P} = \frac{\left[0.4615 \text{ kPa.m}^3/(\text{kg.K})\right](523.15 \text{ K})}{1000 \text{ kPa}} = 0.241 \text{ m}^3/\text{kg}.$$

This involves an error of (0.241 – 0.2327)/0.2327 = 0.036 or 3.6%.

(b) To use the generalized compressibility chart, we need to calculate the reduced temperature and pressure.

$$T_R = \frac{T}{T_{cr}} = 0.808$$

$$P_R = \frac{P}{P_{cr}} = 0.0453.$$

From the chart, the correction factor Z = 0.97.

Thus,

$$v = Zv_{ideal} = (0.97)(0.241 \text{ m}^3/\text{kg}) = 0.234 \text{ m}^3/\text{kg}.$$

This involves an error of (0.234 – 0.2327)/0.2327 = 0.0056 or 0.56%.

Example 2.11

Problem: Determine the pressure of refrigerant-134a at 300°F and 0.7672 ft^3/lb_m, using (a) the computerized tables, (b) the ideal-gas equation of state, and (c) the generalized compressibility chart.

Solution

From the table on critical constants in the Appendix, the relevant properties for refrigerant-134a are obtained:

$$R = 0.1052 \text{ psia.ft}^3/(\text{lb}_m.°\text{R})$$

$$P_{cr} = 588.9 \text{ psia}$$

$$T_{cr} = 673.6°\text{R}.$$

(a) From the computerized tables, P = 100 psia.

(b) Using the ideal-gas equation,

$$P = \frac{RT}{v} = \frac{\left[0.1052 \text{ psia.ft}^3/(\text{lb}_m.°\text{R})\right](759.67°\text{R})}{0.7672 \text{ ft}^3/\text{lb}_m} = 104.2 \text{ psia}.$$

This results in an error of (104.2 – 100)/100 = 0.042 or 4.2%.

(c) To find the correction factor Z from the compressibility chart, we need to calculate the reduced temperature and the pseudoreduced specific volume.

$$T_R = \frac{T}{T_{cr}} = 1.13$$

$$v_R = \frac{v_{actual}}{RT_{cr}/P_{cr}} = \frac{(0.7672\ ft^3/lb_m)(588.9\ psia)}{[0.1052\ psia.ft^3/(lb_m.°R)](673.6°R)} = 6.38$$

$$P_R = 0.175.$$

Thus,

$$P = P_R P_{cr} = 103.1\ psia.$$

This results in an error of $(103.1 - 100)/100 = 0.031$ or 3.1%.

2.9 Other Equations of State

The P-v-T behavior of substances can be represented more accurately by the more complex equations of state. Two of the more well-known ones are the van der Waals equation and the Beattie-Bridgeman equation.

The van der Waals equation is

$$\left(P + \frac{a}{v^2}\right)(v - b) = RT \tag{2.37}$$

where

$$a = \frac{27R^2T_{cr}^2}{64P_{cr}} \quad \text{and} \quad b = \frac{RT_{cr}}{8P_{cr}}.$$

In the van der Waals equation, the term a/v^2 takes care of the intermolecular attractive forces and the term b takes care of the volume occupied by the gas molecules. At ordinary pressures around 1 bar and ordinary room temperatures, the volume occupied by the gas molecules is negligible. However, this volume becomes more significant when pressure is increased. Hence, van der Waals used a correction on the volume with the b term. The two constants are found experimentally.

The Beattie-Bridgeman equation is

$$P = \frac{R_u T}{\bar{v}^2}\left(1 - \frac{c}{\bar{v}\,T^3}\right)(\bar{v} + B) - \frac{A}{\bar{v}^2} \tag{2.38}$$

where

$$A = A_0\left(1 - \frac{a}{\bar{v}}\right) \quad \text{and} \quad B = B_0\left(1 - \frac{b}{\bar{v}}\right).$$

The five constants in the above equation, a, b, c, A, and B, are found by curve-fitting to experimental data. It can be used without much error when ρ is less than 0.75 ρ_{crit}, where ρ_{crit} is the density at critical point.

There are many multiconstant equations of state. Employing computers, equations with many constants have been used.

Example 2.12

Problem: Calculate the pressure of carbon dioxide gas at T = 300 K and v = 0.0040 m^3/kg. Use (a) the ideal-gas equation of state, (b) the van der Waals equation of state, and (c) the Beattie-Bridgeman equation of state. Compare the values obtained to the experimentally determined value of 6600 kPa.

Solution

(a) Use the ideal-gas equation of state.

$$P = \frac{RT}{v} = \frac{[0.1889 \text{ kPa.m}^3/(\text{kg.K})](300 \text{ K})}{0.0040 \text{ m}^3/\text{kg}} = 14{,}170 \text{ kPa}.$$

(b) Use the van der Waals equation of state.
From the table on critical properties in the Appendices,

$$a = \frac{27R^2T_{cr}^2}{64P_{cr}} = 3.654 \text{ bar}\left(\frac{\text{m}^3}{\text{kmol}}\right)^2$$

$$b = \frac{RT_{cr}}{8P_{cr}} = 0.0428 \frac{\text{m}^3}{\text{kmol}}$$

then

$$P = \frac{RT}{v - b} - \frac{a}{v^2} = 6950 \text{ kPa}.$$

(c) Use the Beattie-Bridgeman equation.
From the tables,

$$A = 301.7$$
$$B = 0.0617$$
$$c = 6.60 \times 10^5.$$

Since $\bar{v} = (44\ kg/kmol)(0.0040\ m^3/kg) = 0.176\ m^3/kmol$, substituting these values into the equation

$$P = \frac{R_u T}{\bar{v}^2}\left(1 - \frac{c}{\bar{v}T^3}\right)(\bar{v} + B) - \frac{A}{\bar{v}^2} = 6741\ kPa.$$

The Beattie-Bridgeman equation gives the closest answer to the experimentally determined value.

Problems

2.1 Explain the meanings of the following terms: pure substance, critical point, triple point, saturated vapor, saturation pressure, subcooled liquid, superheated vapor, quality.

Thermodynamic Property Data

2.2 Determine the phase or phases of water at the following conditions.

(a) $P = 4$ bars, $v = 0.32\ m^3/kg$
(b) $T = 30°C$ (303.15 K), $v = 31.0\ m^3/kg$
(c) $P = 1.5$ bars, $T = 200°C$ (473.15 K)
(d) $P = 25$ bars, $T = 200°C$ (473.15 K)
(e) $T = 10°C$ (283.15 K), $P = 0.1$ MPa.

2.3 Determine the phase or phases of water at the following conditions.

(a) $P = 60\ lb_f/in.^2$, $v = 6.1\ ft^3/lb_m$
(b) $T = 80°F$ (439.67°R), $v = 600\ ft^3/lb_m$
(c) $P = 60\ lb_f/in.^2$, $T = 300°F$ (759.67°R)
(d) $P = 500\ lb_f/in.^2$, $T = 300°F$ (759.67°R)
(e) $T = 330°F$ (789.67°R), $v = 30\ ft^3/lb_m$.

2.4 The saturation temperature of ammonia T_s corresponding to a pressure of 0.5 bar is –46.53°C (319.68 K). The specific volumes of the saturated liquid and the saturated vapor, respectively, are $v_f = 0.001433\ m^3/kg$, $v_g = 2.1752\ m^3/kg$. Deduce the state of the ammonia for the following cases and explain:

(i) $P = 0.5$ bar, $v = 2.3\ m^3/kg$
(ii) $P = 0.5$ bar, $v = 0.001\ m^3/kg$
(iii) $T = -46.53°C$ (319.68 K), $P = 0.6$ bar

(iv) $P = 0.5$ bar, $T = -40°C$ (233.15 K)
(v) $P = 0.5$ bar, $T = -50°C$ (223.15 K)
(vi) $P = 0.5$ bar, $v = 1\ m^3/kg$

2.5 The saturation temperature of refrigerant-12, T_s, corresponding to a pressure of 0.4 MPa is 8.15°C (281.3 K). The specific internal energies of the saturated liquid and the saturated vapor, respectively, are $u_f = 43.35$ kJ/kg, $u_g = 173.69$ kJ/kg. Deduce the state of the refrigerant for the following cases and explain:

(i) $P = 0.4$ MPa, $u = 175$ kJ/kg
(ii) $P = 0.4$ MPa, $u = 40$ kJ/kg
(iii) $P = 0.4$ MPa, $T = 10°C$ (283.15 K)
(iv) $T = 8.15°C$ (281.3 K), $P = 0.42$ MPa
(v) $P = 0.4$ MPa, $T = -4°C$ (269.15 K)
(vi) $P = 0.4$ MPa, $u = 100$ kJ/kg

2.6 The saturation temperature of refrigerant-134a, T_s, corresponding to a pressure of 30 $lb_f/in.^2$ is 15.38°F (475.05°R). The specific enthalpies of the saturated liquid and the saturated vapor, respectively, are $h_f = 16.31$ Btu/lb_m, $h_g = 103.96$ Btu/lb_m. Deduce the state of the refrigerant for the following cases and explain:

(i) $P = 30\ lb_f/in.^2$, $h = 105$ Btu/lb_m
(ii) $P = 30\ lb_f/in.^2$, $h = 12$ Btu/lb_m
(iii) $T = 15.38°F$ (475.05°R), $P = 35\ lb_f/in.^2$
(iv) $P = 30\ lb_f/in.^2$, $T = 20°F$ (479.67°R)
(v) $P = 30\ lb_f/in.^2$, $T = 10°F$ (469.67°R)
(vi) $P = 30\ lb_f/in.^2$, $h = 50$ Btu/lb_m

2.7 Calculate the volume, in cubic meters, of 2 kg of ammonia at 0.4 bar, –5°C (268.15 K).

2.8 Calculate the volume, in cubic feet, of 2 lb_m of ammonia at 20 $lb_f/in.^2$, 10°F (469.67°R).

2.9 Find the quality of a two-phase liquid-vapor mixture of

(a) Water at 20°C (293.15 K) with a specific volume of 10 m^3/kg.
(b) Refrigerant-12 at 30°C (303.15 K) with a specific volume of 0.01 m^3/kg.
(c) Refrigerant-134a at 1 bar with a specific volume of 0.1 m^3/kg.
(d) Ammonia at 3.5 bars with a specific volume of 0.2 m^3/kg.
(e) Water at 2 bars with a specific volume of 0.6 m^3/kg.

2.10 Find the quality of a two-phase liquid-vapor mixture of

(a) Water at 90°F (549.67°R) with a specific volume of 400 ft^3/lb_m.
(b) Refrigerant-12 at 100°F (559.67°R) with a specific volume of 0.25 ft^3/lb_m.
(c) Refrigerant-134a at 200°F (659.67°R) with a specific volume of 0.05 ft^3/lb_m.
(d) Ammonia at 100 $lb_f/in.^2$ with a specific volume of 2.0 ft^3/lb_m.
(e) Water at 1000 $lb_f/in.^2$ with a specific volume of 0.4 ft^3/lb_m.

2.11 A substance, called "lambda" behaves somewhat like water. When the pressure is n kPa for substance lambda, the saturation temperature is m K. The specific volumes of the saturated liquid and the saturated vapor, respectively, are $v_f = \alpha m^3/kg$, $v_g = \beta m^3/kg$, where $(\beta - \alpha) = 0.001$ m^3/kg. Stipulate the state of the substance lambda for the following cases and explain:

(i) T = m K, P = (n + 2) kPa
(ii) P = n kPa, v = (α – 0.00001) m^3/kg
(iii) P = n kPa, v = (β + 0.00002) m^3/kg
(iv) P = n kPa, v = (α + 0.0012) m^3/kg
(v) P = n kPa, T = (m – 2) K
(vi) P = n kPa, T = (m + 4) K

2.12 Water is in a closed vessel at 100°C (373.15 K) and 25% quality. What is the volume of the water if its mass is 10 kg?

2.13 Refrigerant-134a is in a closed vessel at 4 bars and 50% quality. Compute the volume of the refrigerant, if its mass is 3 kg.

2.14 Refrigerant-12 is in a closed vessel at 34°C (307.15 K) with a quality of 60%. What is its specific volume?

2.15 Ammonia is in a closed vessel at 3 bars and 20% quality. If the mass of the ammonia is 4 kg, determine the volume of the ammonia.

2.16 Water is in a closed vessel at 212°F (671.67°R) and 25% quality. What is the volume of the water if its mass is 10 lb?

2.17 Refrigerant-134a is in a closed vessel at 80 $lb_f/in.^2$ and 50% quality. Find the volume of the refrigerant, if its mass is 3 lb.

2.18 Refrigerant-12 is in a closed vessel at 60°F with a quality of 60%. What is its specific volume?

2.19 Ammonia is in a closed vessel at 60 $lb_f/in.^2$ and 20% quality. If the mass of the ammonia is 4 lb, determine the volume of the ammonia.

2.20 A closed rigid tank contains 10 kg of water at an initial pressure of 10 bars and a quality of 25%. Heat is added to the system until the tank contains saturated vapor. Find the final pressure, in bars, and the volume of the tank, in cubic meters.

2.21 A closed system comprises a two-phase liquid-vapor mixture of ammonia in equilibrium at 40°C (313.15 K). The ratio of the mass of vapor to that of liquid ammonia is 1:3. Determine the quality of the ammonia and the total volume of the system, in cubic meters, if the mass is 10 kg.

2.22 A closed system comprises a two-phase liquid-vapor mixture of refrigerant-12 in equilibrium at 180°F (639.67°R). The ratio of the mass of vapor to that of liquid ammonia is 1:5. Determine the quality of the ammonia and the total volume of the system, in cubic feet, if the mass is 10 lb.

2.23 A closed rigid tank contains steam. The steam is initially at 10 bars and 320°C (593.15 K). Heat is lost to the surroundings by cooling, and the temperature drops. Determine the pressure at which condensation first takes place, in bars, and the fraction of the total mass that has condensed when the temperature reaches 100°C (373.15 K). What percentage of the volume is occupied by the saturated vapor at the final state?

2.24 A closed rigid tank contains ammonia. The ammonia is initially at 50 $lb_f/in.^2$ and 320°F (779.67°R). Heat is lost to the surroundings by cooling, and the temperature drops. Determine the pressure at which condensation first takes place, in bars, and the fraction of the total mass that has condensed when the temperature reaches –20°F (439.67°R). What percentage of the volume is occupied by the saturated vapor at the final state?

2.25 There are 4 kg of refrigerant-134a contained in a rigid pressure vessel of 1.6 m^3. Heat is added to the refrigerant. It is designed to have a safety valve to be set at a maximum temperature of 120°C (393.15 K); at what pressure should this valve be set?

2.26 A pressure vessel contains water at 162°C (435.15 K) with the vapor volume being one-fifth the total volume. What is its initial quality? It is then heated until the pressure reaches 1 MPa. Find the temperature and quality at this final state.

2.27 A closed rigid tank comprises two equal-volume sections. One is filled with ammonia at a pressure of 8 bars and 20% quality, and the other is empty. The partition between the sections is removed, and the ammonia expands to occupy the whole tank. The final pressure is 6 bars. What is the quality of the ammonia in the final state?

Processes

2.28 Carbon dioxide in a piston-cylinder device is initally at the superheated vapor state. It is then cooled at constant pressure to saturated vapor, at which point the piston is locked. The cooling continues to a lower temperature. Indicate these processes on a T-v diagram. Explain the shapes of the processes.

2.29 In a piston-cylinder equipment, a gas undergoes a polytropic process such that PV^n = constant. The initial pressure is 2 bars, the initial volume is 1 m^3, and the final volume is 1.5 m^3. Calculate the work for the process, in kiloJoules, if (a) n = 2.0, (b) n = 0.0, and (c) n = 1.0.

2.30 PV^n = constant is the equation that describes the polytropic process undergone by a gas. The initial volume is 1 ft^3, the initial pressure is 2 atm, and the final volume is 1.5 ft^3. Compute the work for the process, in kiloJoules, if (a) n = 2.0, (b) n = 0.0, and (c) n = 1.0.

2.31 A piston-cylinder equipment compresses air in a polytropic process from P_1 = 1 atm, T_1 = 25°C (298.15 K) to P_2 = 4 atm. Compute the work involved if n = 1.1.

2.32 In a piston-cylinder device, air is compressed in a polytropic process from P_1 = 1.5 atm, T_1 = 20°C (293.15 K) to P_2 = 3 atm. Calculate the work n = k = 1.4.

2.33 A piston-cylinder device is used to transform air in a polytropic compression from P_1 = 1 atm, T_1 = 80°F (539.67°R) to P_2 = 4 atm. Determine the work involved if n = 1.1.

2.34 A piston-cylinder equipment compresses air in a polytropic process from P_1 = 1.5 atm, T_1 = 70°F (529.67°R) to P_2 = 3 atm. Find the work involved if n = k = 1.4.

2.35 Let 2 kg of steam with a quality of 10% be heated at a constant pressure of 1.5 bars until the temperature reaches 320°C (593.15 K). Determine the work done by the steam.

2.36 A 100-mm-diameter cylinder contains 1 l of ammonia at 0°C (273.15 K). A 40-kg piston is supported by the ammonia. If heat is added until the temperature is 20°C (293.15 K), calculate the work done.

2.37 Let 4 lb_m of steam with a quality of 10% be heated at a constant pressure of 20 $lb_f/in.^2$ until the temperature reaches 500°F (959.67°R). Determine the work done by the steam.

2.38 A 6-in.-diameter cylinder contains 40 $in.^3$ of ammonia at 30°F (489.67°R). A 100-lb_m piston is supported by the ammonia. If heat is added until the temperature is 90°F (549.67°R), calculate the work done.

2.39 A round balloon containing 10 kg of refrigerant-134a as saturated vapor at 30°C (303.15 K) initially, is attached by a valve to a 2 m^3 evacuated tank. The material of the balloon is such that the pressure inside is proportional to its radius. The valve is then opened, letting the refrigerant flow into the tank until the pressure within the balloon has decreased to 500 kPa, at which time the valve is shut. The final temperature in both the tank and the balloon is 30°C (303.15 K). Calculate (a) the final pressure in the vessel and (b) the work done by the refrigerant.

2.40 Ammonia in a piston-cylinder device is initially at 60°C (333.15 K), $x = 1$. It is then expanded in a process such that $Pv^2 =$ constant to a pressure of 3 MPa. Find the final specific volume and temperature.

The Ideal-Gas Equation of State and Compressibility Factor

2.41 Find the specific volume of water vapor at 0.01 MPa and 100°C (373.15 K), using (a) the ideal-gas equation of state and (b) the generalized compressibility chart. Evaluate the values obtained by comparing them to the experimental volume of 17.196 m^3/kg.

2.42 Find the pressure of refrigerant-134a at 200°F (659.67°R) and 0.7239 ft^3/lb_m, using (a) the computerized tables, (b) the ideal-gas equation of state, and (c) the generalized compressibility chart.

2.43 Determine the specific volume of refrigerant-134a at 0.50 MPa and 100°C (373.15 K), using (a) the ideal-gas equation of state and (b) the generalized compressibility chart. Evaluate the values obtained by comparing them to the experimental volume of 0.05805 m^3/kg.

2.44 Determine the pressure of water vapor at 400°F (859.67°R) and 4.934 ft^3/lb_m, using (a) the computerized tables, (b) the ideal-gas equation of state, and (c) the generalized compressibility chart.

2.45 Find the pressure of water vapor at 360°F (819.67°R) and 3.0 ft^3/lb_m, utilizing (a) the computerized tables, (b) the ideal-gas equation of state, and (c) the generalized compressibility chart.

Other Equations of State

2.46 Explain the physical meaning of the two constants in the van der Waals equation of state.

2.47 Find the pressure of oxygen gas at $T = 300$ K and $v = 0.0040$ m^3/kg. Use (a) the ideal-gas equation of state, (b) the van der Waals equation of state, and (c) the Beattie-Bridgeman equation of state.

2.48 Compute the pressure of nitrogen gas at $T = 300$ K and $v = 0.0040$ m^3/kg. Employ (a) the ideal-gas equation of state, (b) the van der Waals equation of state, and (c) the Beattie-Bridgeman equation of state.

2.49 Find the pressure of carbon dioxide gas at $T = 250$ K and $v = 0.0040$ m^3/kg. Employ (a) the ideal-gas equation of state, (b) the van der Waals equation of state, and (c) the Beattie-Bridgeman equation of state.

2.50 Refrigerant-12 at 1.5 MPa has a specific volume of 0.015 m^3/kg. Calculate the temperature of the refrigerant based on (a) the thermodynamic tables and (b) the ideal-gas equation.

Computer, Design, and General Problems

2.51 Write an essay on the selection of working fluids by engineers and scientists in their various inventions.

2.52 Write an essay on the role of working fluids in energy-related equipment and devices.

2.53 Compile a list of the devices used by scientists in recent history to measure work. Discuss the major differences between these devices.

2.54 Select one of the work-measurement devices in the previous question. Modify it using modern tools and devices. Discuss the improvements you have made on the design.

2.55 Design a piece of equipment to measure the power produced by a single person.

2.56 Design a piece of equipment to measure the power produced by an animal, e.g., a horse.

2.57 Design a piece of equipment to observe the sequence of events as a two-phase liquid-vapor mixture is heated at constant volume near its critical point. Discuss the possible observations if the equipment were used for water and for ammonia.

2.58 Design an apparatus to measure the work done against friction for a block sliding down a plane. Discuss the use of different materials to obtain different frictional work quantities.

2.59 Design an apparatus to demonstrate the elasticity of various materials.

2.60 Compile a list of C_v and C_p values for simple gases. Note the atomicity of these gases. Is there a relationship between the atomicity and the ratio of specific heats?

2.61 Write an interactive computer program that determines the work and heat transfer, per unit mass, for an ideal gas undergoing a polytropic process. Inputs could be the gas, initial pressure and temperature, polytropic exponent, and the final pressure. Instead of the final pressure, the final temperature could be used.

3

Mass Conservation and the First Law of Thermodynamics

CONTENTS

3.1 Mass Conservation

Mass can neither be created nor destroyed. Consider a control volume. This is a volume in space that has been arbitrarily defined for the purpose of analysis. Let the mass in the control volume initially be m_{ini} and at the end be m_{fin}, representing the final mass. Let the mass entering the control volume be m_{in} and the mass exiting the control volume be m_e.

Mass conservation for the control volume shown in Figure 3.1 requires that

$$m_{ini} + m_i = m_{fin} + m_e. \tag{3.1}$$

This equation states that the initial mass and the incoming mass should equal the final mass and the exiting mass. In the rate form, it may be written as

$$\dot{m}_{ini} + \dot{m}_i = \dot{m}_{fin} + \dot{m}_e. \tag{3.2}$$

These are the forms of the mass conservation equations most commonly employed in this book because they are most useful that way.

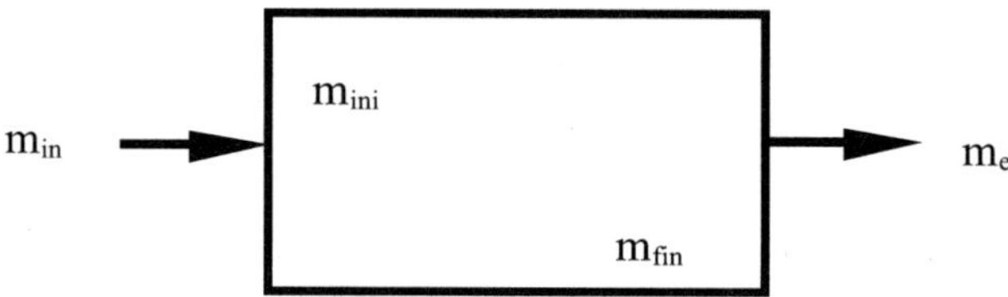

FIGURE 3.1
Control volume for mass conservation.

3.2 The First Law of Thermodynamics

The first law of thermodynamics is also known as the conservation of energy. It states that energy can neither be created nor destroyed.

3.2.1 Stored Forms of Energy

Work and heat are forms of energy transfer, and they are not stored forms of energy in a system. The energy of a system is composed of several forms.

The energy of the molecules of a physical system is represented by the internal energy. The system as a unit possesses kinetic energy and potential energy. The arrangement of the atoms in a substance is a manifestation of the chemical energy of the substance. The nuclear energy of a system is due to the cohesion of the nuclear particles.

Thermodynamics emphasizes the changes of this stored energy rather than the absolute quantities. Any arbitrary value E_1 may be given to the value of the energy of a system at a given state 1, so no particular importance can be associated to the value of E_1 at this or any other state. Only changes in the energy of a system are useful, and thus of significance.

3.2.2 Internal (Thermal) Energy, U

In general, the internal energy of a substance is due to the molecules that make up the substance. The internal energy is associated with the internal, microscopic motions of molecules. Two major contributions to the internal energy of a substance are (1) the kinetic internal energy that is due to the velocity of the molecules; and (2) the potential internal energy that is due to the attractive forces existing between molecules.

The changes in the velocity of the molecules may be manifested by changes in temperature of the substance. A change in phase may be a manifestation of the variation of position of the molecules. Intermolecular forces are very significant in influencing the internal energy of dense gases, liquids, and solids. The kinetic internal energy and potential internal energy are useful simplified terms to describe the internal energy. These should not be confused with the (gravitational) potential energy and kinetic energy possessed by a system as a whole, that are described in the following subsections.

3.2.3 (Gravitational) Potential Energy, P.E.

The energy possessed by the system due to its elevation or position is called its (gravitational) potential energy. This is a macroscopic form of energy. By Newton's law of motion,

$$F = ma = mg \tag{3.3}$$

where a is the acceleration of the mass m and g is the gravitational acceleration.

By the definition of work,

$$\text{P.E.} = W = \int_{z_1}^{z_2} F dz = \int_{z_1}^{z_2} mg\, dz = mg(z_2 - z_1). \tag{3.4}$$

Usually, the datum in Figure 3.2 is taken to be at sea level. Under this assumption, the P.E. term may be represented by mg(z – 0) = mgz, where the coordinate z is the height above sea level.

3.2.4 Kinetic Energy, K.E.

The energy possessed by the system as a result of its velocity is its kinetic energy. It is a function of the macroscopic motion of the system. The kinetic energy is the work that could be done in bringing to rest a medium that is in motion, with a velocity $\vec{V}$, in the absence of gravity. Thus, if

$$F = ma = -m\frac{d\vec{V}}{dt} \tag{3.5}$$

then,

$$
\begin{aligned}
K.E. = W = \int_0^x F\,\mathbf{d}x &= -\int_V^0 m\frac{\mathbf{d}\vec{V}}{\mathbf{dt}}dx = -\int_V^0 m\vec{V}d\vec{V} \\
&= \frac{1}{2}mV^2.
\end{aligned}
\tag{3.6}
$$

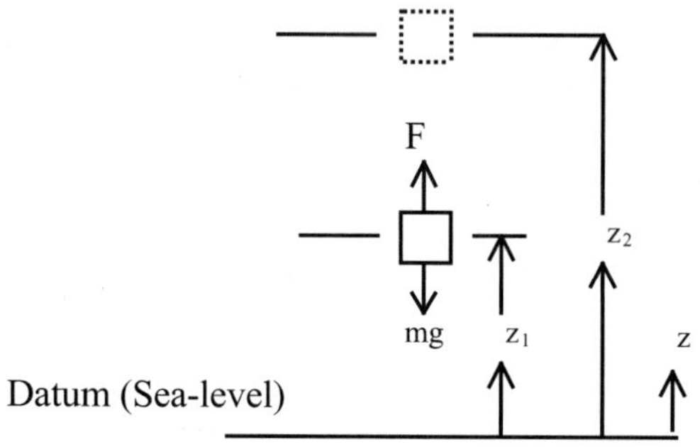

FIGURE 3.2
Concept of (gravitational) potential energy for a body (system).

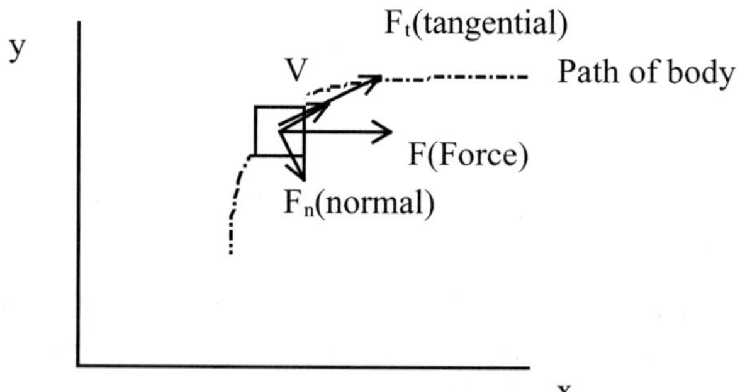

FIGURE 3.3
Concept of kinetic energy for a body (system).

3.2.5 Chemical Energy

The energy possessed by the system due to the arrangement of the atoms comprising the molecule is its chemical energy. This form of energy is not discussed in detail in this introductory book on engineering thermodynamics.

3.2.6 Nuclear Energy

The energy possessed by the system, due to the cohesive forces holding the protons and neutrons together as the nucleus of the atom, is its nuclear energy. Nuclear energy is not further discussed in this book, as it is not in the purview of classical thermodynamics.

The various forms of energy are summarized in Table 3.1. The transfer forms of energy, like work and heat, are path functions and thus can only be identified as they cross the system boundary. The possessed forms of energy are point functions. The difference in values of these point functions at two different states depends only on the two states. The values of the path functions (work and heat) going from one state to the other state, depend on the path taken.

TABLE 3.1

Forms of Energy

Transfer Forms of Energy	
Work	Potential: Force
Heat	Potential: Temperature
Energy Possessed by Substances and Systems	
By substances as entities:	
Potential (gravitational)	Manifested by elevation
Kinetic	Manifested by velocity
Internal or intrinsic:	
Thermal	
Molecular kinetic	Manifested by temperature
Molecular potential	Manifested by phase
Chemical	Manifested by changes in molecular composition
Nuclear	Manifested by changes in atomic composition

3.3 First Law for a Control Volume

An open system is one where there is flow of matter or mass in or out of the boundaries. The boundaries essentially define a control volume, that is, a volume that is defined as the space of interest in which we are concentrating our analysis. We consider the volume initially with a mass m_{ini} [Figure 3.4, Part (a)]. Associated with this mass are the forms of energy we are interested in at this time, namely, internal energy, kinetic energy, and potential energy.

The energies initially in the control volume are thus

$$m_{ini}\left(u_{ini} + \frac{V_{ini}^2}{2} + gz_{ini}\right).$$

The symbol E is often used to represent the total energy. Thus,

$$E_{ini} = m_{ini}\left(u_{ini} + \frac{V_{ini}^2}{2} + gz_{ini}\right)$$

represents the total initial energy of the system.

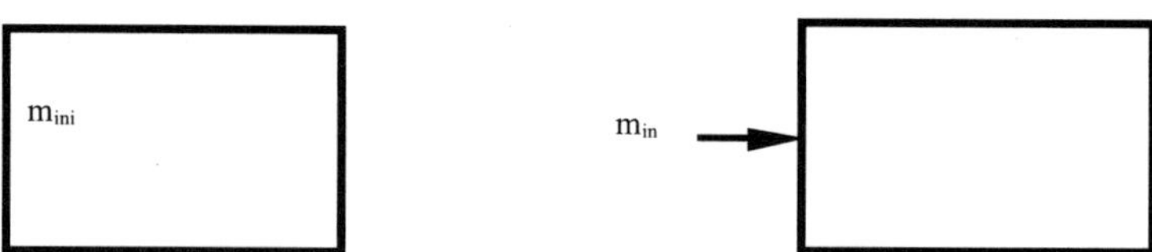

(a) Initial mass in the control volume (b) Incoming mass into the control volume

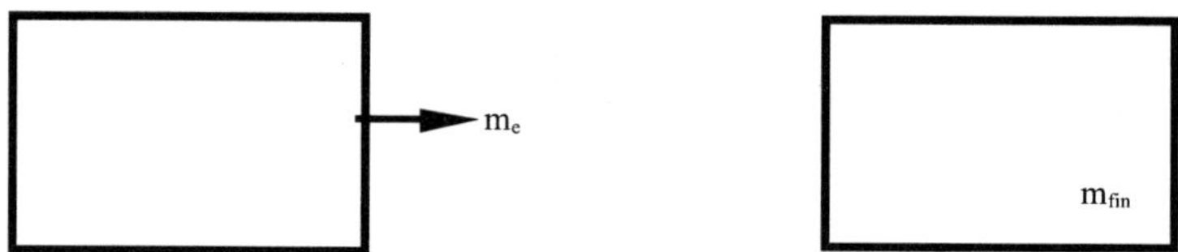

(c) Mass exiting the control volume (d) Final mass in the control volume

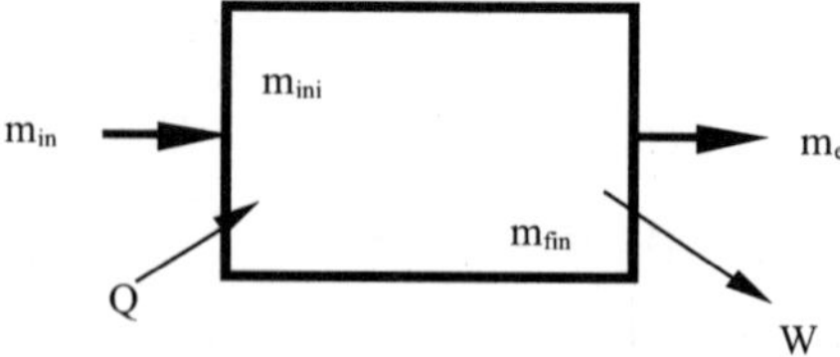

(e) Control volume with one inlet and one exit

FIGURE 3.4
Control volume.

Consider a mass m_{in} entering the control volume, as in Figure 3.4, Part (b). The energies associated with this mass are the internal energy, the kinetic energy, the potential energy, and the energy required to push the mass into the control volume. The internal energy and the energy required to push the mass into the control volume is the enthalpy of m_{in}. This fact may be shown in the following discussion. There is work at an inlet *in* associated with the pressure of the flowing substance. The rate of energy transfer by work can be expressed as the product of a force and the velocity at the point the force is applied. Thus, the rate at which work is done at the inlet by the normal force due to pressure is the product of the normal force, $P_{in} A_{in}$ and the fluid velocity V_{in}. In other words,

$$\begin{pmatrix} \text{rate of energy transfer} \\ \text{by work into the control volume} \\ \text{at inlet } in \end{pmatrix} = (P_{in} A_{in}) V_{in} = P_{in} v_{in}$$

where P_{in} is the pressure, A_{in} is the area, V_{in} is the velocity, and v_{in} is the specific volume at the inlet, respectively. The velocity and pressure are assumed constant with position over the inlet flow area. This rate of energy transfer into the control volume at inlet *in* has been shown to be equal to $P_{in} v_{in}$, and usually is called the incoming *flow work*. The energies associated with m_{in} are

$$m_{in}\left(u_{in} + P_{in} v_{in} + \frac{V_{in}^2}{2} + gz_{in}\right) = m_{in}\left(h_{in} + \frac{V_{in}^2}{2} + gz_{in}\right).$$

Consider a mass m_e exiting a control volume, as depicted in Figure 3.4, Part (c). The energies associated with this mass are the internal energy, the kinetic energy, the potential energy, and the energy required to push the mass out of the control volume. The internal energy and the energy required to push the mass out of the control volume is the enthalpy of m_e. By an argument similar to that used above, it can be shown that the energy required to push the mass out of the control volume is equal to $P_e v_e$, and is usually called the exiting *flow work*.

The energies associated with m_e are

$$m_e\left(u_e + P_e v_e + \frac{V_e^2}{2} + gz_e\right) = m_e\left(h_e + \frac{V_e^2}{2} + gz_e\right).$$

There is mass entering the control volume and mass exiting the control volume, so the mass within the control volume does not remain the same. Let us call the final mass m_{fin} [Figure 3.4, Part (d)]. Associated with this mass are its internal energy, kinetic energy, and potential energy. Hence, the energies left in the control volume are

$$m_{fin}\left(u_{fin}+\frac{V_{fin}^2}{2}+gz_{fin}\right).$$

In addition, there is heat entering the system, and work being done by the system besides the flow work terms discussed above, that are specifically related to the incoming and exiting masses. These are forms of energy that have to be taken into account in the energy balance of the control volume. The final diagram, Figure 3.4, Part (e), shows the various masses, and the heat and work interactions. The conservation of energy equation for the control volume in Figure 3.4, Part (e) is

$$Q+m_{in}\left(h_{in}+\frac{V_{in}^2}{2}+gz_{in}\right)+m_{ini}\left(u_{ini}+\frac{V_{ini}^2}{2}+gz_{ini}\right)$$

$$=m_e\left(h_e+\frac{V_e^2}{2}+gz_e\right)+m_{fin}\left(u_{fin}+\frac{V_{fin}^2}{2}+gz_{fin}\right)+W. \tag{3.7}$$

Note that the internal energies are used for the masses that do not cross the system boundaries, and the enthalpies are used for the masses that do cross the system boundaries. If the number of mass streams entering and exiting the system are more than one in each, all the energies of the masses have to be considered. This is also true if there are more than one heat transfer process and more than one work process going on. Taking into consideration multiple heat transfer and work processes, the first law equation becomes

$$\sum_k {}_kQ+\sum_l {}_l m_{in_l}\left(h_{in}+\frac{V_{in}^2}{2}+gz_{in}\right)+m_{ini}\left(u_{ini}+\frac{V_{ini}^2}{2}+gz_{ini}\right)=$$

$$\sum_n {}_n m_{e_n}\left(h_e+\frac{V_e^2}{2}+gz_e\right)+m_{fin}\left(u_{fin}+\frac{V_{fin}^2}{2}+gz_{fin}\right)+\sum_p {}_pW. \tag{3.8}$$

In a steady-state steady-flow (S.S.S.F.) process, the initial mass is equal to the final mass and the initial state is the same as the final state, i.e.,

$$m_{ini}=m_{fin} \tag{3.9}$$

and

$$\left(u_{ini}+\frac{V_{ini}^2}{2}+gz_{ini}\right)=\left(u_{fin}+\frac{V_{fin}^2}{2}+gz_{fin}\right). \tag{3.10}$$

So the energy conservation equation for a S.S.S.F. process is

$$Q + m_{ini}\left(h_{in} + \frac{V_{in}^2}{2} + gz_{in}\right) = m_e\left(h_e + \frac{V_e^2}{2} + gz_e\right) + W. \tag{3.11}$$

A closed system is one where no mass enters or leaves the system. Under these conditions,

$$m_{in} = m_e = 0 \text{ and } m_{ini} = m_{fin} = m.$$

So the energy conservation equation for a closed system is

$$Q + m\left(u_{ini} + \frac{V_{ini}^2}{2} + gz_{ini}\right) = m\left(u_{fin} + \frac{V_{fin}^2}{2} + gz_{fin}\right) + W. \tag{3.12}$$

It is important to emphasize here that the energy conservation equation for the closed system is embedded in the conservation equation for the open system. It is easier for the student of thermodynamics to appreciate the generality of the open system in this fashion.

The above equations are similarly modified if there is more than one mass stream entering or more than one mass stream exiting the system for the S.S.S.F. process; if there is more than one heat transfer process, or more than one work process in either the S.S.S.F. process or the closed system, the equations are similarly modified to take into account the multiple processes.

The rate equation for the control volume for a single mass stream in and out of the system and single heat and work processes, is

$$\dot{Q} + \dot{m}_{in}\left(h_{in} + \frac{V_{in}^2}{2} + gz_{in}\right) + \dot{m}_{ini}\left(u_{ini} + \frac{V_{ini}^2}{2} + gz_{ini}\right) =$$
$$\dot{m}_e\left(h_e + \frac{V_e^2}{2} + gz_e\right) + \dot{m}_{fin}\left(u_{fin} + \frac{V_{fin}^2}{2} + gz_{fin}\right) + \dot{W}. \tag{3.13}$$

The corresponding equation for the multiple mass streams and boundary processes is

$$\sum_k {}_k\dot{Q} + \sum_l {}_l\dot{m}_{in}\left(h_{in} + \frac{V_{in}^2}{2} + gz_{in}\right)_l + \dot{m}_{ini}\left(u_{ini} + \frac{V_{ini}^2}{2} + gz_{ini}\right) =$$
$$\sum_n {}_n\dot{m}_e\left(h_e + \frac{V_e^2}{2} + gz_e\right)_n + \dot{m}_{fin}\left(u_{fin} + \frac{V_{fin}^2}{2} + gz_{fin}\right) + \sum_p {}_p\dot{W}. \tag{3.14}$$

In terms of the total energy E, it may be written as

$$\sum_k {}_k\dot{Q} + \sum_l {}_l\dot{E}_{in} + \dot{E}_{ini} = \sum_n {}_n\dot{E}_e + \dot{E}_{fin} + \sum_p {}_p\dot{W}. \qquad (3.15)$$

Example 3.1

Problem: Given the open system shown, where water is the working fluid, with three incoming masses and two outgoing masses, calculate the internal energy of the mass finally remaining in the control volume. Specify another property of this final mass.

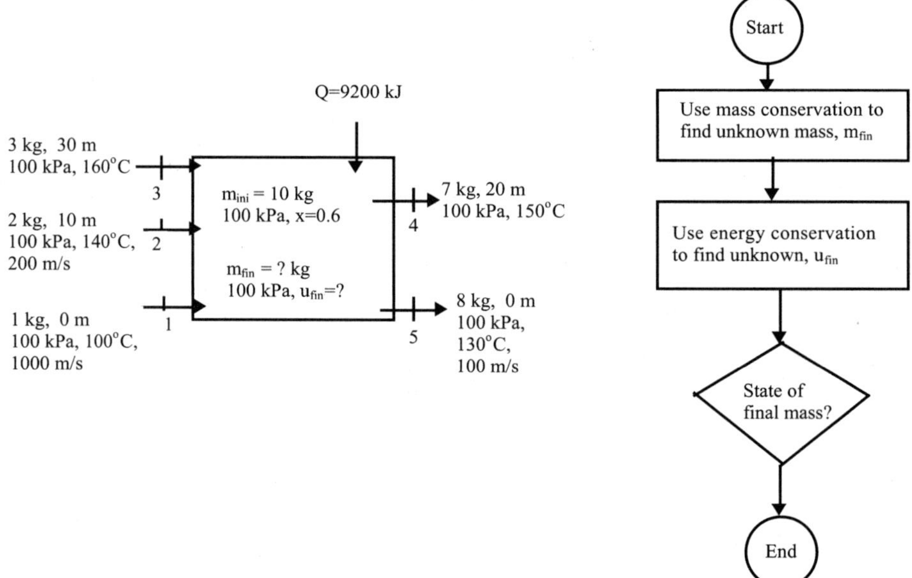

Schematic and logic diagram for Example 3.1.

Solution

Assumptions: (1) If the velocity of the mass is not given, its K.E. is neglected.

Analysis:

From the thermodynamic tables, for the incoming mass at state 1, $h_1 = 2676$ kJ/kg

For the incoming mass at state 2, $h_2 = 2757$ kJ/kg

For the incoming mass at state 3, $h_3 = 2796$ kJ/kg

For the initial mass, $u_{ini} = 1670.6$ kJ/kg

For the outgoing mass at state 4, $h_4 = 2776$ kJ/kg
For the outgoing mass at state 5, $h_5 = 2737$ kJ/kg.

The conservation of mass for the control volume gives

$$m_1 + m_2 + m_3 + m_{ini} = m_{fin} + m_4 + m_5.$$

Substituting values for the known masses,

$$1 + 2 + 3 + 10 = m_{fin} + 7 + 8.$$

Thus, $m_{fin} = 1$ kg.

The first law energy balance for the control volume is

$$Q + m_1\left(h_1 + \frac{V_1^2}{2} + gz_1\right) + m_2\left(h_2 + \frac{V_2^2}{2} + gz_2\right) + m_3\left(h_3 + \frac{V_3^2}{2} = gz_3\right) + m_{ini}u_{ini}$$
$$= m_{fin}u_{fin} + m_4\left(h_4 + gz_4\right) + m_5\left(h_5 + \frac{V_5^2}{2} + gz_5\right).$$

Substituting the values for the various parameters,

$$9200 + 1\left(2676 + \frac{1000 \times 1000}{2 \times 1000} + 0\right) + 2\left(2757 + \frac{200 \times 200}{2 \times 1000} + \frac{10 \times 9.81}{1000}\right)$$
$$+3\left(2796 + \frac{30 \times 9.81}{1000}\right) + 10 \times 1670.6$$
$$= u_{fin} + 7\left(2776 + \frac{20 \times 9.81}{1000}\right) + 8\left(2737 + \frac{100 \times 100}{2 \times 1000} + 0\right).$$

Therefore, $u_{fin} = 1656$ kJ/kg.

From the thermodynamic tables, the internal energies of saturated vapor and saturated liquid at 100 kPa, respectively, are 2506 kJ/kg and 417.3 kJ/kg. Hence, the state of the final mass is a mixture of vapor and liquid. From the thermodynamic tables, the final temperature is 99.62°C.

The logic diagram for solving the above problem may be represented as shown. It is not necessary to draw this logic diagram in solving the problem, but the sketch definitely clarifies the principal steps in the problem-solving process. The mass conservation equation is first used to find the unknown mass. Then, the conservation of energy (first law of thermodynamics) is used to find the unknown internal energy of the final mass in the control volume. After this internal energy has been found, the state of the final mass has to be determined. In establishing the state of the final mass, we have to look at the u_f and u_g at the final pressure of 100 kPa. If the internal energy is greater than u_g, the state is superheated vapor. If the internal energy is equal to u_g, the state is saturated vapor. If the internal energy is between u_f and u_g, the state is a mixture of liquid and vapor. If the internal energy is equal to u_f, the state is saturated liquid. If the internal energy is less than u_f, the state is compressed liquid.

Example 3.2

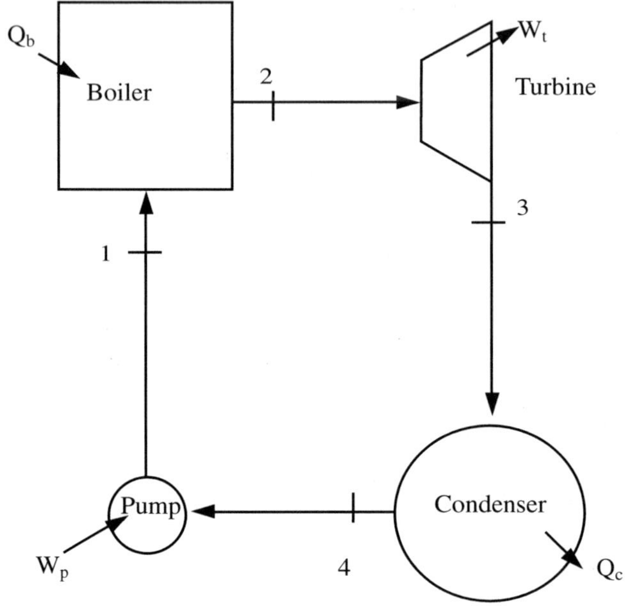

Schematic for Example 3.2.

Problem: The sketch above is a schematic of a simple power plant system. It shows a boiler that heats liquid water to steam. The steam is then expanded in a turbine to do work. The expanded steam then goes to the condenser, where it is condensed into liquid water again. The pump increases the pressure of the water from the condenser, so that the water can enter the boiler, which is at a high pressure. Apply the first law of thermodynamics to the system as a whole, and then individually to each of the four pieces of equipment.

Solution

In the example above, the general first law for the whole system, Equation (3.12), which is a closed system, is

$$Q_b - Q_c + m\left(u_{ini} + \frac{V_{ini}^2}{2} + gz_{ini}\right) = m\left(u_{fin} + \frac{V_{fin}^2}{2} + gz_{fin}\right) + W_t - W_p.$$

The terms in the equation above are canceled because the system undergoes a cycle. The equation reduces to $Q_b - Q_c = W_t - W_p$.

If a boundary is drawn around the boiler, it will define an open system undergoing a S.S.S.F. process. The states of the water are designated 1, 2, 3, and 4, respectively. Neglecting the K.E. and P.E. terms, the first law equation for the boiler is $Q_b + m_1 h_1 = m_2 h_2$. Since $m_1 = m_2 = m$, then $Q_b + m h_1 = m h_2$.

Similarly, each of the other pieces of equipment, the turbine, the condenser, and the pump, may be defined as an open system undergoing a S.S.S.F. system. Neglecting the K.E. and P.E. terms, the first law equation for the turbine is $mh_2 = mh_3 + W_t$.

Neglecting the K.E. and P.E. terms, the first law equation for the condenser is $-Q_c + mh_3 = mh_4$.

Neglecting the K.E. and P.E. terms, the first law equation for the pump is $mh_4 = mh_1 - W_p$.

Example 3.3

Problem: In Example 3.2, the state of the water in various parts of the cycle shown is as follows: $P_1 = 0.3$ MPa, $T_1 = 95°C$; $P_2 = 0.3$ MPa, $T_2 = 600°C$; $P_3 = 0.07$ MPa, $T_3 = 400°C$; $P_4 = 0.07$ MPa, and $x_4 = 0.0$. Consider 1 kg of water undergoing the cycle. Calculate the heat supplied to the boiler q_b, the heat rejected in the condenser q_c, the work output from the turbine w_t, and the work input into the pump, w_p. By considering the system as a whole, calculate the net work output from the closed system. In addition, calculate the ratio of this net work output to the heat supplied to the boiler.

Solution

The enthalpy of water in the compressed liquid region at T′ is approximately equal to the enthalpy of water in the saturated liquid state at T′. The reason is because pressure does not have much effect on the enthalpy of compressed liquid water. Hence, $h_1 = 398$ kJ/kg.

The enthalpy values of the water at the other states are as follows:

$$h_2 = 3703.2 \text{ kJ/kg}$$
$$h_3 = 3279 \text{ kJ/kg}$$
$$h_4 = 376.7 \text{ kJ/kg.}$$

For the boiler, which undergoes a S.S.S.F. process, the energy conservation equation takes the rate form of Equation (3.11),

$$\dot{Q} + \dot{m}_{in}\left(h_{in} + \frac{V_{in}^2}{2} + gz_{in}\right) = \dot{m}_e\left(h_e + \frac{V_e^2}{2} + gz_e\right) + \dot{W}.$$

To simplify, we can assume the changes in K.E. and P.E. are negligible. Work is not done by the working fluid in a boiler. The above equation reduces to

$$\dot{Q}_b = \dot{m}(h_2 - h_1)$$

$$q_b = \frac{\dot{Q}_b}{\dot{m}} = (h_2 - h_1) = 3305 \frac{kJ}{kg}.$$

For the turbine, which undergoes a S.S.S.F. process, the energy conservation equation takes the rate form of Equation (3.11),

$$\cancel{\dot{Q}} + \dot{m}_{in}\left(h_{in} + \frac{V_{in}^2}{2} + gz_{in}\right) = \dot{m}_e\left(h_e + \frac{V_e^2}{2} + gz_e\right) + \dot{W}.$$

To simplify, we can assume the changes in K.E. and P.E. are negligible. Heat transfer is not important for a turbine. The above equation reduces to

$$\dot{W}_t = \dot{m}(h_2 - h_3)$$

$$w_t = \frac{\dot{W}_t}{\dot{m}} = (h_2 - h_3) = 424 \frac{kJ}{kg}.$$

For the condenser, which undergoes a S.S.S.F. process, the energy conservation equation takes the rate form of Equation (3.11),

$$\dot{Q} + \dot{m}_{in}\left(h_{in} + \frac{V_{in}^2}{2} + gz_{in}\right) = \dot{m}_e\left(h_e + \frac{V_e^2}{2} + gz_e\right) + \cancel{\dot{W}}.$$

To simplify, we can assume the changes in K.E. and P.E. are negligible. Work is not done by the working fluid in a condenser. The above equation reduces to

$$\dot{Q}_c = \dot{m}(h_4 - h_3)$$

$$q_c = \frac{\dot{Q}_c}{\dot{m}} = (h_4 - h_3) = -2902 \frac{kJ}{kg}.$$

For the pump, which undergoes a S.S.S.F. process, the energy conservation equation takes the rate form of Equation (3.11),

$$\cancel{\dot{Q}} + \dot{m}_{in}\left(h_{in} + \frac{V_{in}^2}{2} + gz_{in}\right) = \dot{m}_e\left(h_e + \frac{V_e^2}{2} + gz_e\right) + \dot{W}.$$

To simplify, we can assume the changes in K.E. and P.E. are negligible. Heat transfer is not important for a pump. The above equation reduces to

$$\dot{W}_p = \dot{m}(h_4 - h_1)$$

$$w_p = \frac{\dot{W}_p}{\dot{m}} = (h_4 - h_1) = -21\frac{kJ}{kg}.$$

Net work output $= w_t + w_p = 403$ kJ/kg.

$$\text{Ratio} = \frac{\text{Net work output}}{\text{Boiler heat input}} = 0.122.$$

This ratio of net work output to the boiler heat input is a measure of the performance of the power plant system. A further discussion of this ratio is given later in the chapter.

The logic diagram for solving the above problem may be represented as shown. It is similar to that shown for Example 3.1. The mass conservation equation is not used explicitly, but by assuming a unit mass of working fluid undergoing the cycle, mass conservation is automatically satisfied. This particular technique is often used for a system when the actual mass is not known. The energy conservation equation (first law) is used several times in this problem.

3.4 First Law for a Control Mass

As shown in the previous section, the energy conservation equation for a closed system or a control mass system with one heat interaction and one work interaction is

$$Q + m\left(u_{ini} + \frac{V_{ini}^2}{2} + gz_{ini}\right) = m\left(u_{fin} + \frac{V_{fin}^2}{2} + gz_{fin}\right) + W. \tag{3.12}$$

Note that enthalpies are not used since there is no work involved in pushing mass in or out of the system boundaries. When there is more than one heat or work interaction at the system boundaries, the energy conservation equation in terms of total energy is

$$\sum_k {}_kQ + m\left(u_{ini} + \frac{V_{ini}^2}{2} + gz_{ini}\right) = m\left(u_{fin} + \frac{V_{fin}^2}{2} + gz_{fin}\right) + \sum_p {}_pW. \tag{3.16}$$

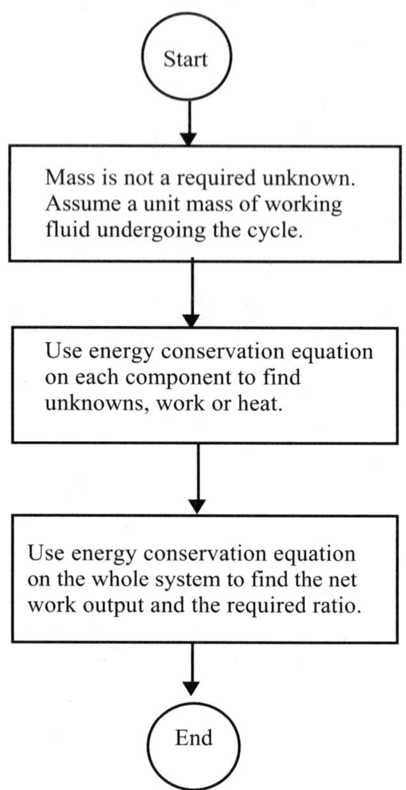

Logic diagram for Example 3.3.

The rate conservation equation in terms of total energy is

$$\sum_k {}_k\dot{Q} + \dot{E}_{ini} = \dot{E}_{fin} + \sum_p {}_p\dot{W}. \tag{3.17}$$

It is easy to see that Equation (3.17) is a special case or subset of Equation (3.15). Since there is no mass moving in or out of a control mass system, there is no need to consider incoming energy with incoming mass or outgoing energy with outgoing mass. The only boundary interactions are heat and work interactions, and care has to be taken to consider all such interactions.

In Example 3.2, a simple power plant system is discussed. Taken as a whole, the system is a control mass system. The conservation equation is simple, like Equation (3.12), but the equation relates only the heat and work interactions at the boundary. There is no difference between initial and final energies of the water (control mass) because the state conditions of the water at any point in the cycle are unchanged with respect to time. It is assumed that at the time of observation, the states of the water are the same as they are on the following day, and the next day, and so on.

If each of the four pieces of equipment in the power plant system are taken individually, the system obtained for each is a control volume system. The energy conservation equation for a simple control volume system takes the form of Equation (3.7), and a relation is obtained between the heat and work interactions, the states of the water, and the K.E. and P.E. terms. Note that it is this later form of analysis that allows us to determine the different states of the water throughout the cycle, whereas the control mass analysis did not. The important fact to be learned is that the understanding of the problem and the choice of analysis are the keys to solving thermodynamic problems.

In many thermodynamic problems, it is required to determine a property of the working fluid (in our example, water). The approach taken often depends on whether it will produce a relationship between the unknown property with the given quantities. It is important to note that for a control mass analysis, no flow goes in or out of the control mass; however, the boundaries of the control mass are not fixed in space and may move. Consider a deflated balloon as an example. One can choose the mass in the balloon as a control mass system. Initially, the system boundary is around the deflated balloon. Then the balloon is inflated from a tank of helium, for instance. The system boundary is now around the inflated balloon. Note that the system boundary has moved in the example, even though the system is a control mass.

Example 3.4

Problem: A rigid, insulated tank is initially evacuated. It is then connected via a valve to a steam line that supplies steam at 5 bars and 240°C. The valve is opened and steam flows slowly into the tank until the pressure reaches 5 bars; then the valve is closed. Determine the final temperature of the steam in the tank.

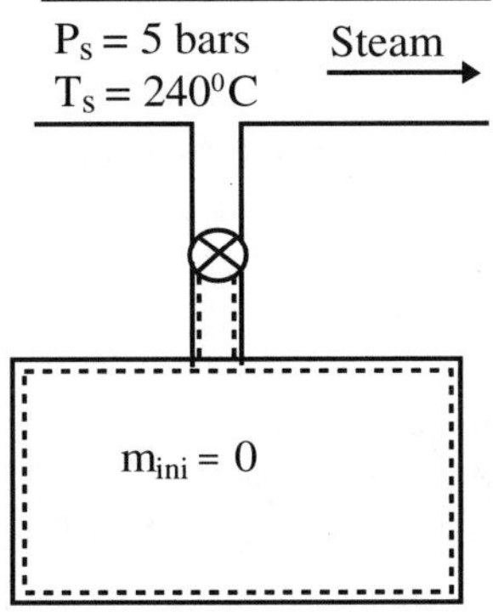

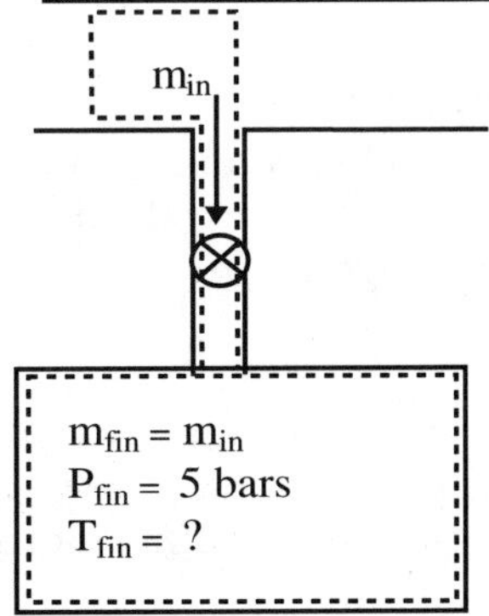

Schematics for Example 3.4.

Solution

Mass conservation equation is

$$\sum m_{in} - \sum m_{e} = m_{fin} - m_{ini}.$$

But $m_{ini} = m_e = 0$, thus $m_{in} = m_{fin}$. There is no heat transfer, Q = 0. Since the tank is rigid, no work is done, W = 0. The general first law equation, Equation (3.7), is

$$\cancel{Q} + m_{in}\left(h_{in} + \frac{V_{in}^2}{2} + gz_{in}\right) + \underbrace{\cancel{m}_{ini}}_{=0}\left(u_{ini} + \frac{V_{ini}^2}{2} + gz_{ini}\right) =$$

$$\underbrace{\cancel{m}_{e}}_{=0}\left(h_{e} + \frac{V_{e}^2}{2} + gz_{e}\right) + m_{fin}\left(u_{fin} + \frac{V_{fin}^2}{2} + gz_{fin}\right) + \cancel{W}.$$

In the system selected, there is no initial mass or $m_{ini} = 0$, and in the final conditions, there is no mass exiting the system or $m_e = 0$. Of the terms remaining in the equation, the K.E. and P.E. terms may be taken to be zero since there is no information provided. The energy conservation equation reduces to $m_{in} h_{in} = m_{fin} u_{fin}$. But $m_{in} = m_{fin}$, so $u_{fin} = h_{in}$.

The final internal energy of the steam in the tank is equal to the enthalpy of the incoming steam. The incoming steam is at P_{in} = 5 bars, and T_{in} = 240°C, hence h_{in} = 2939.9 kJ/kg. From P_{fin} = 5 bars, u_{fin} = 2939.9 kJ/kg, we find from the computerized tables that T_{fin} = 386°C.

In can be seen that the temperature of the steam in the tank is 146°C above that of the supply steam. The reason is that the flow work part (Pv) of the enthalpy (h = u + Pv), is converted to sensible internal energy once the flow is stopped in the tank. This conversion is manifested as an increase in temperature.

The logic diagram for this problem is the same as that for Example 3.1.

Example 3.5

Problem: A 5 kW fan is used in a room for circulation purposes. Assuming a sealed room that is well insulated, determine the internal energy increase after 2 h of operation.

Solution

Assumptions: (1) The K.E. and P.E. terms are negligible.

Analysis: The sealed room is selected as the system and is, by definition, a closed system. Furthermore, the assumption of a well-insulated room implies that it is undergoing an adiabatic process. The first law equation, Equation (3.12), is

$$\not{Q} + m\left(u_{ini} + \frac{V_{ini}^2}{2} + gz_{ini}\right) = m\left(u_{fin} + \frac{V_{fin}^2}{2} + gz_{fin}\right) + W.$$

With the assumption above that Δ K.E. = Δ P.E. = 0, the above equation becomes

$$m(u_{init} - u_{fin}) = U_{ini} - U_{fin} = W.$$

The work input is

$$W = (-5\ \text{kW})(2\ \text{h})(3600\ \text{s/h})\left(\frac{1\ \text{kJ/s}}{1\ \text{kW}}\right) = -36,000\ \text{kJ}.$$

There is a negative sign because work is done on the system. The internal energy rise is

$$U_{fin} - U_{ini} = -W = 36,000\ \text{kJ}.$$

3.5 First Law Applied to Various Processes

3.5.1 Turbines

A turbine is a piece of machinery that converts the energy in a fluid to a rotating shaft via a set of rotating blades. The rotating shaft may be coupled to an electric generator, for instance, to do work. It is usual to neglect any heat transfer or changes in K.E. and P.E., unless otherwise shown to be a poor assumption.

For the turbine, which undergoes a S.S.S.F. process, the energy conservation equation takes the rate form of Equation (3.11),

$$\not{\dot{Q}} + \dot{m}_{in}\left(h_{in} + \not{\frac{V_{in}^2}{2}} + \not{gz_{in}}\right) = \dot{m}_e\left(h_e + \not{\frac{V_e^2}{2}} + \not{gz_e}\right) + \dot{W}.$$

Heat transfer is not important for a turbine. The above equation reduces to

$$\dot{W}_t = \dot{m}(h_{in} - h_e) \quad \text{or} \quad w_t = (h_{in} - h_e). \tag{3.18}$$

Example 3.6

Problem:

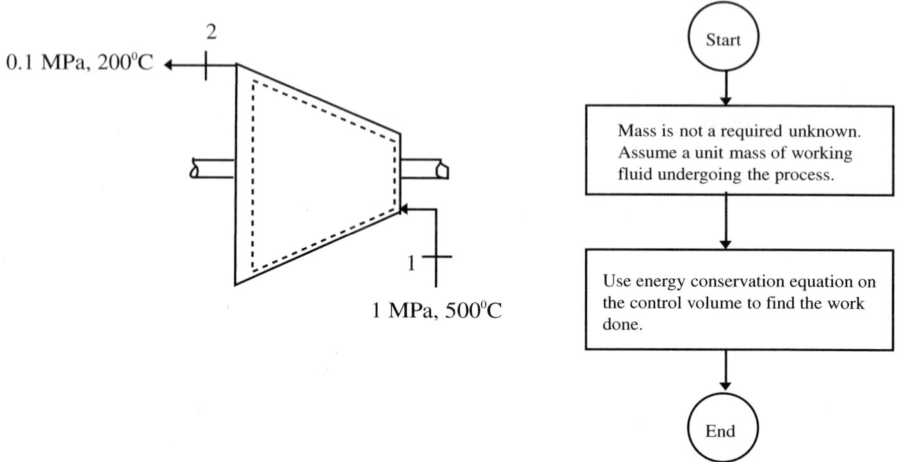

Schematic and logic diagram for Example 3.6.

A turbine converts energy to work from steam supplied at 500°C and 1 MPa. The steam leaves at 200°C and 0.1 MPa. Calculate the work output from the turbine.

Solution

The energy conservation equation, Equation (3.18), gives $w_t = (h_1 - h_2)$. The tables give these values for the steam:

$$h_1 = 3478 \text{ kJ/kg}$$
$$h_2 = 2875 \text{ kJ/kg}.$$

Note that the specific volume has increased from 0.3541 m^3/kg to 2.172 m^3/kg. The schematic of the turbine shows the increased specific volume of the steam between the entrance and the exit. Thus, the work output from the turbine is $w_t = 603$ kJ/kg of steam.

The logic diagram for solving the problem may be represented as shown. The mass conservation equation is not used explicitly. We assume a unit mass of working fluid undergoing the process, so the conservation of mass is automatically satisfied. The conservation of energy (first law) is used once to find the unknown boundary quantity, work, in this case.

Example 3.7

Problem: Steam enters a turbine at 3 MPa and 600°C with a velocity of 180 m/s, and leaves saturated at an outlet pressure of 70 kPa with a velocity of 150 m/s. Neglect any heat transfer. The steam flow rate is 10 kg/s. Calculate (a) the change in kinetic energy and (b) the turbine output. What is the percentage of the quantity (a) divided by the quantity (b)?

Solution

From the data tables for steam,

$$h_1 = 3682.3 \text{ kJ/kg}$$
$$h_2 = 2660.0 \text{ kJ/kg}.$$

(a)
$$\Delta \text{K.E.} = \dot{m}\left(\frac{v_2^2 - v_1^2}{2}\right)$$
$$= \left(10\frac{\text{kg}}{\text{s}}\right)\left(\frac{150^2 - 180^2}{2}\right)\frac{\text{m}^2}{\text{s}^2} = -49{,}500 \text{ J/s} = -49.5 \text{ kJ/s}.$$

(b) The turbine is a S.S.S.F. system. The energy conservation equation is

$$\dot{W}_t = \dot{m}(h_1 - h_2)$$
$$\dot{W}_t = (3682.3 - 2660)\frac{\text{kJ}}{\text{kg}}\left(10\frac{\text{kg}}{\text{s}}\right) = 10{,}223\frac{\text{kJ}}{\text{s}} = 10{,}223 \text{ kW}.$$

The kinetic energy change is just 0.5% of the turbine power output. In most practical situations, the kinetic energy change is neglected for the steam turbine.

The logic diagram for this problem is very similar to that for Example 3.6. The energy conservation equation is used once.

3.5.2 Compressors and Pumps

A compressor or a blower increases the energy in a gas, which is manifested as increased pressure. A pump provides the same effect to a liquid. It is usual to neglect heat transfer, as well as changes in K.E. and P.E. For the compressor or pump, which undergoes a S.S.S.F. process, the energy conservation equation takes the rate form of Equation (3.11),

$$\cancel{\dot{Q}} + \dot{m}_{in}\left(h_{in} + \cancel{\frac{V_{in}^2}{2}} + \cancel{gz_{in}}\right) = \dot{m}_e\left(h_e + \cancel{\frac{V_e^2}{2}} + \cancel{gz_e}\right) + \dot{W}.$$

Under steady-state conditions, the energy conservation equation gives

$$-\dot{W}_p = \dot{m}(h_e - h_{in}) \quad \text{or} \quad -w_p = (h_e - h_{in}). \tag{3.19}$$

For incompressible fluids such as liquid water, the equation may be written as

$$-w_p = \frac{P_e - P_{in}}{\rho}. \tag{3.20}$$

Example 3.8

Problem:

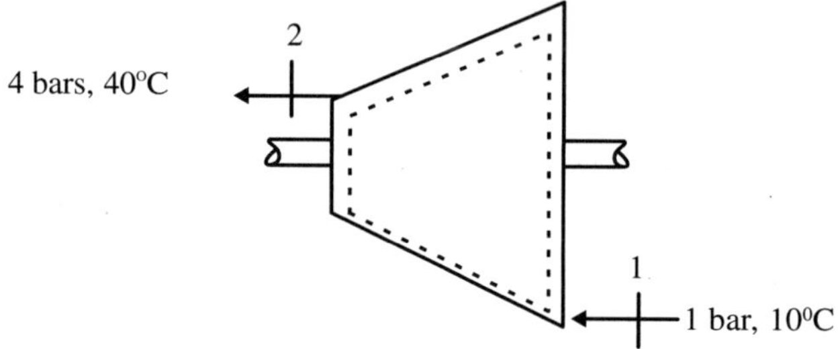

Schematic for Example 3.8.

A compressor increases the pressure of refrigerant R-134a from 1 bar and 10°C to 4 bars and 40°C. Calculate the amount of electricity input into the compressor.

Solution

The energy conservation equation, Equation (3.19), gives $-w_c = (h_2 - h_1)$. The tables give these values for the refrigerant:

$$h_1 = 411.7 \text{ kJ/kg}$$
$$h_2 = 432.5 \text{ kJ/kg}.$$

Note that the specific volume has been reduced from 0.2253 m^3/kg to 0.0594 m^3/kg. The schematic of the compressor shows an entrance that is larger than the exit because the specific volume of the vapor is reduced by the compressor. Thus, $-w_c = 20.8$ kJ/kg. The amount of electricity input into the compressor is 20.8 kJ/kg of refrigerant flow.

The logic diagram for this problem is very similar to that for Example 3.6. The energy conservation equation is used once.

3.5.3 Throttling Devices

A throttling device (Figure 3.5) may be analyzed as a steady-flow adiabatic process that causes a pressure drop with negligible kinetic or potential energy change.

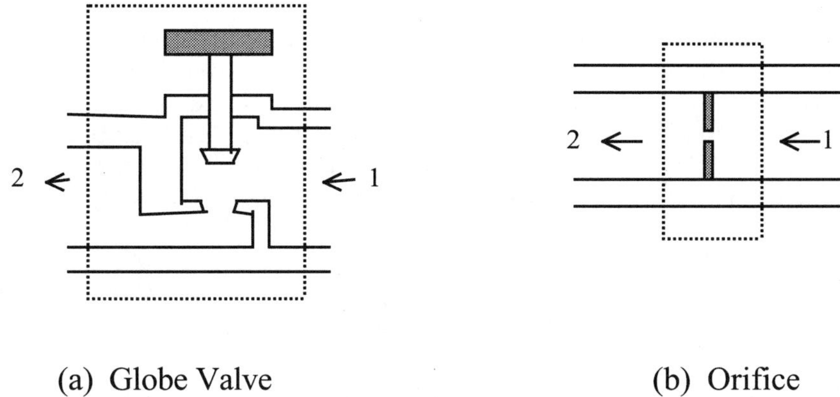

FIGURE 3.5
Throttling devices.

For a control volume that comprises this device, the energy conservation equation gives

$$0 = \dot{m}_1\left(h_1 + \frac{v_1^2}{2} + gz_1\right) - \dot{m}_2\left(h_2 + \frac{v_2^2}{2} + gz_2\right).$$

Since $\dot{m}_1 = \dot{m}_2$, and there is hardly any change in K.E. or P.E.,

$$h_1 = h_2. \tag{3.21}$$

When the flow through a valve or a restriction (e.g., orifice) is analyzed this way, the process is called a throttling process.

Example 3.9

Problem: Before a throttling valve, the steam is at 4000 kPa and 300°C and after the valve, its pressure has been reduced to 1000 kPa. Determine the temperature and specific volume of the steam after the valve.

Solution

The steam tables indicate that the steam before entering the throttling valve is in the superheated region. The tables give the enthalpy as $h_1 = 2961$ kJ/kg. Thus the enthalpy after the valve is also equal to this value, $h_2 = h_1$. From the computerized tables, it is found that $T_2 = 258.5°C$, $v_2 = 0.2366$ m³/kg.

The logic diagram for this problem is very similar to that for Example 3.6. The energy conservation equation is used once.

3.5.4 Nozzles and Diffusers

A nozzle (Figure 3.6) is a device that increases the velocity of a flowing fluid. The pressure is reduced in the process. A diffuser is a device that reduces the velocity of a flowing fluid. The pressure is increased in the process. There is no work output from these devices and normally negligible heat transfer takes place. For the nozzle or diffuser, which undergoes a S.S.S.F. process, the energy conservation equation takes the rate form of Equation (3.11),

$$\not{\dot{Q}} + \dot{m}_{in}\left(h_{in} + \frac{V_{in}^2}{2} + gz_{in}\right) = \dot{m}_e\left(h_e + \frac{V_e^2}{2} + gz_e\right) + \not{\dot{W}}.$$

With no significant changes in P.E., and $\dot{m}_{in} = \dot{m}_e$, the above equation becomes

$$h_2 - h_1 = \frac{V_1^2}{2} - \frac{V_2^2}{2}. \tag{3.22}$$

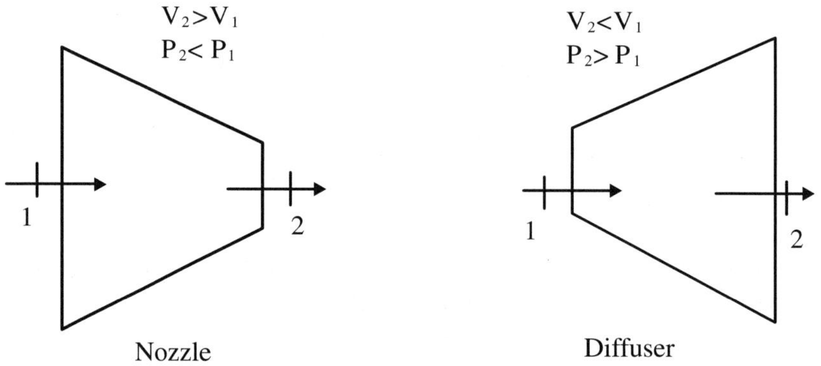

FIGURE 3.6
Nozzle and diffuser (subsonic flow).

For subsonic flow, the nozzle has a decreasing area in the direction of flow and a diffuser has an increasing area. For supersonic flow, the nozzle has an increasing area in the direction of flow and a diffuser has a decreasing area. Supersonic flow is treated in a later chapter.

In general, the volumetric flow rate is given by the mass flow rate multiplied by the specific volume, $\dot{m} \times v$. Assuming uniform velocity, the volumetric flow rate is also given by the area of flow multiplied by the velocity, $A \times V$. Hence,

$$\dot{m}v = AV. \tag{3.23}$$

This relationship is useful in general in nozzle problems and flow problems.

Example 3.10

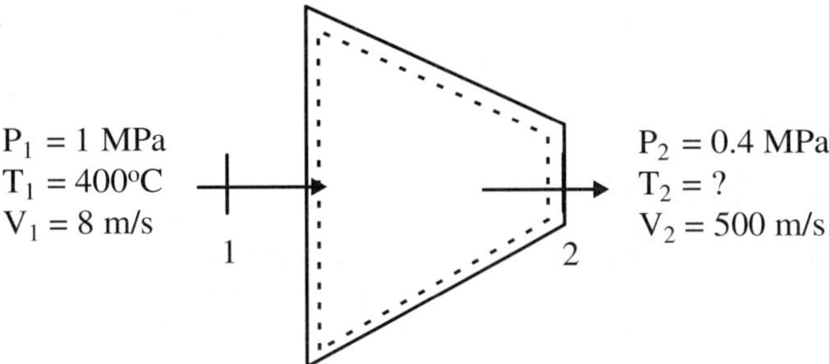

Schematic for Example 3.10.

Problem: Steady-state steady-flow steam enters a nozzle at $P_1 = 1$ MPa and 400°C, and a velocity of 8 m/s. At the exit, $P_2 = 0.4$ MPa, and the velocity is 500 m/s. The mass flow rate is 10 kg/s. Determine the exit temperature and the exit area of the nozzle, in square meters. Assume that the heat transfer and change in P.E. are negligible.

Solution

At the exit, from Equation (3.23),

$$\dot{m}\,v_2 = A_2 V_2.$$

Thus,

$$A_2 = \frac{\dot{m}v_2}{V_2}.$$

The energy conservation equation for the nozzle gives us

$$h_2 = h_1 + \frac{V_1^2}{2} - \frac{V_2^2}{2}.$$

From the tables, $h_1 = 3264$ kJ/kg.
Substituting this value in the expression for h_2,

$$h_2 = 3264\frac{\text{kJ}}{\text{kg}} + \left[\frac{8^2 - 500^2}{2}\right]\left(\frac{\text{m}^2}{\text{s}^2}\right)\left(\frac{1\text{N}}{1\text{kg.m}/\text{s}^2}\right)\left(\frac{1\text{kJ}}{10^3\text{ N.m}}\right)$$
$$= 3264 - 125 = 3139 \text{ kJ/kg}.$$

From the tables, $T_2 = 335°C$, $v_2 = 0.696\ m^3/kg$.
The exit area is

$$A_2 = \frac{(10\ kg/s)(0.696\ m^3/kg)}{500\ m/s} = 0.0139\ m^2.$$

3.5.5 Heat Exchangers

Heat exchangers are important devices in engineering. They are used to transfer energy from a hot substance/body to a cooler substance/body or to the surroundings, via heat transfer. Energy is transferred from the hot combustion gases to the water in the tubes of a boiler, an essential component of a power plant. Energy is also transferred from the hot water of the radiator in an automobile engine to the atmosphere.

In many heat exchangers, flow enters at one temperature and leaves at a different temperature. The pressure drop is usually neglected and the K.E. and P.E. changes are also neglected. No work is done in a heat exchanger.

In the schematic of a heat exchanger given in Figure 3.7, the energy conservation equation for the single control volume is

$$0 = \dot{m}_{hf}(h_{hf2} - h_{hf1}) + \dot{m}_{cf}(h_{cf2} - h_{cf1}). \quad (3.24)$$

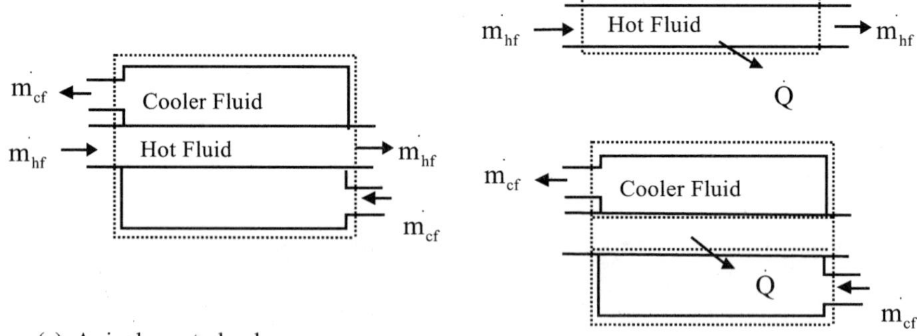

FIGURE 3.7
A heat exchanger.

The heat energy that leaves the hot fluid is transferred to the cooler fluid. For the two separate control volumes shown in Figure 3.7, Part (b),

$$\dot{Q} = \dot{m}_{cf}\left(h_{cf2} - h_{cf1}\right)$$
$$\dot{Q} = \dot{m}_{hf}\left(h_{hf2} - h_{hf1}\right). \tag{3.25}$$

Example 3.11

Problem: Steam enters a heat exchanger at 100°C and 0.01 MPa, and leaves as saturated liquid at 0.01 MPa. The steam mass flow rate is 10 kg/s. Water enters to cool the steam at 20°C and 0.01 MPa. Determine the minimum mass flow rate of the cooling water so that the water does not completely vaporize.

Solution

The energy conservation equation gives

$$\dot{m}_{hf}\left(h_{hf2} - h_{hf1}\right) = \dot{m}_{cf}\left(h_{cf1} - h_{cf2}\right).$$

We assume a saturated vapor state for the exiting cooling water at 0.01 MPa to obtain the maximum allowable exiting enthalpy. From the tables,

$$h_{hf1} = 2687 \text{ kJ/kg}$$
$$h_{hf2} = 191.8 \text{ kJ/kg}, \quad T_{hf2} = 45.8°\text{C}$$
$$h_{cf2} = 2585 \text{ kJ/kg}.$$

The value of h_{cf1} is taken to be approximately the same as the enthalpy of saturated liquid water at 20°C, since the pressure does not affect the enthalpy values very much

$$h_{cf1} = 83.95 \text{ kJ/kg}.$$

Substituting the values into the equation above,

$$10(191.8 - 2687) = \dot{m}_{cf}(83.95 - 2585);$$
$$\dot{m}_{cf} = 9.98 \text{ kg/s}.$$

The heat transfer for the process is found by using one of the energy equations for the two separate control volumes.

$$\dot{Q} = \dot{m}_{cf}\left(h_{cf2} - h_{cf1}\right)$$

$$\dot{Q} = (9.98 \text{ kg/s})(2585 - 83.95)(\text{kJ/kg})$$
$$= 24{,}952 \text{ kJ/s}$$
$$= 24.95 \text{ MW}.$$

3.6 Thermodynamic Cycles

Thermodynamic cycles occur when the substance or working fluid have the same initial and final states. These cycles may be considered as a special case of closed systems. The energy balance for a closed system was given in Equation (3.12) as

$$Q + m\left(u_{ini} + \frac{V_{ini}^2}{2} + gz_{ini}\right) = m\left(u_{fin} + \frac{V_{fin}^2}{2} + gz_{fin}\right) + W.$$

Since there is no change in energy of the system undergoing a cycle, the initial and final states of m are the same. So

$$W_{cycle} = Q_{cycle}. \tag{3.26}$$

A system may deliver a net amount of work during each cycle. Such a system is said to undergo a power cycle [Figure 3.8, Part (a)]. The net work output is equal to the net heat transfer to the cycle.

$$W_{cycle} = Q_{in} - Q_{out} \quad \text{(power cycle)}. \tag{3.27}$$

The performance of the cycle may be evaluated by the extent to which the heat energy added to the system, Q_{in}, is converted to net work output. This is called the *thermal efficiency*, and is defined as

$$\eta = \frac{W_{cycle}}{Q_{in}} \quad \text{(power cycle)}. \tag{3.28}$$

This thermal efficiency is the ratio calculated in Example 3.3.

Using Equation (3.27), another form is

$$\eta = \frac{Q_{in} - Q_{out}}{Q_{in}} = 1 - \frac{Q_{out}}{Q_{in}} \quad \text{(power cycle)}. \tag{3.29}$$

The value of the thermal efficiency is limited by the second law of thermodynamics, discussed in Chapter 4.

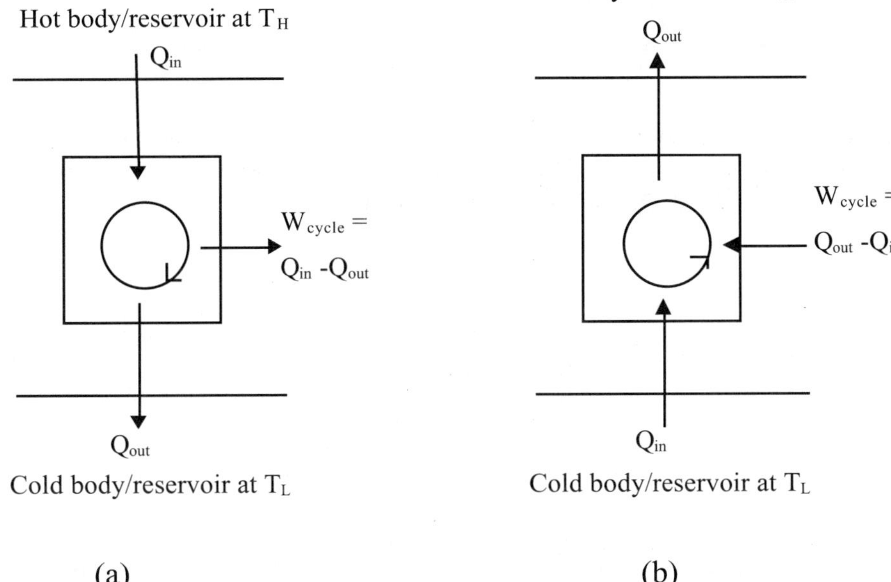

FIGURE 3.8
Schematic diagrams: (a) power cycles, (b) refrigeration and heat pump systems.

The thermal efficiency η is defined as the ratio of the required work output to the cost in terms of heat energy input. From such a definition, the thermal efficiency should be as large as possible. However, there are restrictions imposed by the second law of thermodynamics.

A system may absorb a certain quantity of work, in the process of transferring heat from a cold body to a hot body. Such a system is called a refrigeration system or a heat pump system, shown in Figure 3.8, Part (b). A refrigeration system, for instance, transfers heat from the cold food to the hot surroundings while using electricity. A heat pump system, for example, transfers heat from the cold surroundings during winter to the warm interior of a house while consuming electricity. For such cycles,

$$W_{cycle} = Q_{out} - Q_{in} \quad \text{(refrigeration and heat pump cycles).} \tag{3.30}$$

The performance of the refrigeration cycle may be evaluated by the quantity of heat energy removed from the cold body, Q_{in}, compared to work input. This is called the *coefficient of performance*, and is defined as

$$\beta = \frac{Q_{in}}{W_{cycle}} \qquad \left(\text{refrigeration cycle}\right). \tag{3.31}$$

Using Equation (3.27), another form is

$$\beta = \frac{Q_{in}}{Q_{out} - Q_{in}} \qquad \text{(refrigeration cycle)}. \qquad (3.32)$$

The performance of the heat pump cycle may be evaluated by the quantity of heat energy transferred to the hot body, Q_{out}, compared to work input. This is called the *coefficient of performance*, and is defined as

$$\gamma = \frac{Q_{out}}{W_{cycle}} \qquad \text{(heat pump cycle)}. \qquad (3.33)$$

Using Equation (3.27), another form is

$$\gamma = \frac{Q_{out}}{Q_{out} - Q_{in}} \qquad \text{(heat pump cycle)}. \qquad (3.34)$$

It can be seen that the value of γ cannot be less than unity.

The coefficient of performance β and γ are defined as the ratios of required heat transfer effect to the cost in terms of work necessary to achieve that effect. From such a definition, these ratios should be as large as possible. However, there are restrictions imposed by the second law of thermodynamics.

Example 3.12

Problem: If the maximum thermal efficiency of a cycle is 80%, and the maximum net work output is 1 kJ / s, what is the corresponding heat input into the cycle?

Solution

$$\text{Thermal efficiency } \eta = 0.8 = \frac{W_{cycle}}{Q_{in}}$$

$$Q_{in} = \frac{W_{cycle}}{0.8} = 1.25\ \frac{kJ}{s}.$$

Example 3.13

Problem: In Example 3.3, show that $W_{cycle} = Q_{cycle}$. Calculate the thermal efficiency η.

Solution

$$Q_{cycle} = 3305.1 - 2901.9 = 403.2\ kJ/kg$$

$$W_{cycle} = 424.6 - 21.4 = 403.2\ kJ/kg.$$

Hence,

$$W_{cycle} = Q_{cycle},$$

and

$$\eta = \frac{W_{cycle}}{Q_{in}} = 12.2\%.$$

Example 3.14

Problem: In a refrigeration cycle, the $Q_{in} = 100$ kJ/s and $Q_{out} = 175$ kJ/s. Calculate the work input into the cycle and the coefficient of performance.

Solution

$$W_{cycle} = Q_{out} - Q_{in} = 75 \text{ kJ/s}$$

$$\beta = \frac{Q_{in}}{W_{cycle}} = 1.33.$$

Example 3.15

Problem: In a heat pump cycle, the $Q_{out} = 100$ kJ/s and $Q_{in} = 50$ kJ/s. Calculate the work input into the cycle, and the coefficient of performance.

Solution

$$W_{cycle} = Q_{out} - Q_{in} = 50 \text{ kJ/s}$$

$$\gamma = \frac{Q_{out}}{W_{cycle}} = 2.0.$$

Problems

3.1 Water is boiling in a kettle, and steam is leaving from the spout. Apply the first law of thermodynamics to the kettle alone. Then apply the first law of thermodynamics to the kettle and the element used to heat the kettle. Write the energy conservation equations.

3.2 Given the open system above where water is the working fluid, with three incoming masses and two outgoing masses. Calculate the internal energy of the mass finally remaining in the control volume. Specify another property of this final mass.

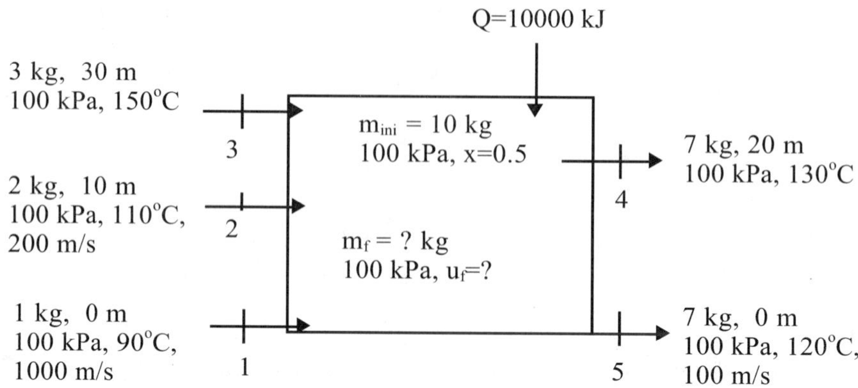

Schematic for Problem 3.2.

3.3 In Example 3.2, the state of the water in various parts of the cycle shown are as follows:

$$P_1 = 0.5 \text{ MPa} \quad T_1 = 120°\text{C (393.15 K)}$$
$$P_2 = 0.5 \text{ MPa} \quad T_2 = 700°\text{C (973.15 K)}$$
$$P_3 = 0.1 \text{ MPa} \quad T_3 = 440°\text{C (713.15 K)}$$
$$P_4 = 0.1 \text{ MPa} \quad x_4 = 0.0$$

Consider 1 kg of water undergoing the cycle. Find the heat supplied to the boiler q_b, the heat rejected in the condenser q_c, the work output from the turbine w_t, and the work input into the pump, w_p.

By considering the system as a whole, find the net work output from the closed system. In addition, calculate the ratio of this net work output to the heat supplied to the boiler.

3.4 In Example 3.2, the state of the water in various parts of the cycle shown are as follows:

$$P_1 = 80 \text{ psia} \quad T_1 = 250°\text{F (709.67°R)}$$
$$P_2 = 80 \text{ psia} \quad T_2 = 1200°\text{F (1659.67°R)}$$
$$P_3 = 10 \text{ psia} \quad T_3 = 700°\text{F (1159.67°R)}$$
$$P_4 = 10 \text{ psia} \quad x_4 = 0.0$$

Compute the heat supplied to the boiler q_b, the heat rejected in the condenser q_c, the work output from the turbine w_t, and the work input into the pump, w_p. Consider 1 lb_m of water undergoing the cycle.

If the system is taken as a whole, compute the net work output from the closed system. What is the ratio of this net work output to the heat supplied to the boiler?

3.5 An open system has an inlet fluid stream of 0.52 kg/s at a velocity of 28 m/s at an elevation of 28 m. At the outlet, which is at 60 m, the fluid leaves at a velocity of 8.9 m/s. The enthalpies of the incoming and exiting fluid are 2560 kJ/kg and 2576 kJ/kg, respectively. If the work done by the system is 2.08 kW, determine the heat supplied to the system.

3.6 A rigid, insulated tank is initially evacuated. It is then connected via a valve to a steam line that supplies steam at 3 bars and 320°C (593.15 K). The valve is opened and steam flows slowly into the tank until the pressure reaches 3 bars; then the valve is closed. Determine the final temperature of the steam in the tank.

3.7 A balloon is flat, and may be assumed to contain no air. It is then connected via a valve to a tank of air at 300°C (573.15 K) and at a certain pressure above atmospheric pressure. The valve is opened and air flows slowly into the balloon until the pressure reaches the same pressure as in the tank; then the valve is closed. Determine the final temperature of the air in the balloon. (Hint: the internal energy and the enthalpy of air may be treated as functions of temperature only. Neglect work done by the balloon.)

3.8 A rigid, insulated tank is initially evacuated. It is then connected via a valve to a steam line that supplies steam at 40 psia and 600°F (1059.67°R). The valve is opened and steam flows slowly into the tank until the pressure reaches 40 psia; then the valve is closed. Determine the final temperature of the steam in the tank.

3.9 A balloon is flat, and may be assumed to contain no air. It is then connected via a valve to a tank of air at 520°F (979.67°R) and at a certain pressure above atmospheric pressure. The valve is opened and air flows slowly into the balloon until the pressure reaches the same pressure as in the tank; then the valve is closed. Determine the final temperature of the air in the balloon. (Hint: the internal energy and the enthalpy of air may be treated as functions of temperature only. Neglect work done by the balloon.)

3.10 Consider a turbine through which steam flows at 5 kg/s. The entry conditions are 3 MPa and 700°C (973.15 K) with a velocity of 160 m/s, and the exit conditions are a saturated state at 100 kPa with a velocity of 140 m/s. Neglect any heat transfer. The steam flow rate is 5 kg/s. Determine (a) the change in kinetic energy and (b) the turbine output. What is the percentage of the quantity (a) divided by the quantity (b)?

3.11 Steam at 20 lb/s enters a turbine at 350 $lb_f/in.^2$ MPa and 1200°F (1659.67°R) with a velocity of 450 ft/s, and leaves saturated at an outlet pressure of 20 $lb_f/in.^2$ with a velocity of 400 ft/s. Compute (a) the change in kinetic energy and (b) the turbine output. What is the percentage of the quantity (a) divided by the quantity (b)? Assume no heat transfer.

3.12 A 13.5 kW fan is used in a room for circulation purposes. Assuming a sealed room that is well insulated, determine the internal energy increase after 1.5 h of operation.

3.13 A 10 hp fan is used in a room for circulation purposes. Assuming a sealed room that is well insulated, determine the internal energy increase after 2 h of operation.

3.14 A piston of mass 50 kg holds a mass of water vapor. Heat is added in the amount of 1 kJ. Calculate the change in internal energy of the water vapor.

3.15 A mass of water vapor is held by a piston of mass 100 lb_m. Add 1000 Btu of heat to the fluid. Determine the change in internal energy of the water vapor.

First Law Applied to Various Processes

3.16 A turbine converts energy to work from steam supplied at 600°C (873.15 K) and 1.5 MPa. The steam leaves at 280°C (653.15 K) and 0.1 MPa. Calculate the work output from the turbine.

3.17 A turbine converts energy to work from ammonia supplied at 200°C (473.15 K) and 0.7 MPa. The ammonia leaves at 15°C (288.15 K) and 0.06 MPa. Calculate the work output from the turbine.

3.18 A turbine converts energy to work from steam supplied at 1000°F (1459.67°R) and 100 psia. The steam leaves at 500°F (959.67°R) and 14.7 psia. Calculate the work output from the turbine.

3.19 A turbine converts energy to work from ammonia supplied at 380°F (839.67°R) and 100 psia. The ammonia leaves at 70°F (529.67°R) and 10 psia. Calculate the work output from the turbine.

3.20 A compressor changes the state of refrigerant R-134a from 1.4 bar and 20°C (293.15 K) to 5 bars and 80°C (353.15 K). Determine the amount of electricity consumed by the compressor.

3.21 The pressure of refrigerant R-12 is increased by a compressor from 1 bar and 20°C (293.15 K) to 4 bars and 100°C (373.15 K). Compute the amount of electricity input into the compressor.

3.22 The state of ammonia is changed from 0.5 bar and 10°C (283.15 K) to 4 bars and 200°C (473.15 K). Find the amount of electricity consumed by the compressor.

3.23 A compressor changes the state of refrigerant R-134a from 20 psia and 80°F (539.67°R) to 80 psia and 180°F (639.67°R). Determine the amount of electricity input into the compressor.

3.24 A compressor increases the pressure of refrigerant R-12 from 15 psia and 60°F (519.67°R) to 70 psia and 180°F (639.67°R). Determine the electricity consumption of the compressor.

3.25 The state of ammonia is changed from 10 psia and 40°F (499.67°R) to 50 psia and 300°F (759.67°R). Compute the amount of electricity consumed by the compressor.

3.26 Ammonia expands through a valve from a pressure of 700 kPa to a pressure of 150 kPa. The exiting ammonia has a quality of 67%. Determine the inlet temperature.

3.27 A valve is used to reduce the pressure of refrigerant-12 from 200 $lb_f/in.^2$ to half. The exiting refrigerant has a quality of 50%. Determine the inlet temperature.

3.28 The conditions of the steam before a throttling valve are 3000 kPa and 300°C (573.15 K). After the valve, its pressure is 2000 kPa. Find the temperature and specific volume of the steam after the valve.

3.29 A throttling valve causes steam at 450 psia and 570°F (1029.67°R) to expand to 290 psia. Find the temperature and specific volume of the steam after the valve.

3.30 At steady state, a nozzle causes steady-flow steam at $P_1 = 0.1$ MPa and 500°C (773.15 K), and a velocity of 12 m/s, to transform to $P_2 = 0.05$ MPa with a velocity of 450 m/s. The mass flow rate is 8 kg/s. Find the exit temperature and the exit area of the nozzle, in square meters. Heat transfer and change in P.E. are negligible.

3.31 Steady-flow steam at 20 lb/s enters a nozzle at $P_1 = 200$ psia and 600°F (1059.67°R), and a velocity of 10 ft/s. At steady state, the exit conditions are $P_2 = 100$ psia and velocity = 600 ft/s. Find the exit temperature and the exit area of the nozzle, in square feet. Assume that the heat transfer and change in P.E. are negligible.

3.32 Steam enters a heat exchanger at 100°C (373.15 K) and 0.01 MPa, and leaves as saturated liquid at 0.01 MPa. The steam mass flow rate is 10 kg/s. Ammonia enters to cool the steam at 50°C (323.15 K) and 0.1 MPa. Determine the minimum mass flow rate of the ammonia so that the ammonia does not completely vaporize.

3.33 Steam enters a heat exchanger at 250°F (709.67°R) and 14.7 psia, and leaves as saturated liquid at 14.7 psia. The steam mass flow rate is 20 lb_m/s. Ammonia enters to cool the steam at 32.8°F (426.87°R) and 14.504 psia. Determine the minimum mass flow rate of the ammonia so that the ammonia does not completely vaporize.

3.34 In an open heat exchanger, one stream of 55 kg/s water enters at 540 kPa and 35°C (308.15 K), and another stream enters at 540 kPa, and $x = 0.39$. The exit water is in the saturated liquid state. Determine the mass flow rate at the second inlet stream.

3.35 Steam of 0.98 quality enters the condenser of a steam power plant at 10 kPa and a mass flow rate of 36,000 kg/h. The water for cooling the steam comes from the ocean; but to avert thermal pollution, the cooling water is not allowed to increase more than 6°C (279.15 K). If the steam exits the condenser as saturated liquid at 10 kPa, find the mass flow rate of the cooling water from the ocean.

3.36 A rigid vessel whose volume is 1.7 m^3 contains steam at 200°C (473.15 K) and 60% quality. Determine the heat transferred if the vessel is cooled to 50°C (323.15 K).

3.37 Six kilograms of liquid water (no vapor) at 25°C (298.15 K) is contained in a piston-cylinder device; the weight of the piston is such that the pressure of the liquid is 750 kPa. An unexpected fire results in the water heating up and expanding, causing the piston to rise frictionlessly until it lodges at a position where the volume enclosed is 1.25 m^3. More heat from the fire causes the water to convert entirely into saturated vapor. Calculate the heat transferred to the water and the work done by the water.

3.38 A rigid tank contains 3 kg of water at 1200 kPa and 15% quality. The tank is then heated, and saturated vapor is allowed to leave the tank so that a constant pressure is kept within the tank until the liquid has decreased to 25% its initial mass. Calculate (a) the heat transferred and (b) the mass of vapor that left the tank.

Thermodynamic Cycles

3.39 A closed system executes a process from state 1 to state 2 where its internal energy is increased by 88 Btu, and the heat added to the system is 156 Btu. The amount of work done on the system to restore it to its initial state (state 2 to state 1) is 47 Btu. Evaluate the heat transfer of the process 2 → 1.

3.40 A closed system undergoes a cycle which comprises three processes, as listed in the table below. Calculate the unknown values. All quantities are in kiloJoules.

Process	Q	W	Δ(U + K.E. + P.E.)
1→2	50	a	30
2→3	–20	b	c
3→1	d	50	40

3.41 A closed system undergoes a cycle which comprises three processes, as listed in the table below. Calculate the unknown values. All quantities are in British thermal units.

Process	Q	W	Δ(U + K.E. + P.E.)
1→2	150	a	130
2→3	–80	b	c
3→1	d	70	50

3.42 The maximum thermal efficiency of a cycle is 95%, and the maximum net work output is 2 kJ/s. Calculate the corresponding heat input into the cycle.

3.43 If the minimum thermal efficiency of a cycle is 60%, and the heat input into the cycle is 10 kJ/s, what is the minimum work output?

3.44 If the maximum net work output of a cycle is 2 Btu/s and the maximum thermal efficiency is 95%, what is the corresponding heat input into the cycle?

3.45 The minimum thermal efficiency of a cycle is 55%, and the heat input into the cycle is 10 Btu/s. Compute the minimum work output.

3.46 In Problem 3.3, show that $W_{cycle} = Q_{cycle}$. Calculate the thermal efficiency η.

3.47 In Problem 3.4, show that $W_{cycle} = Q_{cycle}$. Calculate the thermal efficiency η.

3.48 In a refrigeration cycle, the $\dot{Q}_{in} = 100$ kJ/s and $\dot{Q}_{out} = 250$ kJ/s. Calculate the work input into the cycle, and the coefficient of performance.

3.49 The power consumption of a refrigerator is 400 W. If the coefficient of performance is 1.5, calculate $\dot{Q}_{in}$.

3.50 In a refrigeration cycle, the $\dot{Q}_{in} = 100$ Btu/s and $\dot{Q}_{out} = 200$ Btu/s. Calculate the work input into the cycle, and the coefficient of performance.

3.51 The power consumption of a refrigerator is 0.35 Btu/s. If the coefficient of performance is 1.35, calculate $\dot{Q}_{in}$.

3.52 The quantities $\dot{Q}_{out} = 200$ kJ/s and $\dot{Q}_{in} = 150$ kJ/s apply to a heat pump cycle. Compute the work input into the cycle, and the coefficient of performance.

3.53 The power consumption of a heat pump for a residential building is 3 kW. If the coefficient of performance is 2.0, calculate $\dot{Q}_{out}$.

3.54 In a heat pump cycle, the $\dot{Q}_{out} = 200$ Btu/s and $\dot{Q}_{in} = 125$ Btu/s. Calculate the work input into the cycle, and the coefficient of performance.

3.55 A heat pump for a residential building consumes power at a rate of 3 Btu/s. If the coefficient of performance is 1.8, calculate $\dot{Q}_{out}$.

Computer, Design, and General Problems

3.56 Water is boiling in a kettle, and steam is leaving from the spout. With a thermometer and a ruler, explain how you can determine the average velocity of the steam leaving the kettle.

3.57 Research the experimental work related to the first law of thermodynamics, including that done by James Prescott Joule. Discuss Joule's contribution and, in particular, the events that were pivotal to the acceptance of the first law.

3.58 A man is exercising on a stationary bicycle in an exercise room. The room is insulated from its surroundings, and is supplied with electricity for lights and air-conditioning. Consider a system consisting of the man and the bicycle. Apply the conservation of energy principle or first law of thermodynamics to this system. Now, consider an expanded system that comprises the whole room. Apply the first law of thermodynamics to this expanded system.

3.59 Hydroelectricity is based on the conversion of the hydraulic head of water (usually from a river or lake) to electricity via a hydraulic turbine-generator. Write an essay on the major hydroelectric projects in the world, concentrating on the figures of available hydraulic head, water flow rate, and the seasonal variability of these quantities.

3.60 Consider your own home as a control volume. Set up the energy balance of your home. Take into consideration the heat gained by the home in the daytime when the home may be cooler than the surroundings, and the heat lost by the home at night when the home may be warmer than the surroundings. How do you account for yourself and your parents as you and they go in and out of the house?

4

Second Law of Thermodynamics and Entropy

CONTENTS

4.1 Introduction

If we look around us, we see many examples of work that is completely converted into heat. However, we do not see any cycling system that completely converts heat into work. Such a complete conversion would not be a violation of the first law of thermodynamics. The fact that heat cannot be completely converted into work is the basis for the second law of thermodynamics. The justification for the second law is empirical, that is, it is based on experimental evidence.

When we consider a mass of hot water in a room at a lower temperature, experience tells us that the water will cool to the room temperature eventually; the hot water will not gain heat from the lower-temperature room. The second law helps to predict the direction of this spontaneous process and other processes like it.

The second law of thermodynamics has many uses. They include the following:

1. Means to establish the conditions for equilibrium;
2. Means to predict the direction of processes;
3. Means to set the conditions for the ideal or reversible process, which can then be used as a standard of comparison;
4. Allowing the establishment of the best theoretical performance of engines, refrigerators, heat pumps, and other devices;
5. Means to quantitatively account for factors that prevent the best theoretical performance;
6. Allowing properties like h to be determined in terms of other properties that are more easily measured;
7. Allowing a temperature scale to be defined that is independent of the thermometric fluid.

Associated with item 3 above, reversible processes were first introduced in Chapter 2, Section 2.6.2. It is emphasized here that the concept of a reversible process is to establish a standard for the comparison of actual processes. All actual processes are irreversible. The reversible process provides the maximum accomplishment, e.g., it provides the greatest amount of work or requires the least amount of work to bring about a specified change.

The factors that make processes irreversible include friction, viscosity, free expansion, inelastic deformation, heat transferred across a finite temperature difference, mixing of substances, all chemical reactions, sudden change of phase, losses in electrical resistors, and hysteresis effects. It is often useful in analyses to classify irreversibilities into two categories. *Internal irreversibilities* are those within the system boundaries. *External irreversibilities* are those in the surroundings of the system, often the immediate surroundings.

There are many different ways of stating the second law. For instance, Caratheodory's statement is as follows: "There exist arbitrarily close to any given state of a system other states which cannot be reached from it by reversible adiabatic processes."

The Kelvin-Planck statement of the second law is as follows: "It is impossible for any cycling device to exchange heat with only a single reservoir and produce positive work." Heat cannot be continuously and completely converted into work; a fraction of the heat must be rejected to another reservoir at a lower temperature. The second law of thermodynamics thus places a restriction on the first law in the relation to the way energy is transferred. Work can be continuously and completely converted into heat, but not vice versa.

If the Kelvin-Planck statement was not true and heat could be completely converted into work, the heat might be obtained from a low-temperature source, converted into work, and the work converted back into heat in a region of higher temperature. The net result of this series of events would be the flow of heat from a low-temperature region to a high-temperature region with no other effect. This phenomenon has never been observed and is contrary to all our experience.

The Clausius statement of the second law is as follows: "No process is possible whose sole result is the removal of heat from a reservoir at one temperature and the absorption of an equal quantity of heat by a reservoir at a higher temperature." This statement does not say that it is impossible to transfer heat from a lower-temperature body to a higher-temperature body. This is exactly what a refrigerator does when it receives an energy input, usually in the form of work. This transfer of energy from the surroundings constitutes an effect other than the transfer of heat from the lower-temperature body to the higher-temperature body; thus, the "sole effect" of the Clausius statement includes effects within the refrigerating device itself.

Every relevant experiment performed either directly or indirectly verifies the second law. No experiment has produced results that contradict the second law. Therefore, the second law is empirical or based on experimental results.

It is timely to point out that the second law of thermodynamics has been stated as the impossibility of building a perpetual-motion machine of the second kind. This second kind of perpetual-motion machine would take heat from a source and then completely convert this heat into other forms of energy, thus violating the second law. A perpetual-motion machine of the first kind violates the first law. This first kind of perpetual-motion machine would create work from nothing, or create energy. The third kind of perpetual-motion machine would have no friction, and consequently produce no heat while running indefinitely. Note that this third kind of perpetual-motion machine is not related to the third law of thermodynamics. These perpetual motion machines (P.M.M.s) are illustrated in Figure 4.1.

4.1.1 The Kelvin-Planck and the Clausius Statements

In general, two statements are equivalent if the truth of each statement results in the truth of the other, or if the violation of each statement results in the violation of the other. The rest of this section sets out to prove the equivalence of the Clausius and the Kelvin-Planck statements of the second law of thermodynamics.

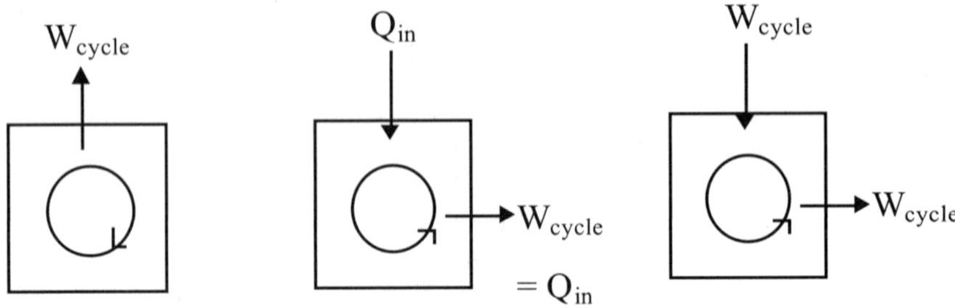

FIGURE 4.1
Perpetual motion machines (P.M.M.s).

In Figure 4.2 is shown a system (inside the dashed rectangle) that violates the Clausius statement. It is composed of a normal engine and a refrigerator that violates the Clausius statement.

The net heat from the low-temperature reservoir is zero ($Q_2 - Q_2 = 0$). The heat transferred into the high-temperature reservoir is $Q_1 - Q_2$. The net work done by the system is $Q_1 - Q_2$. Therefore, the Kelvin-Planck statement has been violated. In other words, a system that violates the Clausius statement has been shown to also violate the Kelvin-Planck statement.

In Figure 4.3 is shown a system (inside the dashed rectangle) that violates the Kelvin-Planck statement. It is composed of a normal refrigerator and an engine that violates the Kelvin-Planck statement.

The net heat from the low-temperature reservoir is Q_2. The net heat transferred into the high-temperature reservoir is Q_2. The net work done by the system is zero. Therefore, the Clausius statement has been violated. In other words, a system that violates the Kelvin-Planck statement has been shown to also violate the Clausius statement.

4.2 Statements of the Second Law

Ever since scientists have been working independently, they have come up with many theories that were only later shown to be similar. Hence, many corollaries of the second law of thermodynamics exist. This section lists various statements of the second law.

Corollary 1: (Carnot Principle or Carnot Theorem.) No engine operating between two given reservoirs can have a greater efficiency than a reversible engine operating between the same two reservoirs.

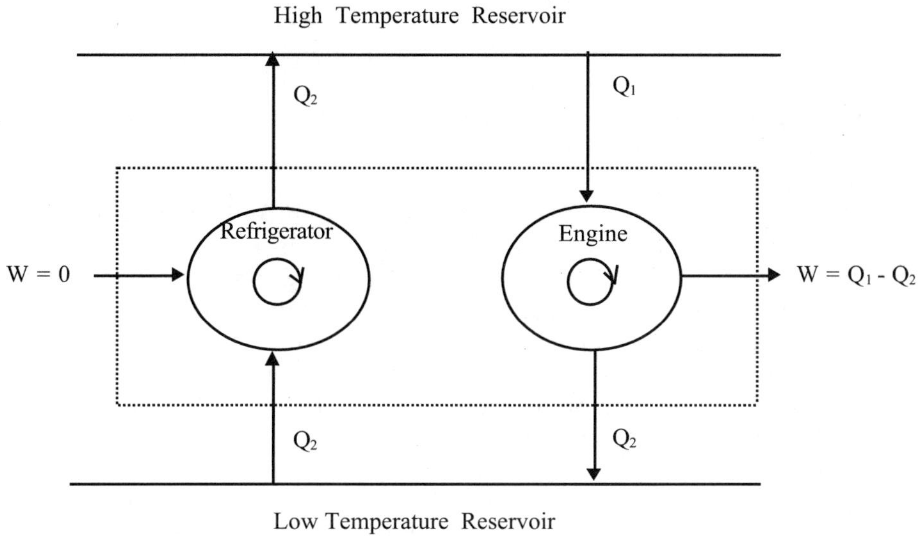

High Temperature Reservoir
Q2
Q1
Refrigerator
Engine
W = 0
W = Q1 - Q2
Q2
Q2
Low Temperature Reservoir

FIGURE 4.2
A combination that violates the Clausius statement.

High Temperature Reservoir
Q2
Q1 + Q2
Engine
W = Q1
Refrigerator
Q2 = 0
Q2
Low Temperature Reservoir

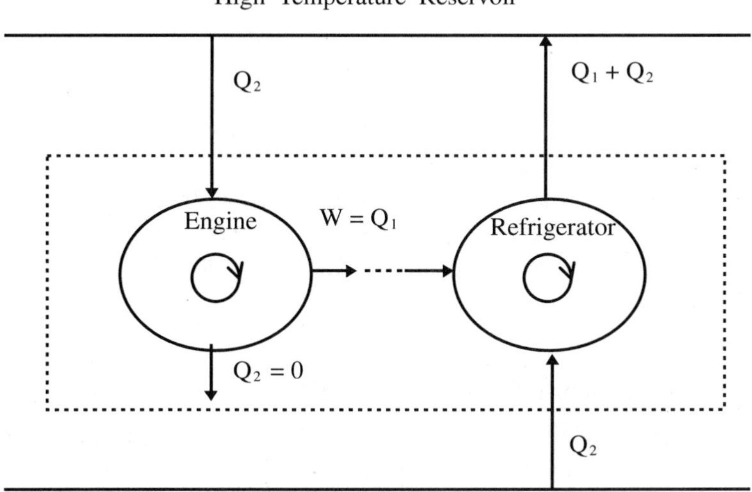

FIGURE 4.3
A combination that violates the Kelvin-Planck statement.

Corollary 2: (Carnot Corollary.) All reversible engines operating between the same temperature limits have the same efficiency.

Corollary 3: The efficiency of any reversible engine operating between two reservoirs is independent of the nature of the working fluid and depends only on the temperature of the reservoirs, for any reversible process undergone by the system between states 1 and 2. This property is called the entropy.

Corollary 4: It is theoretically impossible to reduce the temperature of a system to absolute zero by a series of finite processes.

Corollary 5: (The Thermodynamic Temperature Scale.) Define the ratio of two temperatures as the ratio of the heat absorbed by a Carnot engine to the heat rejected when the engine is operated between reservoirs at these temperatures. Thus the equality $Q_L / Q_H = T_L / T_H$ becomes a matter of definition, and the fundamental problem of thermometry, that of establishing a temperature scale, reduces to a problem in calorimetry.

Corollary 6: (The Inequality of Clausius.) When a system is carried around a cycle and the heat δQ added to it at every point is divided by its temperature, the sum of all such quotients is less than zero for irreversible cycles and in the limit is equal to zero for reversible cycles:

$$\oint \delta Q / T \leq 0 \tag{4.1}$$

Corollary 7: There exists a property (denoted by s) of a system such that a change in its value is equal to

$$s_2 - s_1 = \int_1^2 \delta Q / T \tag{4.2}$$

for any reversible process undergone by the system between states 1 and 2. This property is called the entropy.

Corollary 8: (Principle of the Increase of Entropy.) In any process whatever between two equilibrium states of a system, the increase in entropy of the system plus the increase in entropy of its surroundings is equal to or greater than zero.

4.2.1 Thermodynamic Temperature Scale

Corollary 5, above, is the basis of the thermodynamic temperature scale. Consider a system undergoing an ideal power cycle operating between reservoirs at temperatures τ_H and τ_L on a scale that has yet to be defined. Based on Carnot's Corollary (Corollary 2, above), the thermal efficiency of the cycle is

$$\eta = \eta(\tau_H, \tau_L). \tag{4.3}$$

Since the thermal efficiency is defined as

$$\eta = \frac{W_{net}}{Q_H} = \frac{Q_H - Q_L}{Q_H},$$

then

$$\eta = (\tau_H, \tau_L) = 1 - \frac{Q_L}{Q_H}. \tag{4.4}$$

Thus,

$$\frac{Q_L}{Q_H} = 1 - \eta(\tau_H, \tau_L) \tag{4.5}$$

$$\left(\frac{Q_L}{Q_H}\right)_{\text{reversible cycle}} = \phi(\tau_H, \tau_L) \tag{4.6}$$

where $\phi(\tau_H, \tau_L)$ is a function of τ_H and τ_L.

The above equation provides a basis for defining a thermodynamic temperature scale, that is, a scale that is independent of the thermometric substance. The Kelvin temperature scale is obtained by making the choice that $\phi(\tau_H, \tau_L) = T_L/T_H$, where T is the letter representing a temperature on the Kelvin scale. Therefore,

$$\left(\frac{Q_L}{Q_H}\right)_{\text{reversible cycle}} = \frac{T_L}{T_H}. \tag{4.7}$$

In other words, the two temperatures on the Kelvin scale are in the same ratio as the magnitudes of the heat transfer gained and that rejected, respectively, by a system undergoing a reversible cycle between two reservoirs at these temperatures. The temperatures on the Kelvin scale are called absolute temperatures.

The triple point of water (the state at which all three phases of water are in equilibrium) was given the value of 273.16 K. The magnitude of a kelvin is defined as 1/273.16 of the temperature interval between absolute zero temperature and the triple-point temperature of water.

4.3 Entropy of a Pure, Simple Compressible Substance

Equation (4.2), which is essentially the equation from one of the corollaries of the second law, can be used to assign a value to the entropy. The value of entropy at any state y relative to the value at the reference state x is calculated by integration

$$S_y = S_x + \left(\int_x^y \frac{\delta Q}{T} \right)_{\text{int rev}}. \tag{4.8}$$

Both the entropy value and the reference state can be chosen arbitrarily.

The use of an arbitrary reference state is satisfactory only when entropy differences are of relevance since the reference value cancels. When there are chemical reactions, it is necessary to calculate the absolute values of entropy determined by the third law of thermodynamics.

Tables of thermodynamic data are available for entropy values, together with values of v, u, h, T, and P. Computerized thermodynamic tables are becoming more commonplace because of their ease of use, and these computerized tables are highly recommended.

4.3.1 Tds Equations

The entropy values can also be found by Tds equations, which are developed in this section. These equations are used beyond finding entropy values; they can be used for deriving many important relations for pure, simple compressible systems, including ways for building the property tables for u, h, and s.

For a pure, simple compressible system undergoing an internally reversible process in the absence of system motion and gravitational effects, an energy balance in differential form is

$$(\delta Q)_{\text{int rev}} = dU + (\delta W)_{\text{int rev}}. \tag{4.9}$$

For a simple compressible system, the work is given by

$$(\delta W)_{\text{int rev}} = PdV. \tag{4.10}$$

On a differential basis, the entropy change equation is

$$dS = \left(\frac{\delta Q}{T} \right)_{\text{int rev}} \tag{4.11}$$

or

$$(\delta Q)_{\text{int rev}} = TdS. \tag{4.12}$$

Substituting Equations (4.10) and (4.12) into Equation (4.9), the first TdS equation is obtained:

$$TdS = dU + PdV. \tag{4.13}$$

The second Tds equation is derived from Equation (4.13) using H = U + PV. The differential form is

$$dH = dU + d(PV) = dU + PdV + VdP.$$

On rearranging,

$$dU + PdV = dH - VdP.$$

Putting this into Equation (4.6) provides the second TdS equation:

$$TdS = dH - VdP. \tag{4.14}$$

The TdS equations can be written as a per unit mass basis as

$$Tds = du + Pdv \tag{4.15}$$

$$Tds = dh - vdP \tag{4.16}$$

or as a per mole basis as

$$Td\bar{s} = d\bar{u} + Pd\bar{v} \tag{4.17}$$

$$Td\bar{s} = d\bar{h} - \bar{v}dP. \tag{4.18}$$

Regardless of the derivation assumptions, the entropy change for a system obtained by integrating these equations is the change of any process of the system, reversible or irreversible, between two equilibrium states. Since entropy is a property, the change in entropy between two states is not dependent on the details of the process joining the two states.

4.3.2 Entropy Change of an Ideal Gas

From Equations (4.15) and (4.16)

$$ds = \frac{du}{T} + \frac{P}{T}dv \tag{4.19}$$

$$ds = \frac{dh}{T} - \frac{v}{T}dp. \tag{4.20}$$

In addition, for an ideal gas

$$\begin{aligned} du &= C_v\,(T)dT \\ dh &= C_p\,(T)dT \\ Pv &= RT. \end{aligned}$$

Thus, Equations (4.19) and (4.20) become, respectively,

$$ds = C_v(T)\frac{dT}{T} + R\frac{dv}{v} \tag{4.21}$$

and

$$ds = C_p(T)\frac{dT}{T} - R\frac{dP}{P}. \tag{4.22}$$

Since $C_P(T) = C_v(T) + R$ where R = gas constant. On integration,

$$s(T_2, v_2) - s(T_1, v_1) = \int_{T_1}^{T_2} C_v(T)\frac{dT}{T} + R\ln\left(\frac{v_2}{v_1}\right) \tag{4.23}$$

$$s(T_2, P_2) - s(T_1, P_1) = \int_{T_1}^{T_2} C_p(T)\frac{dT}{T} - R\ln\left(\frac{P_2}{P_1}\right). \tag{4.24}$$

Write

$$s^\circ(T) = \int_0^T C_p(T)\frac{dT}{T} \tag{4.25}$$

which is the specific entropy at a pressure of 1 atm and temperature T; $s^\circ(T)$ is a function only of temperature T.

$$\begin{aligned} \int_{T_1}^{T_2} C_P\frac{dT}{T} &= \int_0^{T_2} C_P\frac{dT}{T} - \int_0^{T_1} C_P\frac{dT}{T} \\ &= s^\circ(T_2) - s^\circ(T_1) \end{aligned} \tag{4.26}$$

Therefore,

$$s(T_2, P_2) - s(T_1, P_1) = s^\circ(T_2) - s^\circ(T_1) - R\ln\left(\frac{P_2}{P_1}\right). \tag{4.27}$$

Or on a per mole basis,

$$\bar{s}(T_2, P_2) - \bar{s}(T_1, P_1) = \bar{s}^{\circ}(T_2) - \bar{s}^{\circ}(T_1) - \bar{R}\ln\left(\frac{P_2}{P_1}\right). \tag{4.28}$$

The above are relations obtained from an exact analysis when the specific heats are variable. When the specific heats are approximately constant, by carrying out the integration of Equations (4.23) and (4.24) we obtain,

$$s(T_2, v_2) - s(T_1, v_1) = C_{v,av} \ln\frac{T_2}{T_1} + R\ln\left(\frac{v_2}{v_1}\right) \tag{4.29}$$

$$s(T_2, P_2) - s(T_1, P_1) = C_{p,av} \ln\frac{T_2}{T_1} - R\ln\left(\frac{P_2}{P_1}\right). \tag{4.30}$$

By multipying these relations by molar mass, we obtain the entropy changes on a unit-mole basis.

$$\bar{s}(T_2, v_2) - \bar{s}(T_1, v_1) = \bar{C}_{v,av} \ln\frac{T_2}{T_1} + R_u \ln\left(\frac{v_2}{v_1}\right) \tag{4.31}$$

$$\bar{s}(T_2, P_2) - \bar{s}(T_1, P_1) = \bar{C}_{p,av} \ln\frac{T_2}{T_1} - R_u \ln\left(\frac{P_2}{P_1}\right). \tag{4.32}$$

4.3.3 Entropy Change of an Incompressible Substance

An incompressible substance may be modelled by assuming that the specific volume is constant, $dv = 0$, and that the specific heat is a function of temperature only, $C_{v\,=}\,C(T)$. Thus, the differential change in specific internal energy is $du = C(T)\,dT$ and Equation (4.19) reduces to

$$ds = \frac{C(T)dT}{T} + \frac{Pdv}{T} = \frac{C(T)dT}{T}. \tag{4.33}$$

On integration, the specific entropy change is

$$s_2 - s_1 = \int_{T_1}^{T_2} \frac{C(T)}{T} dT \qquad \text{(incompressible)}. \tag{4.34}$$

When the specific heat is assumed constant, this gives

$$s_2 - s_1 = C \ln \frac{T_2}{T_1} \qquad \text{(incompressible, constant c).} \qquad (4.35)$$

4.3.4 Important Facts about Entropy

The following are important facts about entropy:

1. The entropy of a system is defined for equilibrium states only.
2. Only entropy differences may be computed.
3. The entropy of a system in an equilibrium state is independent of the history of the system. It is a property. It is a point function, not a path function.
4. Entropy changes can be computed from the heat transfer for reversible processes only.
5. Entropy changes for an irreversible process from one equilibrium state to another may be determined by the following methods.
   (a) Devising a reversible process connecting the same two end states.
   (b) Using thermodynamic tables (subtract the two end-state values).
   (c) Using s = s(T,P) if this functional relationship is known.
   (d) Using computer-aided thermodynamic tables.
6. For reversible processes, the heat transfer may be computed directly from the properties and is just the area under the process curve represented on the (T-s) diagram.

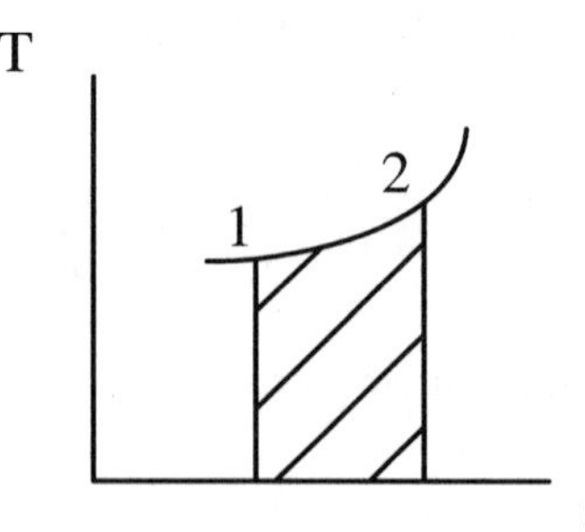

FIGURE 4.4
(T-s) diagram.

Since for reversible processes, $\delta Q = Tds$, then

$$_1Q_2 = \int_1^2 Tds. \tag{4.36}$$

Even after all the foregoing treatment, a student of thermodynamics may reasonably inquire about the physical nature of entropy. The answer is that both energy and entropy are abstract concepts. Energy, however, is a part of our everyday language and we all were familiar with the term before seeing it in early science courses. Entropy is not part of our common vocabulary. Entropy is introduced in this chapter, and shown to be very useful in thermodynamic analysis. Energy and entropy are important concepts in the remainder of the book.

For a solid at equilibrium with its vapor at a specified temperature and pressure, the entropy of the solid is less than that of the vapor. Since the popular scientific models of solids and gases provide us with a mental physical picture of the solid as consisting of molecules which are less random than those of a gas, we are led to one interpretation that entropy is a measure of randomness. The above statements may also be made for a liquid at equilibrium with its vapor at a specified temperature and pressure. The entropy of the liquid is less than that of the vapor of the same substance.

Example 4.1

Problem: A piston-cylinder device contains 5 lb_m of refrigerant-12 at 15 $lb_f/in.^2$ and $x = 0.1$. The refrigerant is then heated at constant pressure by the addition of 500 Btu of heat. Determine the entropy change of the refrigerant.

Solution

From the tables, state 1: $h_1 = 11.17$ Btu/lb_m, and $s_1 = 0.0256$ Btu/(lb_m.°R)

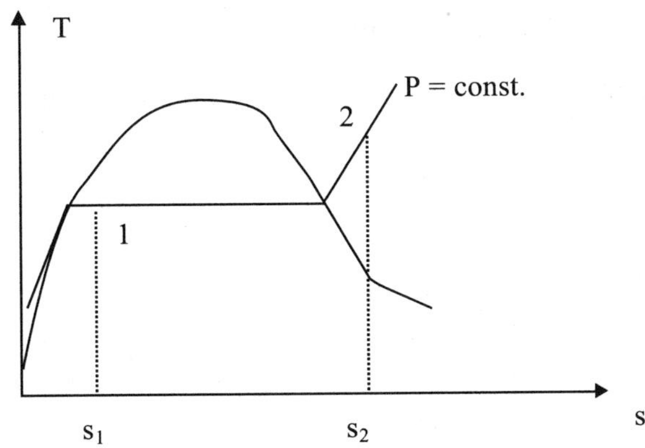

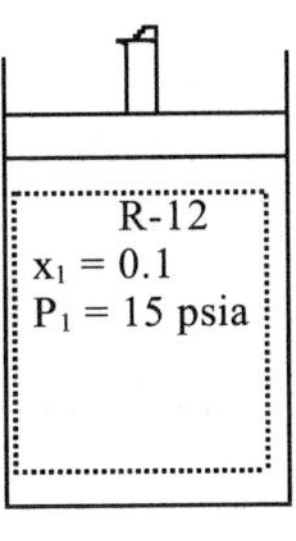

Schematic for Example 4.1.

From the first law of thermodynamics applied to the system,

$$Q + mu_1 = mu_2 + P(v_2 - v_1)$$
$$Q = m(h_2 - h_1).$$

Thus,

$$500 = 5(h_2 - h_1)$$

and

$$h_2 = 111.17 \text{ Btu/lb}_m.$$

Since

$$P_2 = P_1 = 15 \text{ lb}_f/\text{in.}^2,$$

from the thermodynamic tables it may be found that

$$T_2 = 220.7°F \text{ and } s_2 = 0.2365 \text{ Btu/(lb}_m°R).$$

The entropy change in the refrigerant is

$$\Delta S = m(s_2 - s_1) = (5 \text{ lb}_m)(0.2365 - 0.0256) = 1.05 \text{ Btu/°R}.$$

Example 4.2

Problem: A fixed mass of oxygen undergoes a thermodynamic cycle comprising the following three processes consecutively:

Process 1→2, constant pressure cooling from $P_1 = 0.5$ MPa, $V_1 = 0.01$ m³.
Process 2→3, isothermal heating to $P_3 = 0.1$ MPa, $T_3 = 25°C$, $V_3 = 0.01$ m³.
Process 3→1, constant volume heating back to state 1.

Use the ideal gas equations with $C_p = 0.9$ kJ/(kg.K) to determine the entropy change in each process.

Solution

From the ideal gas law,

$$m = \frac{P_3 V_3}{RT_3} = \frac{\left(0.1 \times 10^6 \frac{N}{m^2}\right)\left(0.01\, m^3\right)}{\left(\frac{8314}{36} \frac{N.m}{kg.K}\right)(298.15\, K)} = 0.0145 \text{ kg}.$$

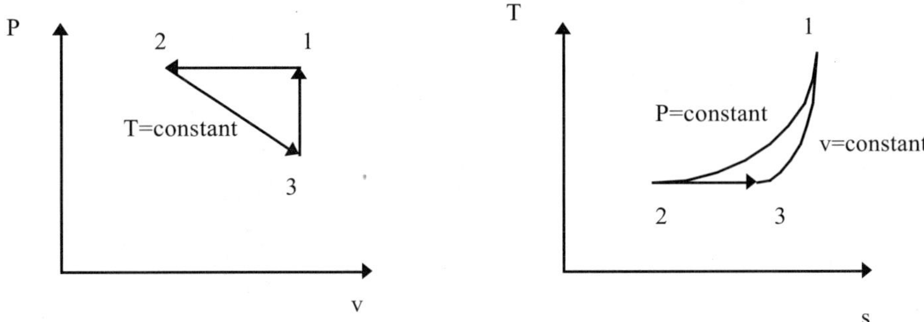

Schematic for Example 4.2.

Since

$$V_3 = V_1,\ P_1V = mRT_1, \text{ and } P_3V = mRT_3,$$

then

$$T_1 = \frac{P_1}{P_3}T_3 = \left(\frac{0.5}{0.1}\right)(298.15) = 1491\ \text{K}.$$

From Equation (4.22),

$$C_v = C_P - R = 0.9\frac{\text{kJ}}{(\text{kg.K})} - \frac{8.314}{36}\frac{\text{kJ}}{(\text{kg.K})} = 0.669\frac{\text{kJ}}{(\text{kg.K})}.$$

Process 1→2: Since $P_1 = P_2$, Equation (4.30) becomes

$$s(T_2, P_2) - s(T_1, P_1) = C_{P,av}\ln\frac{T_2}{T_1} = C_{P,av}\ln\frac{T_3}{T_1}$$

$$S(T_2, P_2) - S(T_1, P_1) = mC_{P,av}\ln\frac{T_3}{T_1}$$

$$= (0.0145\ \text{kg})\left(0.9\frac{\text{kJ}}{\text{kg.K}}\right)\ln\frac{298.15}{1491} = -0.0210\frac{\text{kJ}}{\text{K}}.$$

Process 2→ 3: Since $T_2 = T_3$, Equation (4.30) becomes

$$s(T_3, P_3) - s(T_2, P_2) = -R\ln\frac{P_3}{P_2} = -R\ln\frac{P_3}{P_1}$$

$$S(T_3, P_3) - S(T_2, P_2) = -mR\ln\frac{P_3}{P_1} = -(0.0145)\left(\frac{8.314}{36}\right)\ln\frac{0.1}{0.5} = 0.0054\frac{\text{kJ}}{\text{kg}}.$$

Process 3→1: Since $V_1 = V_3$, Equation (4.29) becomes

$$s(T_1, P_1) - s(T_3, P_3) = C_{v,av} \ln \frac{T_1}{T_3}$$

$$S(T_1, P_1) - S(T_3, P_3) = mC_{v,av} \ln \frac{T_1}{T_3}$$

$$= (0.0145\,\text{kg})\left(0.669 \frac{\text{kJ}}{\text{kg.K}}\right) \ln \frac{1491}{298.15} = 0.0156 \frac{\text{kJ}}{\text{K}}.$$

4.4 Carnot Cycle

In this section, we are introducing the topic of the most efficient cycle for a heat engine. Consider a heat engine, which runs in a cycle between a given high-temperature reservoir and a low-temperature reservoir. From the concept of a reversible process introduced in Chapter 2, we know that the heat engine will have maximum efficiency if each of its processes is reversible. Since each process is reversible, the engine cycle is also reversible. When the cycle is reversed, the heat engine becomes a refrigerator. Such a cycle is named after the French engineer, Nicolas Leonard Sadi Carnot (1792–1832).

Figure 4.5 is a schematic of a simple power plant operating on a Carnot cycle. The four basic processes are as follows:

1→2 Reversible isothermal process — heat is transferred from the high-temperature reservoir.

2→3 Reversible adiabatic process — temperature of the working fluid decreases from the high temperature to the low temperature.

3→4 Reversible isothermal process — heat is transferred to the low-temperature reservoir.

4→1 Reversible adiabatic process — temperature of the working fluid increases from the low temperature to the high temperature.

The system undergoing a reversed Carnot cycle is a refrigerator. This is shown by the schematic in Figure 4.6. The four basic processes are as follows:

2→1 Reversible isothermal process — heat is transferred to the high-temperature reservoir.

1→4 Reversible adiabatic process — temperature of the working fluid decreases from the high temperature to the low temperature.

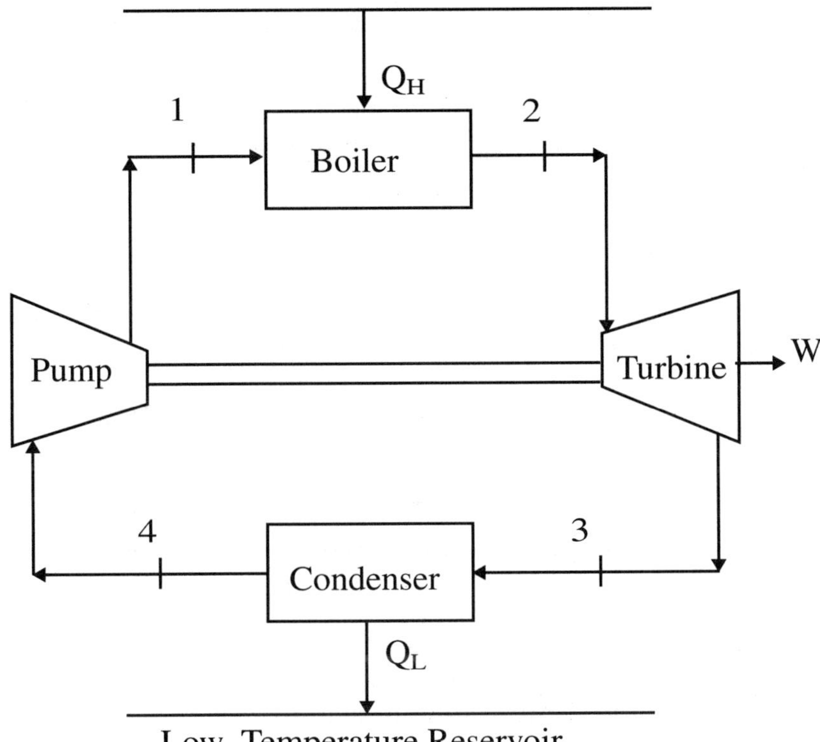

FIGURE 4.5
Heat engine operating on a Carnot cycle.

4→3 Reversible isothermal process — heat is transferred from the low-temperature reservoir.

3→2 Reversible adiabatic process — temperature of the working fluid increases from the low temperature to the high temperature.

Since the Carnot cycle is a reversible cycle with two isothermal processes and two adiabatic processes, the cycle appears as a rectangle on the T-s diagram of the working fluid. For a system undergoing the forward Carnot cycle, a sketch on the T-s diagram is shown in Figure 4.7.

The reversible adiabatic processes of the cycle require that

$$s_2 - s_3 = s_4 - s_1 = 0. \tag{4.37}$$

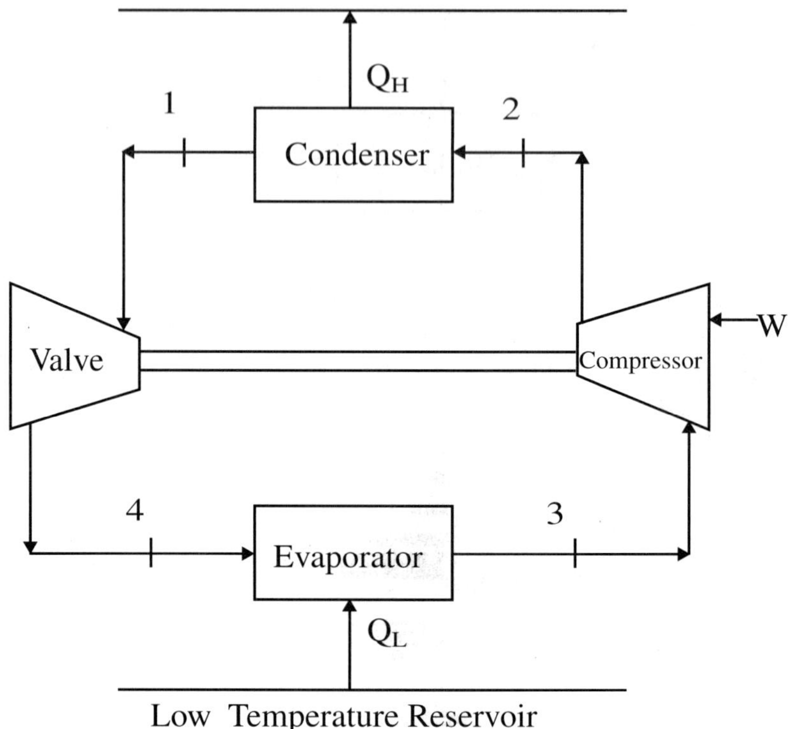

FIGURE 4.6
Refrigerator operating on a reversed Carnot cycle.

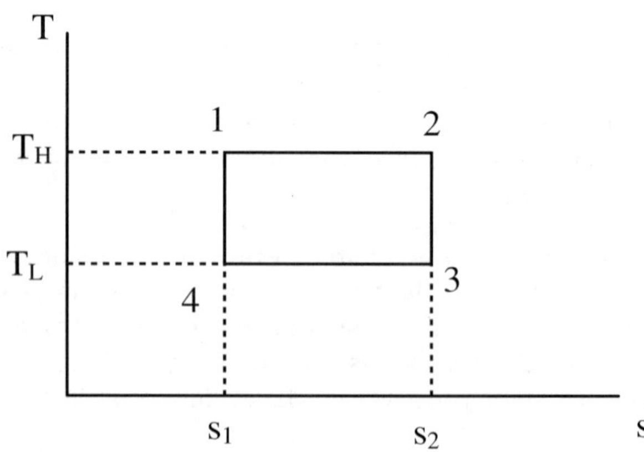

FIGURE 4.7
(T-s) diagram of a Carnot cycle.

It follows that ${}_2Q_3 = {}_4Q_1 = 0$. Since the heat transferred in reversible processes are the areas under the curve in the T-s diagram, it follows that

$$ {}_1Q_2 = T_H(s_2 - s_1). \tag{4.38} $$

$$ {}_3Q_4 = T_L(s_4 - s_3) = -T_L(s_3 - s_4) = -T_L(s_2 - s_1). \tag{4.39} $$

The thermal efficiency of the cycle is defined as the net work of the cycle divided by the heat input. From the first law for a closed system, the net work done by the cycle is

$$ W_{net} = {}_1Q_2 + {}_3Q_4 = (T_H - T_L)(s_2 - s_1). \tag{4.40} $$

Thus, the thermal efficiency is

$$ \eta = \frac{W_{net}}{{}_1Q_2} = \frac{(T_H - T_L)}{T_H}. \tag{4.41} $$

This is the thermal efficiency of an engine operating on a Carnot cycle. Since the Carnot cycle has the maximum efficiency, this is the maximum efficiency of any cycle operating between two temperatures T_H and T_L. This maximum efficiency is independent of the working fluid. In other words, no matter which fluid is used, any engine operating on a Carnot cycle between two temperatures will have the same efficiency.

Example 4.3

Problem: A Carnot engine operates between two reservoirs maintained at 300°C and 10°C, respectively. The heat transfer from the high-temperature reservoir is 10 kW. What is the output of the engine? Determine the heat rejected at the low-temperature reservoir.

Solution

Efficiency of the Carnot engine is

$$ \eta = \frac{\dot{W}_{net}}{\dot{Q}_H} = \frac{(T_H - T_L)}{T_H}. $$

Thus,

$$ \dot{W}_{net} = \left(1 - \frac{283.15}{573.15}\right) 10 = 5.06 \text{ kW}. $$

The first law gives us $\dot{Q}_L = \dot{Q}_H - \dot{W} = 10 - 5.06 = 4.94$ kW.

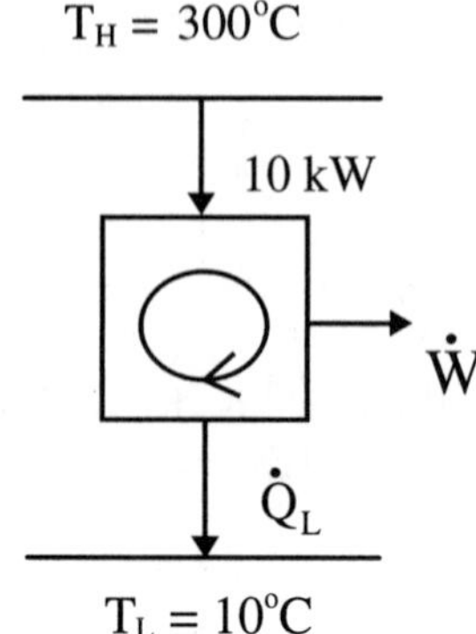

Schematic for Example 4.3.

Example 4.4

Problem: A refrigerator is cooling a space to –20°C by transferring heat to the atmosphere at 10°C. By assuming a Carnot refrigerator, calculate the COP. If there was a heat pump heating a living space to 10°C from the environment at –20°C, determine the COP of the heat pump.

Solution

For the refrigerator,

$$\beta = \frac{Q_L}{W} = \frac{1}{T_H / T_L - 1},$$

thus,

$$\beta = \frac{1}{\frac{283.15}{253.15} - 1} = 8.44.$$

For the heat pump,

$$\gamma = \frac{Q_H}{W} = \frac{1}{1 - T_L / T_H},$$

thus,

$$\gamma = \frac{1}{1 - \frac{253.15}{283.15}} = 9.44.$$

4.5 Second Law in Entropy for a Control Volume

Recall that an open system is one where there is flow of matter or mass in or out of the boundaries. The boundaries essentially define a control volume, that is, a volume that is defined as the space of interest in which we are concentrating our analysis. Figure 4.8 is identical to Figure 3.4. We consider the volume initially with a mass m_{ini} as in Figure 4.8, Part (a). Associated with this mass is its entropy.

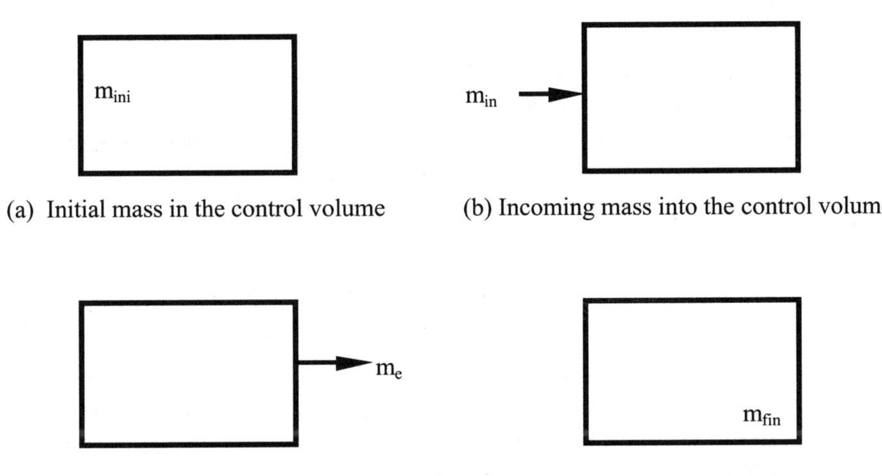

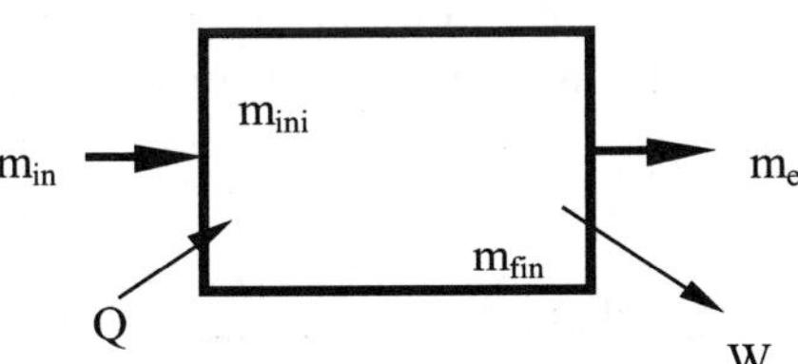

FIGURE 4.8
Control volume.

The entropy initially in the control volume is thus $m_{ini}\ s_{ini}$.

Consider a mass m_{in} entering the control volume, as in Figure 4.8, Part (b). The entropy associated with m_{in} is $m_{in}\ s_{in}$.

Consider a mass m_e exiting a control volume, as in Figure 4.8, Part (c). The entropy associated with m_e is $m_e\ s_e$.

There is mass entering the control volume and mass exiting the control volume, so the mass within the control volume does not remain the same. Let us call the final mass remaining in the control volume, m_{fin}, as in Figure 4.8, Part (d). Hence, the entropy left in the control volume is $m_{fin}\, s_{fin}$.

In addition, there is heat entering the system, and work being done by the system. Work does not change the entropy in the system. Heat Q_j brings with it an entropy of

$$\frac{Q_j}{T_j},$$

where T_j is the temperature at which the heat is being transferred. This entropy has to be taken into account in the entropy balance of the control volume. The final diagram in the series, as in Figure 4.8, Part (e), shows the various masses, and the heat and work interactions.

The entropy equation for the control volume in Figure 4.8, Part (e) is

$$\sigma_{cv} + \frac{Q_j}{T_j} + m_{in}s_{in} + m_{ini}s_{ini} = m_e s_e + m_{fin}s_{fin}. \tag{4.42}$$

Note that the sum of the entropy due to the incoming heat transfer, the initial entropy of the system, and the incoming entropy will be less than the sum of the final entropy in the control volume and the outgoing entropy. The difference is made up in the entropy balance equation by σ_{cv}, which is the entropy production term for the control volume. The second law equation, otherwise called the creation of entropy, is not a conservation equation. However, the equation above is in a form similar to a conservation equation, with the entropy production term included to make up the difference. Taking into consideration multiple heat transfer and work processes, the second law in entropy equation becomes

$$\sigma_{cv} + \sum_k \frac{Q_j}{T_j} + \sum_l m_{in}s_{in} + m_{ini}s_{ini} = \sum_n m_e s_e + m_{fin}s_{fin}. \tag{4.43}$$

In a steady-state steady-flow (S.S.S.F.) process, the initial mass is equal to the final mass and the initial state is the same as the final state, i.e.,

$$m_{ini} = m_{fin} \tag{4.44}$$

and

$$s_{ini} = s_{fin}. \tag{4.45}$$

So the second law in entropy equation for a S.S.S.F. process is

$$\sigma_{cv} + \frac{Q_j}{T_j} + m_{in}s_{in} = m_e s_e. \tag{4.46}$$

A closed system is one where no mass enters or leaves the system. Under these conditions, $m_{in} = m_e = 0$ and $m_{ini} = m_{fin} = m$. So the second law in entropy equation for a closed system is

$$\sigma + \frac{Q_j}{T_j} + ms_{ini} = ms_{fin}. \tag{4.47}$$

The above equations are similarly modified if there is more than one mass stream entering or more than one mass stream exiting the system for the S.S.S.F. process; if there is more than one heat transfer process, or more than one work process in either the S.S.S.F. process or the closed system, the equations are similarly modified to take into account the multiple processes.

The rate equation for the control volume for a single mass stream in and out of the system and single heat and work processes, is

$$\dot{\sigma}_{cv} + \frac{\dot{Q}_j}{T_j} + \dot{m}_{in}\, s_{in} + \dot{m}_{ini}\, s_{ini} = \dot{m}_e\, s_e + \dot{m}_{fin}\, s_{fin}. \tag{4.48}$$

The corresponding equation for the multiple mass streams and boundary processes is

$$\dot{\sigma}_{cv} + \sum_k {}_k \frac{\dot{Q}_j}{T_j} + \sum_1 {}_1\, \dot{m}_{in}\, s_{in} + \dot{m}_{ini}\, s_{ini} = \sum_n {}_n \dot{m}_e s_e + \dot{m}_{fin}\, s_{fin}. \tag{4.49}$$

It is instructive to compare Equation (4.49) with Equation (3.14), which is repeated here for convenience:

$$\begin{aligned} &\sum_k {}_k\dot{Q} + \sum_1 {}_1\dot{m}_{in}\left(h_{in} + \frac{V_{in}^2}{2} + gz_{in}\right)_1 + \dot{m}_{ini}\left(u_{ini} + \frac{V_{ini}^2}{2} + gz_{ini}\right) = \\ &\sum_n {}_n\, \dot{m}_e\left(h_e + \frac{V_e^2}{2} + gz_e\right)_n + \dot{m}_{fin}\left(u_{fin} + \frac{V_{fin}^2}{2} + gz_{fin}\right) + \sum_p {}_p\dot{W}. \end{aligned} \tag{3.14}$$

The similarities between the two equations are emphasized here. Note that the work term does not enter into Equation (4.49). The heat transfer term is appropriately modified. Associated with each mass is its entropy. Since the entropy is not conserved, the entropy production term is introduced on the left-hand side to balance the entropy equation. Entropy production is either positive or zero. It can never be negative.

Example 4.5

Problem: Steam at 10 MPa and 500°C is throttled through a valve to a pressure of 5 MPa in a steady-flow process. Determine the entropy production for this process. Is the increase of entropy principle satisfied?

Solution

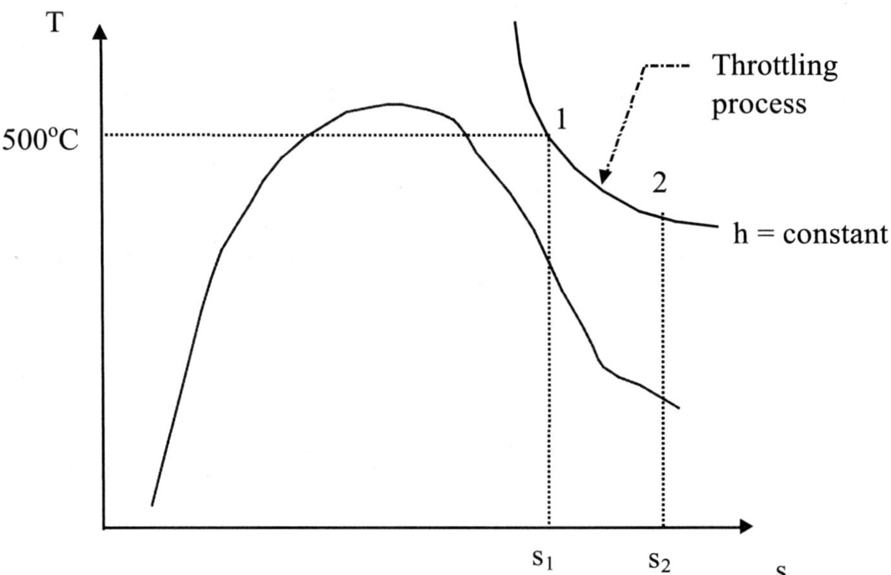

Schematic for Example 4.5.

Assumptions: (1) Heat transfer is negligible in the throttling process.

Analysis:
From the thermodynamic tables,

State 1: $P_1 = 10$ MPa, $T_1 = 500°C$
$s_1 = 6.597$ kJ/(kg.K)
State 2: $P_2 = 5$ MPa, $h_2 = h_1$
$s_2 = 6.897$ kJ/(kg.K)

The throttling process is a S.S.S.F. process. The second law in entropy equation for this throttling process, Equation (4.46), is

$$\sigma_{cv} + \frac{Q_j}{T_j} + m_{in}s_{in} = m_e s_e.$$

Since there is no heat transfer, the equation becomes

$$\sigma_{cv} + m_{in}S_{in} = m_e s_e$$

$$\sigma_{cv} = m\ (s_e - s_{in}) = m\ (s_2 - s_1).$$

The entropy generated per unit mass of steam as it is throttled from the inlet state to the outlet pressure is

$$s_2 - s_1 = (6.897 - 6.597)\ \text{kJ}/(\text{kg.K}) = 0.3\ \text{kJ}/(\text{kg.K}).$$

The increase of entropy principle is satisfied during this process since the total entropy change is positive.

4.5.1 Application to the Power Cycle

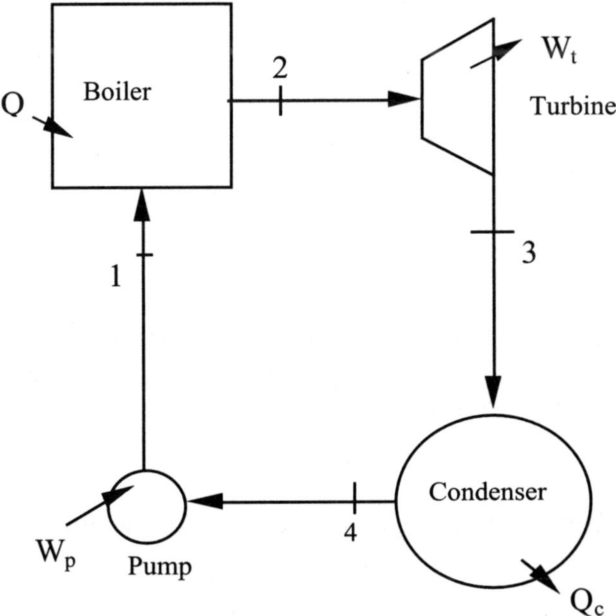

FIGURE 4.9
Schematic of a simple power plant.

Consider a simple power plant with the schematic shown in Figure 4.9 above. The second law in entropy for the whole system which is a control mass or closed system, Equation (4.47), is

$$\sigma_{cy} + \frac{Q_b}{T_b} - \frac{Q_c}{T_c} = m\left(s_{fin} - s_{ini}\right).$$

Since the system undergoes a cycle, $s_{fin} = s_{ini}$ and the equation becomes

$$\sigma_{cy} + \frac{Q_b}{T_b} - \frac{Q_c}{T_c} = 0. \tag{4.50}$$

If a boundary is drawn around the boiler, it will define a control volume or open system undergoing a S.S.S.F. process. The states of the water are designated 1, 2, 3, and 4, respectively. The second law in entropy equation for the boiler, Equation (4.46), is

$$\sigma_{cv} + \frac{Q_j}{T_j} + m_{in}s_{in} = m_e s_e.$$

Substituting the notation for the appropriate states,

$$\sigma_b + \frac{Q_b}{T_b} + m_1 s_1 = m_2 s_2. \tag{4.51}$$

Since $m_1 = m_2 = m$,

$$\sigma_b + \frac{Q_b}{T_b} + m s_1 = m s_2. \tag{4.52}$$

Similarly, each of the other equipment—the turbine, the condenser, and the pump—may be defined as an open system undergoing a S.S.S.F. process. The second law in entropy equation for the turbine is

$$\sigma_t + m s_2 = m s_3. \tag{4.53}$$

The second law in entropy equation for the condenser is

$$\sigma_c - \frac{Q_c}{T_c} + m s_3 = m s_4. \tag{4.54}$$

The second law in entropy equation for the pump is

$$\sigma_p + m s_4 = m s_1. \tag{4.55}$$

Note that the addition of the four entropy equations for the four pieces of equipment gives us the entropy equation for the whole system. In addition, it can be seen that

$$\sigma_{cy} + \sigma_b + \sigma_t + \sigma_c + \sigma_p. \tag{4.56}$$

4.6 Second Law in Entropy for a Control Mass

As shown in the previous section, the second law in entropy equation for a closed system or control mass system with one heat interaction, is

$$\sigma + \frac{Q_j}{T_j} + ms_{ini} = m_{fin}s_{fin}. \tag{4.57}$$

The rate second law in entropy equation for the control mass is

$$\dot{\sigma} + \frac{\dot{Q}_j}{T_j} + \dot{m}s_{ini} = \dot{m}_{fin}s_{fin}. \tag{4.58}$$

The corresponding equation for multiple boundary processes is

$$\dot{\sigma} + \sum_k \frac{\dot{Q}_j}{{}_k T_j} + \dot{m}s_{ini} = \dot{m}_{fin}s_{fin}. \tag{4.59}$$

It is easy to see that Equation (4.59) is a special case or subset of Equation (4.49). Since there is no mass moving in or out of a control mass system, there is no need to consider incoming entropy with incoming mass or outgoing entropy with outgoing mass. The only boundary-interaction contributions to the entropy equation come from heat interactions, and care has to be taken to consider all such interactions.

Example 4.6

Problem: Given the open system shown, which is similar to Example 3.1, with three incoming masses and two outgoing masses, calculate the entropy production.

Solution

Assumptions: (1) If the velocity of the mass is not given, its K.E. is neglected.

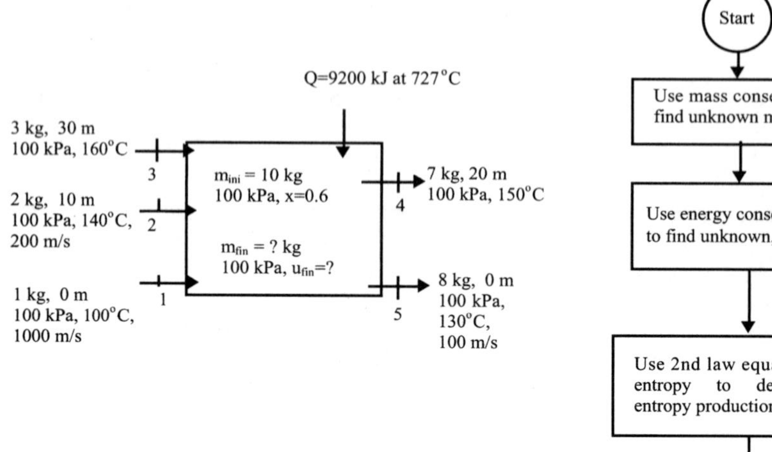

Schematic for Example 4.6.

Analysis:

From the thermodynamic tables, for the incoming mass at state 1, $s_1 = 7.361$ kJ/(kg.K).

For the incoming mass at state 2, $s_2 = 7.563$ kJ/(kg.K).

For the incoming mass at state 3, $s_3 = 7.66$ kJ/(kg.K).

For the initial mass, $s_{ini} = 4.937$ kJ/(kg.K).

For the outgoing mass at state 4, $s_4 = 7.613$ kJ/(kg.K).

For the outgoing mass at state 5, $s_5 = 7.513$ kJ/(kg.K).

For the final mass in the system and the solution given in Example 3.1, $s_{fin} = 4.892$ kJ/(kg.K).

The second law entropy equation for the control volume, Equation (4.43), is

$$\sigma_{cv} + \sum_k \frac{Q_j}{T_j}_k + \sum_1 {}_1 m_{in} s_{in} + m_{ini} s_{ini} = \sum_n {}_n m_e s_e + m_{fin} s_{fin}.$$

Substituting the values for the masses, entropies, heat transfer, and temperature of heat transfer,

$$\sigma_{cv} + \frac{9200}{1000} + 1.(7.361) + 2.(7.563) + 3.(7.66) + 10.(4.937)$$
$$= 7.(7.613) + 8.(7.513) + 1.(4.892)$$
$$\sigma_{cv} + 9.2 + 7.361 + 15.126 + 22.98 + 49.37$$
$$= 53.291 + 60.104 + 4.892$$
$$\sigma_{cv} = 14.3 \text{ kJ / K}.$$

The logic diagram for solving the above problem may be represented as shown. It is an extension to that for Example 3.1. The mass conservation equation is used explicitly in Example 3.1 to find the unknown mass. The energy conservation equation (first law) is used to determine the unknown internal energy of the final mass in the control volume. Then, the second law in entropy equation is used to find the entropy generation term.

Example 4.7

Problem: Saturated liquid ammonia at 0°C is initially contained in a piston-cylinder device. The ammonia converts, finally, to the corresponding saturated vapor state in an adiabatic process. The stirrer causes the change of state, and the piston is free to move during the process. Calculate the work per unit mass of ammonia, and the amount of entropy produced per unit mass.

Solution

Assumptions: (1) The intial and final states are equilibrium states, and (2) there is no change in K.E. or P.E.

Analysis: The volume of the ammonia expands in the process. Thus, there is work done by the control mass (or closed system) during the expansion. Work is done by the stirrer on the ammonia. The first law applied to the control mass gives

$$U_1 = U_2 + W.$$

On a unit mass basis,

$$\frac{W}{m} = -(u_2 - u_1).$$

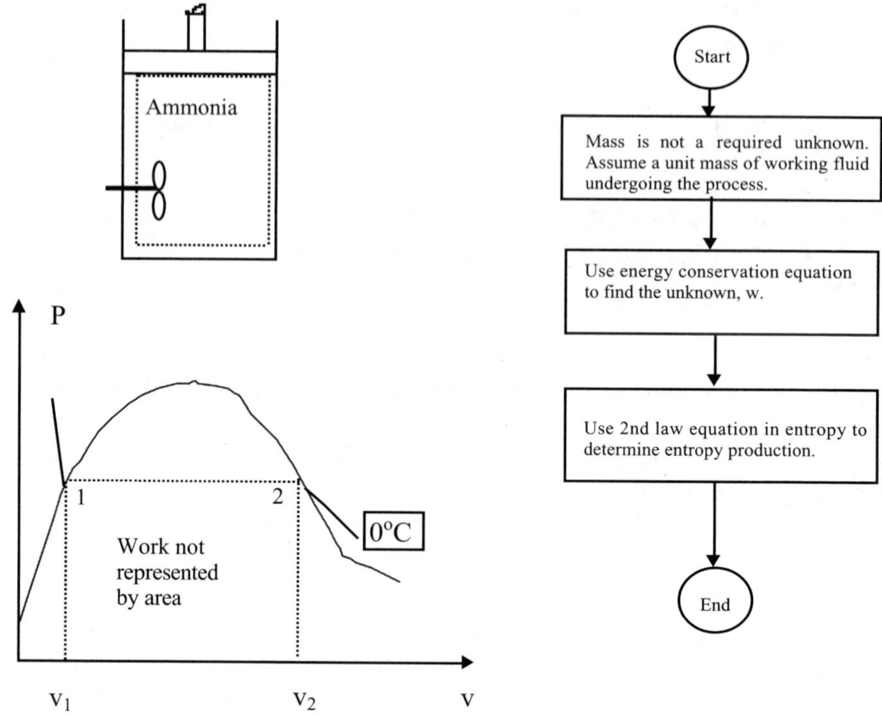

Schematic for Example 4.7.

From the thermodynamic tables,

$$\frac{W}{m} = -1138\frac{kJ}{kg}.$$

The minus sign shows that the work done by the stirrer is more than that done by the ammonia as it expands.

The second law in entropy equation for the system which is a control mass, Equation (4.47), is

$$\sigma + \frac{Q_j}{T_j} + S_1 = S_2.$$

Since there is no heat transfer, this equation becomes

$$\sigma + S_1 = S_2.$$

On a unit mass basis,

$$\frac{\sigma}{m} = s_2 - s_1.$$

From the thermodynamic tables,

$$\frac{\sigma}{m} = 4.62 \frac{kJ}{kg.K}.$$

Note that the entropy production term is positive. Since the process is irreversible, the work is not represented by the area in the P-v diagram. The process is represented by a dashed line, rather than a continuous line. As the process is adiabatic, there is also no significance to the area in the T-s diagram.

The logic diagram for solving the above problem may be represented as shown. It is similar to that for Example 4.6. The mass conservation equation is not used explicitly; a unit mass of the working fluid is considered as undergoing the process. The energy conservation equation (first law) is used to determine the unknown energy quantity, which is work in this case. Then, the second law in entropy equation is used to find the entropy generation term.

Example 4.8

Problem: Refrigerant-134a is compressed adiabatically in a piston-cylinder device from saturated vapor at 0°F to a final pressure of 60 $lb_f/in.^2$ Determine the minimum input work required per unit mass of refrigerant-134a.

Solution

Assumptions: (1) The initial and final states are equilibrium states and (2) there is no change in K.E. or P.E.

Analysis:

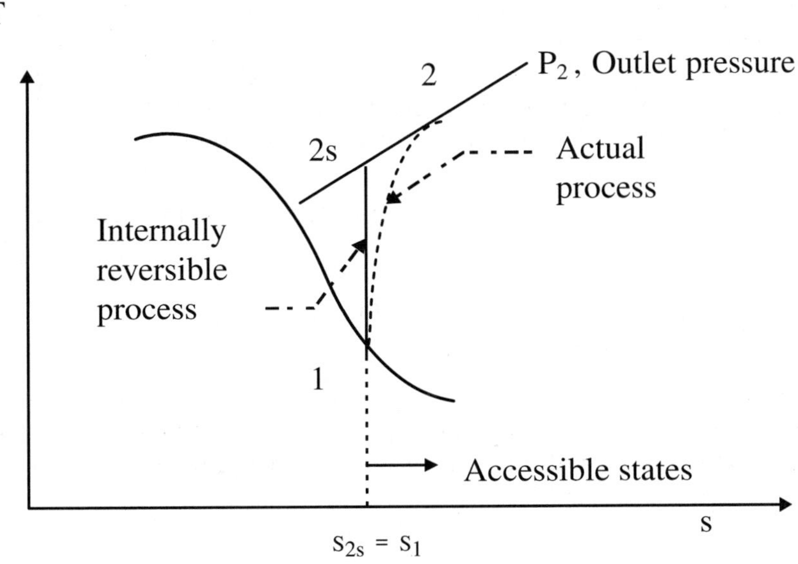

Schematic for Example 4.8.

The first law applied to the control mass gives

$$U_1 = U_2 + W.$$

On a unit mass basis,

$$\frac{W}{m} = -(u_2 - u_1).$$

The minimum input work corresponds to the smallest allowed value for u_2, determined by the second law.

The second law in entropy equation for the system which is a control mass, Equation (4.47), is

$$\sigma + \frac{Q_j}{T_j} + S_1 = S_2.$$

Since there is no heat transfer, this equation becomes

$$\sigma + S_1 = S_2.$$

On a unit mass basis,

$$s_2 - s_1 = \frac{\sigma}{m} \geq 0.$$

Since σ is positive, only states with $s_2 > s_1$ are accessible adiabatically. In the limit as irreversibilities are reduced to zero, $\sigma = 0$ and $s_2 = s_1$. This limiting state is labeled 2s on the diagram, and corresponds to an isentropic compression.

Also by inspection of the thermodynamic tables, the specific internal energy decreases as the temperature decreases when the pressure is fixed. The smallest allowable value for u_2 corresponds to state 2s.

$$\text{With } s_{2s} = s_1 = 0.415 \text{ Btu}/(\text{lb}_m\,°\text{R}),\ P_2 = 60 \text{ lb}_f/\text{in.}^2,$$

$$u_{2s} = 166.61 \text{ Btu}/\text{lb}_m$$

$$\left(-\frac{W}{m}\right)_{min} = u_{2s} - u_1 = 8.6 \text{ Btu}/\text{lb}_m.$$

Note that a greater work input is required for the actual adiabatic compression process as compared to an internally reversible adiabatic process between the same initial state and any given final pressure.

4.7 Isentropic Processes

When an ideal gas undergoes a constant entropy or isentropic process, several relations may be obtained by setting the entropy-change relations developed in Section 4.3 to zero. Consider first the case when a constant specific heat assumption is valid. Setting Equation (4.29) to zero,

$$\ln\frac{T_2}{T_1} = -\frac{R}{C_v}\ln\frac{v_2}{v_1} = \ln\left(\frac{v_1}{v_2}\right)^{R/C_v}. \tag{4.60}$$

Since $R = C_P - C_v$, $k = C_P/C_v$, $R/C_v = k - 1$. Thus,

$$\left(\frac{T_2}{T_1}\right)_{s=constant} = \left(\frac{v_1}{v_2}\right)^{k-1} \quad \text{(ideal gas)}. \tag{4.61}$$

Setting Equation (4.30) to zero,

$$\left(\frac{T_2}{T_1}\right)_{s=constant} = \left(\frac{P_2}{P_1}\right)^{(k-1)/k} \quad \text{(ideal gas)}. \tag{4.62}$$

Substitute Equation (4.61) into Equation (4.62) and simplifying,

$$\left(\frac{P_2}{P_1}\right)_{s=constant} = \left(\frac{v_1}{v_2}\right)^{k} \quad \text{(ideal gas)}. \tag{4.63}$$

These three isentropic relations for ideal gases may also be written as:

$$Tv^{k-1} = \text{constant} \tag{4.64}$$

$$TP^{(1-k)/k} = \text{constant} \qquad \text{(ideal gas)}. \tag{4.65}$$

$$Pv^{k} = \text{constant} \tag{4.66}$$

An average k value for the given temperature range should be used.

When the specific heats are not constant, we obtain the isentropic relation from Equation (4.27). Setting this equation to zero,

$$0 = s°(T_2) - s°(T_1) - R\ln\left(\frac{P_2}{P_1}\right)$$

$$s°(T_2) = s°(T_1) + R\ln\left(\frac{P_2}{P_1}\right) \tag{4.67}$$

where $s^\circ(T_1)$ is the value of s° at the beginning of the process, and $s^\circ(T_2)$ is the value of s° at the end of the process.

Example 4.9

Problem: Hydrogen gas is compressed reversibly in an adiabatic compressor from an initial state of 14.7 $lb_f/in.^2$ and 70°F to a final temperature of 400°F. Determine the outlet pressure of the hydrogen.

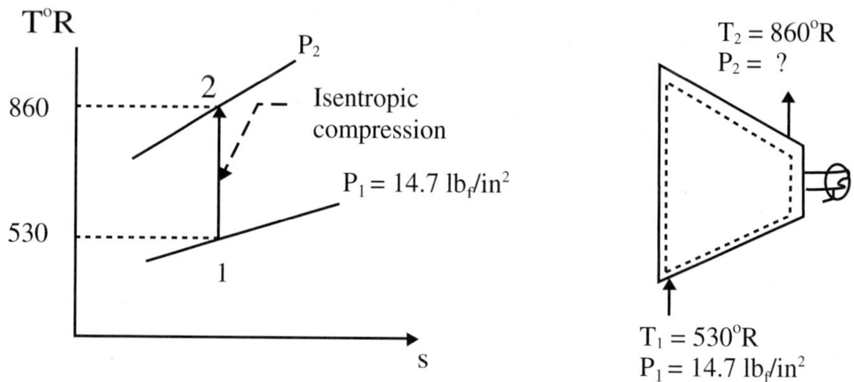

Schematic for Example 4.9.

Solution

Assumptions: (1) Since its initial temperature is very high relative to its critical temperature (T_{cr} = –400°F for hydrogen), the hydrogen may be treated as an ideal gas.

Analysis: The specific heat ratio k of hydrogen is 1.405. The outlet pressure of hydrogen may be determined from Equation (4.55) as

$$P_2 = P_1\left(\frac{T_2}{T_1}\right)^{k/(k-1)} = \left(14.7\ lb_f / in.^2\right)\left(\frac{859.67°R}{529.67°R}\right)^{1.405/0.405} = 78.9\ lb_f / in.^2$$

4.7.1 Relative Pressure and Relative Specific Volume

To simplify calculations when the volume ratio is given instead of the pressure ratio, we define two new dimensionless quantities, the *relative pressure* P_r and the *relative specific volume* v_r. From Equation (4.67),

$$\frac{P_2}{P_1} = \exp\frac{s_2^\circ - s_1^\circ}{R} = \frac{\exp(s_2^\circ / R)}{\exp(s_1^\circ / R)}. \tag{4.68}$$

The quantity $\exp(s^o/R)$ is defined as the relative pressure P_r. Thus,

$$\left(\frac{P_2}{P_1}\right)_{s=constant} = \frac{P_{r2}}{P_{r1}}. \tag{4.69}$$

Since s^o is a function of temperature only, P_r is a function of temperature only. Values of P_r can be tabulated against temperature, and this is done for air in the Appendix.

Using the ideal-gas relation and Equation (4.69),

$$\frac{v_2}{v_1} = \frac{T_2}{T_1}\frac{P_1}{P_2} = \frac{T_2}{T_1}\frac{P_{r1}}{P_{r2}} = \frac{T_2/P_{r2}}{T_1/P_{r1}}. \tag{4.70}$$

The quantity T/P_r is defined as the relative specific volume v_r. Like the relative pressure, it is a function of temperature only. Thus,

$$\left(\frac{v_2}{v_1}\right)_{s=constant} = \frac{v_{r2}}{v_{r1}}. \tag{4.71}$$

Equations (4.70) and (4.71) are only valid for ideal gases undergoing isentropic processes. When the specific heats vary with temperature, they give more accurate data than Equations (4.61) through (4.66). The values of v_r are also listed for air in the Appendix.

Example 4.10

Problem: Air is compressed reversibly in an adiabatic piston-cylinder device from 12°C and 90 kPa. If the compression ratio V_1/V_2 is 10, determine the final temperature of the air.

Solution

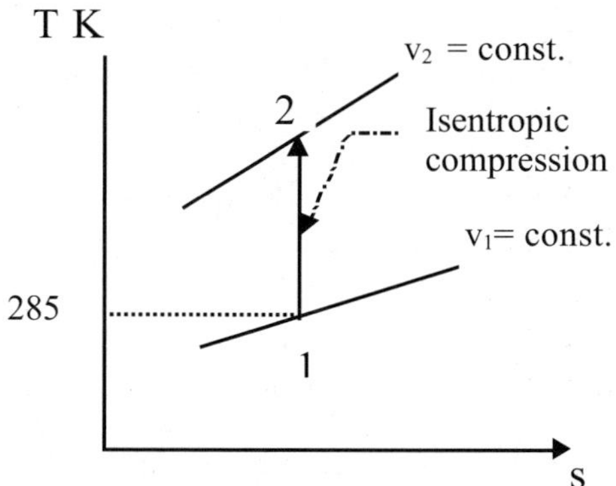

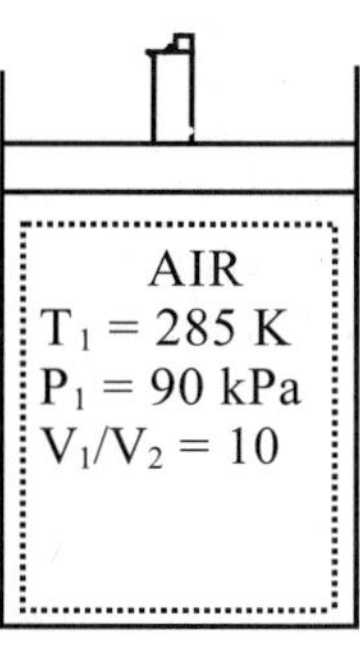

Schematic for Example 4.10.

Assumptions: (1) Since the air is at a high temperature and low pressure relative to its critical values ($T_{cr} = -147°C$ and $P_{cr} = 3390$ kPa for nitrogen, which is the principal component of air), it may be considered as an ideal gas.

Analysis: The process is isentropic because it is both adiabatic and reversible. The final temperature may be calculated from the relative specific value data in the tables as shown below.

For a control mass, $\frac{V_2}{V_1} = \frac{v_2}{v_1}$.

At 285.15 K, $v_{r1} = 204.3$.

From Equation (4.71), $v_{r2} = v_{r1}\left(\frac{v_2}{v_1}\right)$

Thus, $v_{r2} = 20.43$.

From the tables, $T_2 = 697$ K.

4.8 Isentropic Efficiencies

The Carnot cycle is an ideal cycle, to which other practical cycles may be compared. The Carnot cycle is made up of ideal or reversible processes. In the last section, we considered ideal processes. We can use the ideal processes as models for comparison with practical processes. There are engineering devices such as turbines, pumps, and compressors, which may be examined with respect to their performance. We can compare such devices with idealized devices which undergo ideal or reversible processes. The *isentropic efficiency* has been defined as a quantitative measure of this performance.

The isentropic efficiencies of different devices have been defined differently. This is logical because the purposes of different devices are not the same.

4.8.1 Isentropic Efficiency of Turbines

The purpose of a turbine is to do work while the working fluid expands from a high inlet pressure to a low outlet pressure. The desired output is the work produced by the turbine. Hence, the isentropic efficiency of a turbine is defined as the ratio of the actual work of the turbine to the work output if the process between the inlet state and the outlet pressure were isentropic:

$$\eta_{isen} = \frac{\text{actual turbine work}}{\text{isentropic turbine work}} = \frac{w_{actual}}{w_{isen}}. \tag{4.72}$$

The changes in kinetic and potential energies between the inlet and outlet are usually small compared to the change in enthalpy, and thus can be neglected. In an adiabatic situation, there is no heat transfer. The work output can then be expressed in terms of the change in enthalpy. The relation above may be expressed as

$$\eta_{isen} = \frac{h_1 - h_{2a}}{h_1 - h_{2s}} \tag{4.73}$$

where h_1 is the enthalpy at the inlet state, while h_{2a} and h_{2s} are the enthalpy values at the outlet for actual and isentropic expansion processes.

The isentropic efficiencies of most practical turbines range from less than 70 to over 90%. The isentropic efficiency of a turbine is determined by measuring the actual work output, and computing the isentropic work output from the measured inlet state and the outlet pressure.

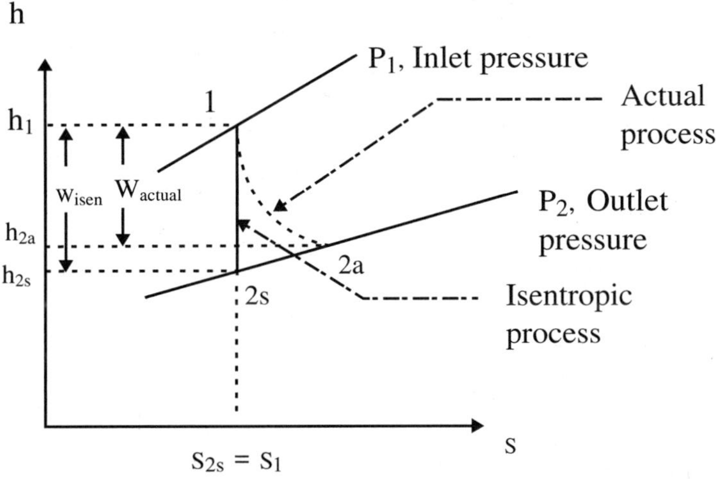

FIGURE 4.10
H-s diagram for the actual and isentropic processes of a turbine.

Example 4.11

Problem: The outlet pressure of a steam turbine operating at steady state is 1 bar. The inlet state of the steam is $P_1 = 7$ bars, $T_1 = 360°C$. If the isentropic efficiency of the turbine is 70%, determine the enthalpy of the steam at the outlet. Calculate the turbine work, per unit mass of steam, in kilojoules per kilogram.

Solution

Assumptions: (1) The system is undergoing an adiabatic process. Any changes in K.E. and P.E. are negligible.

Analysis: From the thermodynamic tables,

$$h_1 = 3184.7 \text{ kJ/kg}, \qquad s_1 = 7.506 \text{ kJ/(kg.K)}$$
$$h_{2s} = 2732 \text{ kJ/kg}$$

From the definition of the isentropic efficiency,

$$\frac{h_1 - h_{2a}}{h_1 - h_{2s}} = 70\%$$

$$h_{2a} = h_1 - 70\%(h_1 - h_{2s})$$
$$= 2867.8 \text{ kJ/kg.}$$

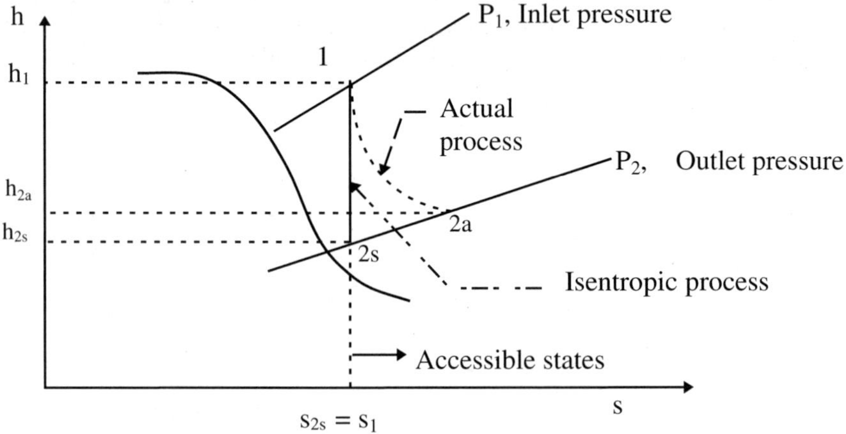

Schematic for Example 4.11.

The turbine work $= h_1 - h_{2a} = 317$ kJ/kg.

4.8.2 Isentropic Efficiency of Pumps and Compressors

The purpose of a pump or a compressor is to increase the pressure of a working fluid from a low inlet pressure to a high outlet pressure, while external work in the form of electricity is supplied. The input is the pump work or the compressor work. Hence, the isentropic efficiency of a pump or a compressor is defined as the ratio of the work input required to raise the pressure of a working fluid to a specified value in an isentropic manner to the actual work input:

$$\eta_{isen} = \frac{\text{isentropic compressor work}}{\text{actual compressor work}} = \frac{w_{isen}}{w_{actual}} \tag{4.74}$$

or

$$\eta_{isen} = \frac{\text{isentropic pump work}}{\text{actual pump work}} = \frac{w_{isen}}{w_{actual}}. \tag{4.75}$$

The changes in kinetic and potential energies between the inlet and outlet are usually small compared to the change in enthalpy, and thus can be neglected. In an adiabatic situation, there is no heat transfer. The work input can then be expressed in terms of the change in enthalpy. The relation above may be expressed as

$$\eta_{isen} = \frac{h_{2s} - h_1}{h_{2a} - h_1} \tag{4.76}$$

where h_1 is the enthalpy at the inlet state, while h_{2a} and h_{2s} are the enthalpy values at the outlet for actual and isentropic compression processes. Note that the isentropic pump or compressor efficiency is defined with the isentropic work input in the numerator instead of in the denominator. Since w_{actual} is larger than w_{isen}, this definition prevents η_{isen} from becoming larger than 100%, which might imply that actual pumps and compressors performed better than the isentropic ones.

When the changes in kinetic and potential energies of a liquid may be neglected, the isentropic efficiency of a pump may be defined as

$$\eta_{isen} = \frac{w_{isen}}{w_{actual}} = \frac{v(P_2 - P_1)}{h_{2a} - h_1} \tag{4.77}$$

where v is the specific volume of the incompressible liquid at P_1 or P_2. Efficiencies of pumps depend on their design and can range from below 70% to above 85%.

Example 4.12

Problem: The outlet pressure of a water pump operating at steady state is 5 bars. The inlet state of the water is 1 bar and 20°C. If the isentropic efficiency of the pump is 80%, determine the enthalpy of the water at the outlet. Calculate the pump work input per unit mass of water, in kilojoules per kilogram.

Solution

Assumption: (1) The system is undergoing an adiabatic process. Any changes in K.E. and P.E. are negligible.

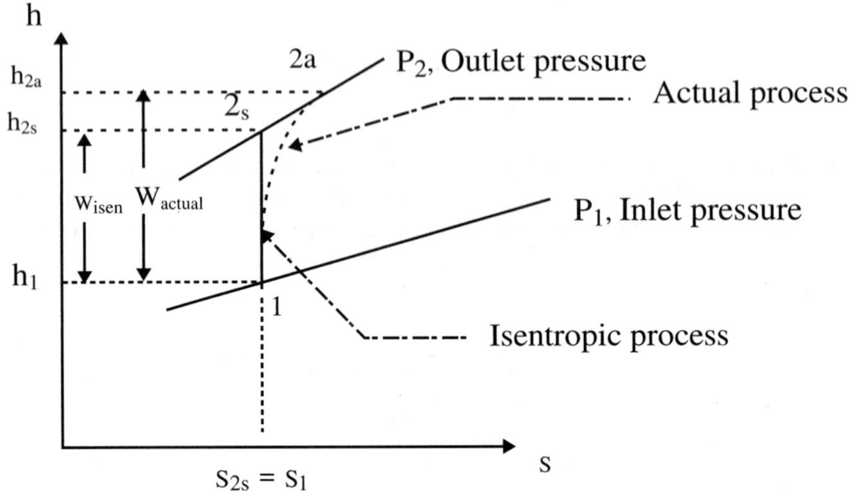

FIGURE 4.11
H-s diagram for the actual and isentropic processes of a pump or a compressor.

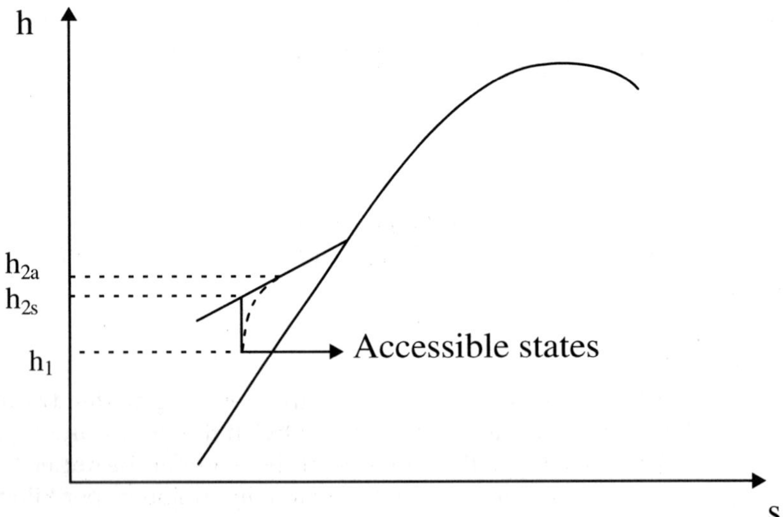

Schematic for Example 4.12.

Analysis: The properties at state 1 are as follows:

$$h_1 = 84.03\ \text{kJ/kg}, \qquad s_1 = 0.2965\ \text{kJ/(kg.K)}$$
$$v_1 = 0.0010017\ \text{m}^3/\text{kg} \approx v_2$$
$$\eta_{isen} = 80\% = \frac{v(P_2 - P_1)}{h_{2a} - h_1}$$
$$h_{2a} = h_1 + \frac{1}{\eta} v_1 (P_2 - P_1)$$
$$= 84.03 + 0.501 = 84.53\ \text{kJ/kg}.$$

The actual pump work is equal to $h_{2a} - h_1 = 0.5$ kJ/kg.

4.8.3 Isentropic Efficiency of Nozzles

Nozzles are used to increase the kinetic energy of the fluid, and the pressure drops in the process. Thus, it is similar to the turbine. The isentropic efficiency of a nozzle is defined as the ratio of the actual specific kinetic energy of the fluid leaving the nozzle, to the kinetic energy at the outlet if an isentropic expansion had taken place between the same inlet state and the same outlet pressure:

$$\eta_{isen} = \frac{\left(v_2^2/2\right)_{actual}}{\left(v_2^2/2\right)_{isen}}. \tag{4.78}$$

Nozzles with isentropic efficiencies of 95% or more are common. It is practical to design nozzles to be nearly internally reversible.

Example 4.13

Problem: Steam enters a nozzle at $P_1 = 70\ \text{lb}_f/\text{in.}^2$, $T_1 = 700°\text{F}$ with a velocity of 150 ft/s. The pressure and temperature at the outlet are $P_2 = 14.504\ \text{lb}_f/\text{in.}^2$ and $T_2 = 500°\text{F}$. These are steady-state conditions, and there is no significant heat transfer. Calculate the nozzle efficiency.

Solution

Assumption: (1) For the nozzle, change in P.E. between the inlet and outlet can be neglected.

Analysis: From the thermodynamic tables,

$$h_1 = 1380.8\ \text{Btu/lb}_m \qquad s_1 = 1.8436\ \text{Btu/(lb}_m.°\text{R)}$$
$$h_{2a} = 1287\ \text{Btu/lb}_m$$

Since $s_{2s} = s_1$, $h_{2s} = 1214\ \text{Btu/lb}_m$.

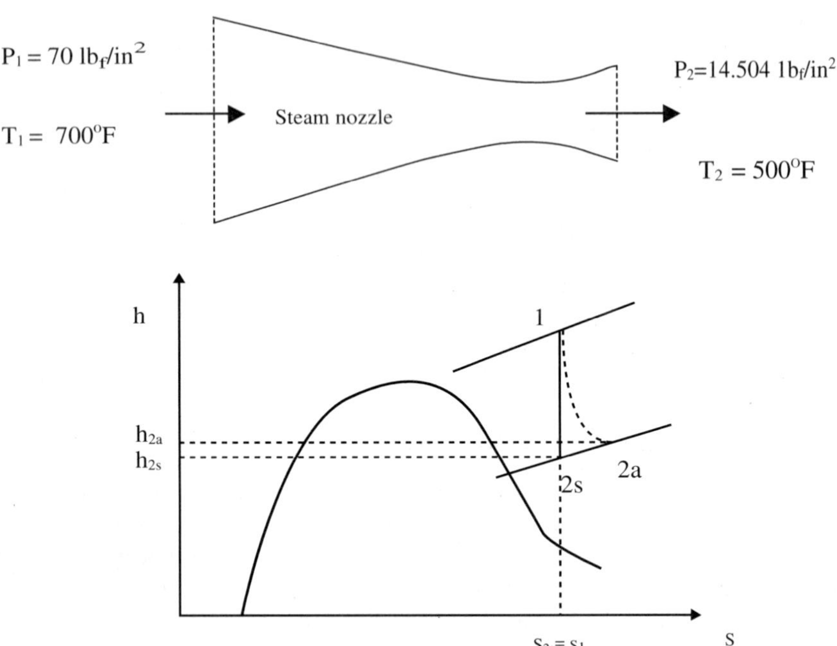

Schematic for Example 4.13.

The first law for the control volume system gives

$$\frac{v_2^2}{2} = h_1 - h_2 + \frac{v_1^2}{2}.$$

Thus,

$$\frac{v_2^2}{2} = 1380.8\,\frac{\text{Btu}}{\text{lb}_\text{m}} - 1287\,\frac{\text{Btu}}{\text{lb}_\text{m}} + \frac{(150\,\text{ft}/\text{s})^2}{2\left(\dfrac{32.2\,\text{lb.ft}/\text{s}^2}{1\,\text{lb}_\text{f}}\right)\left(\dfrac{778\,\text{ft.lb}_\text{f}}{1\,\text{Btu}}\right)} = 94.2\ \text{Btu}/\text{lb}_\text{m}.$$

For the isentropic case,

$$\left(\frac{v_2}{2}\right)^2_{\text{isen}} = 1380.8 - 1214 + \frac{(150)}{2(32.2)(778)} = 167.2\,\frac{\text{Btu}}{\text{lb}_\text{m}}$$

$$\eta_{\text{isen}} = \frac{(v_2^2/2)_{\text{actual}}}{(v_2^2/2)_{\text{isen}}} = \frac{94.2}{167.2} = 0.56\,(56\%).$$

Example 4.14

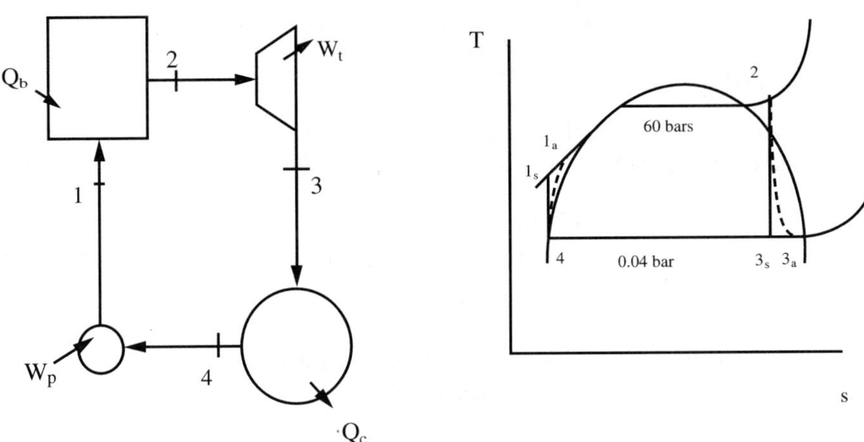

Schematic for Example 4.14.

Problem: The operating parameters of the ideal power cycle shown above, with water as the working fluid, are 60 bars for the boiler, 0.04 bar for the condenser, and the highest temperature in the boiler is 500°C. Calculate the first law efficiency of the ideal power cycle.

The isentropic efficiency of the turbine is actually 85%, and that of the pump is actually 70%. Calculate the entropy production in the turbine and in the pump. In addition, calculate the actual first law efficiency of the power cycle.

Solution

Assumptions: (1) Changes in kinetic and potential energy can be neglected.

Analysis: First, identify the state points on the temperature-entropy T-S diagram. Find the properties for each state, starting at state 2 since both T_2 and P_2 are known.

$$h_2 = 3422 \text{ kJ/kg} \qquad s_2 = 6.88 \text{ kJ/(kg.K)}$$
$$h_{3s} = 2073 \text{ kJ/kg} \qquad s_{3s} = 6.88 \text{ kJ/(kg.K)}$$
$$h_4 = 121.4 \text{ kJ/kg} \qquad s_4 = 0.4226 \text{ kJ/(kg.K)}$$
$$h_{1s} = 128 \text{ kJ/kg} \qquad s_{1s} = 0.4226 \text{ kJ/(kg.K)}$$

The thermal efficiency of the ideal power cycle, or first law efficiency is

$$\eta_I = \frac{W_t - W_p}{Q_b} = \frac{(h_2 - h_{3s}) - (h_{1s} - h_4)}{(h_2 - h_{1s})} = 40.0\%.$$

Since the isentropic efficiency of the turbine is 85%,

$$(h_2 - h_{3a}) = 0.85\ (h_2 - h_{3s}),$$
$$h_{3a} = 2275.4\ \text{kJ/kg, and}$$
$$s_{3a} = 7.6\ \text{kJ/(kg.K)}.$$

The increase in entropy in the turbine is 0.7 kJ/(kg.K).
Since the isentropic efficiency of the pump is 70%,

$$0.7\ (h_{1a} - h_4) = (h_{1s} - h_4),$$
$$h_{1a} = 130.8\ \text{kJ/kg, and}$$
$$s_{1a} = 0.432\ \text{kJ/(kg.K)}.$$

The increase in entropy in the pump is 0.01 kJ/(kg.K).
The actual first law efficiency of the cycle is

$$\eta_I = \frac{W_{t'} - W_{p'}}{Q_b} = \frac{(h_2 - h_{3a}) - (h_{1a} - h_4)}{(h_2 - h_{1a})} = 35\%.$$

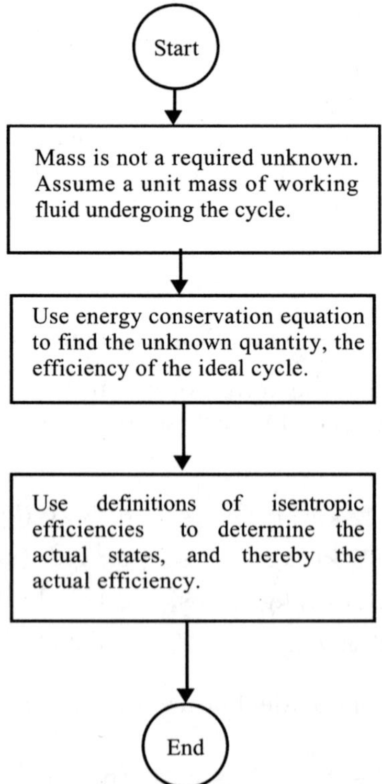

Logic diagram for Example 4.14.

The logic diagram for solving the above problem may be represented as shown. It is similar to that for Example 4.6. The mass conservation equation is not used explicitly; a unit mass of the working fluid is considered as undergoing the cycle. The energy conservation equation (first law) is used to determine the unknown quantity, which is thermal efficiency of the ideal cycle in this case. Then, the definitions of isentropic efficiencies (second law) are used to find the actual states, and thereby the thermal efficiency of the actual cycle.

4.9 Reversible Steady-Flow Processes

This section seeks to study reversible steady-flow processes. The objective is to derive expressions for heat transfer and the work in the absence of irreversibilities (internal irreversibilities).

Consider a control volume around a device with one inlet and one outlet at steady state. The flow is both reversible and isothermal. The second law in entropy equation is

$$\dot{\sigma}_{cv} + \frac{\dot{Q}_{cv}}{T} + \dot{m}s_1 = \dot{m}s_2 \tag{4.79}$$

where 1 and 2 are the inlet and outlet states, respectively, and $\dot{m}$ is the mass flow rate. Since the flow is reversible, $\dot{\sigma}_{CV} = 0$.

$$\frac{\dot{Q}_{cv}}{\dot{m}} = T(s_2 - s_1) \tag{4.80}$$

If the temperature is not constant, we can consider it to change in a series of infinitesimal steps. In such a case, we can integrate to obtain the heat transfer.

$$\left(\frac{\dot{Q}_{CV}}{\dot{m}}\right)_{rev} = \int_1^2 T ds \tag{4.81}$$

In other words, for a reversible process, the magnitude of the heat transfer per unit mass of fluid can be represented by the area under the curve indicating the reversible process on a T-s diagram.

To obtain the expression for work, we apply the first law to the control volume around the device. At steady state this gives

$$\frac{\dot{W}_{CV}}{\dot{m}} = \frac{\dot{Q}_{CV}}{\dot{m}} + (h_1 - h_2) + \left(\frac{v_1^2 - v_2^2}{2}\right) + g(z_1 - z_2). \tag{4.82}$$

For the reversible case, we can substitute Equation (4.81) to obtain

$$\left(\frac{\dot{W}_{CV}}{\dot{m}}\right)_{rev} = \int_1^2 Tds + (h_1 - h_2) + \left(\frac{v_1^2 - v_2^2}{2}\right) + g(z_1 - z_2). \tag{4.83}$$

Since the process is reversible, the working fluid undergoes a sequence of equilibrium states as it passes from inlet to outlet. From Equation (4.16), Tds = dh – vdP, which on integration gives

$$\int_1^2 Tds = (h_1 - h_2) - \int_1^2 vdP. \tag{4.84}$$

Substituting this equation into Equation (4.83) gives

$$\left(\frac{\dot{W}_{CV}}{\dot{m}}\right)_{rev} = -\int_1^2 vdP + \left(\frac{v_1^2 - v_2^2}{2}\right) + g(z_1 - z_2). \tag{4.85}$$

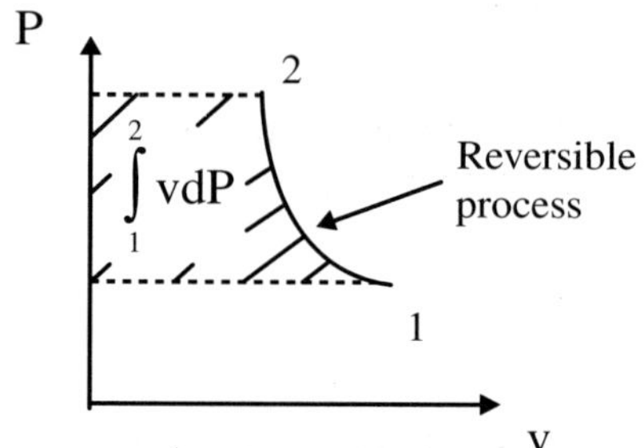

FIGURE 4.12
Interpretation of ∫vdP.

When the reversible process is shown on a P-v diagram, as in Figure 4.12, the magnitude of the integral ∫vdP is represented by the area behind the curve.

In nozzles and diffusers, work $\dot{W}_{CV}$ is zero. For such cases, the first law becomes

$$\int_1^2 vdP + \left(\frac{v_1^2 - v_2^2}{2}\right) + g(z_1 - z_2) = 0. \tag{4.86}$$

If the specific volume is not a function of pressure, as in liquids and other incompressible fluids,

$$v(P_2 - P_1) + \left(\frac{v_1^2 - v_2^2}{2}\right) + g(z_1 - z_2) = 0. \tag{4.87}$$

This is known as the Bernoulli equation and is used in fluid mechanics. It is important to stress this link between thermodynamics and fluid mechanics. The equation is applicable when there are no irreversibilities such as friction. However, the equation can be modified to include irreversibilities.

In turbines, compressors, and pumps there is often no significant change in K.E. or P.E. Using this fact in Equation (4.85),

$$\left(\frac{\dot{W}_{CV}}{\dot{m}}\right)_{rev} = -\int_1^2 vdP. \tag{4.88}$$

With liquids, the specific volume is approximately constant, so

$$\left(\frac{\dot{W}_{CV}}{\dot{m}}\right)_{rev} = -v(P_2 - P_1). \tag{4.89}$$

It is important to note that the larger the specific volume of the working fluid, the larger the reversible work produced or consumed by the steady-flow device. This is also true for actual devices. It is necessary to minimize the specific volume of a fluid during a compression process to minimize the work input and to maximize the specific volume during an expansion process to maximize the work output.

In steam or gas power plants, the pressure drop in the turbine is equal to the pressure rise in the pump or compressor if the pressure losses in the other components are disregarded. In steam power plants, the specific volume of the working fluid is small in the pump and very large in the turbine. Hence, the turbine work output is much greater than the pump work input. This accounts for the success and popularity of this steam cycle.

In contrast, the working fluid (usually air) in gas power plants is compressed in the gas phase. The compressor thus consumes a large portion of the turbine work output. The gas power plant gives less net work per unit mass of the working fluid.

4.9.1 Reversible Steady-Flow Polytropic Processes

A polytropic process is represented by the relation

$$Pv^n = \text{constant}. \tag{4.90}$$

Substituting this in Equation (4.88) and integrating,

$$\left(\frac{\dot{W}_{CV}}{\dot{m}}\right)_{rev} = -\int_1^2 vdP$$

$$= -(\text{constant})^{1/n}\int_1^2 \frac{dP}{P^{1/n}} \quad (4.91)$$

$$= -\frac{n}{n-1}(P_2v_2 - P_1v_1) \qquad (\text{polytropic, } n \neq 1).$$

When n = 1, Pv = constant, and the work is

$$\left(\frac{\dot{W}_{CV}}{\dot{m}}\right)_{rev} = -\int_1^2 vdP$$

$$= -(\text{constant})\int_1^2 \frac{dP}{P} \quad (4.92a)$$

$$= -(P_1v_1)\ln\frac{P_2}{P_1} \qquad (\text{polytropic, } n = 1).$$

The above equations apply to any gas in general. For an ideal gas, Equation (4.91) becomes

$$\left(\frac{\dot{W}_{CV}}{\dot{m}}\right)_{rev} = -\frac{nR}{n-1}(T_2 - T_1) \qquad (\text{ideal gas, } n \neq 1). \quad (4.92b)$$

For a polytropic process of an ideal gas (see Chapter 2, Section 2.6.1),

$$\frac{T_2}{T_1} = \left(\frac{P_2}{P_1}\right)^{(n-1)/n}.$$

Equation (4.92) can be expressed as

$$\left(\frac{\dot{W}_{CV}}{\dot{m}}\right)_{rev} = -\frac{nRT_1}{n-1}\left[\left(\frac{P_2}{P_1}\right)^{(n-1)/n} - 1\right] \qquad (\text{ideal gas, } n \neq 1). \quad (4.93)$$

For an ideal gas, Equation (4.92) becomes

$$\left(\frac{\dot{W}_{CV}}{\dot{m}}\right)_{rev} = -RT\ln\frac{P_2}{P_1} \qquad (\text{ideal gas, } n = 1). \quad (4.94)$$

Example 4.15

Problem: An air compressor operates at steady state with the air inlet state at $P_1 = 1$ bar, $T_1 = 25°C$, and the outlet pressure $P_2 = 10$ bars. The air undergoes a polytropic process with n = 1.2. Determine the work and heat transfer per unit mass of air, in kilojoules per kilogram.

Solution

Assumptions: (1) Air is treated as an ideal gas, and (2) neglect changes in K.E. and P.E.

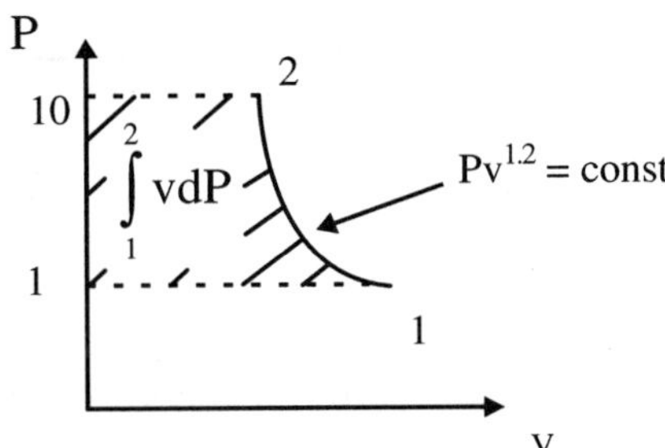

Schematic for Example 4.15.

Analysis: The outlet temperature can be found from

$$T_2 = T_1\left(\frac{P_2}{P_1}\right)^{(n-1)/n} = 298.15\left(\frac{10}{1}\right)^{0.2/1.2} = 437.62 \text{ K}.$$

Substituting values into Equation (4.92),

$$\left(\frac{\dot{W}_{CV}}{\dot{m}}\right)_{rev} = -\frac{nR}{n-1}(T_2 - T_1)$$

$$\left(\frac{\dot{W}_{CV}}{\dot{m}}\right)_{rev} = -\frac{1.2}{1.2-1}\left(\frac{8.314}{28.97}\frac{kJ}{kg \bullet K}\right)(437.62 - 298.15)K = -240\,kJ/kg.$$

Note that this is represented by the area behind the curve on the P-v diagram.

From the first law of thermodynamics,

$$\frac{\dot{Q}_{CV}}{\dot{m}} = \frac{\dot{W}_{CV}}{\dot{m}} + h_2 - h_1.$$

From the thermodynamic tables, $h_1 = 299$ kJ/kg.

Since the enthalpy of an ideal gas is not affected very much by the pressure, the enthalpy at state 2 may be taken to be that of the air at temperature $T_2 = 437.62$ K and a pressure of 0.1 MPa (as listed in the tables). Hence, $h_2 = 439$ kJ/kg.

Thus,

$$\frac{\dot{Q}_{CV}}{\dot{m}} = -240 + (439 - 299) = -100 \text{ kJ / kg.}$$

Problems

Introduction

4.1 A weekend garage tinker claims to have developed a machine that produces work at 15 kW, while consuming only 14 kW of electricity. Evaluate such a claim, or has he developed a perpetual-motion machine?

4.2 A tinker claims to have developed an engine that produces work at 8 kW, while absorbing heat at 8 kW. Evaluate such a claim.

Entropy of a Pure, Simple Compressible Substance

4.3 A tank of constant volume contains 5 kg of ammonia initially at 0.2 MPa and 25°C (298.15 K). The ammonia is then cooled until its pressure drops to 0.08 MPa. Calculate the difference in entropy of the ammonia between its initial and final states.

4.4 A constant-volume tank contains 10 lb_m of ammonia initially at 30 $lb_f/in.^2$ and 80°F (539.67°R). The ammonia is then cooled until its pressure drops to 10 $lb_f/in.^2$ Determine its entropy change from the intial state to the final state.

4.5 There are 2.5 kg of refrigerant 12 at 0.1 MPa and –35°C (238.15 K) contained in a piston-cylinder device. Then 600 kJ of heat are added isobarically to the refrigerant. Determine the entropy of the refrigerant at the initial and final states.

4.6 A piston-cylinder device contains 4 lb_m of refrigerant-12 at 20 $lb_f/in.^2$ and –50°F (409.67°R). The refrigerant is then heated at constant pressure by the addition of 400 Btu of heat. Determine the entropy change of the refrigerant.

4.7 A fixed mass of air undergoes a thermodynamic cycle comprising the following three processes consecutively:

Process 1→2, constant pressure cooling from $P_1 = 0.5$ MPa, $V_1 = 0.01$ m^3.

Process 2→3, isothermal heating to $P_3 = 0.1$ MPa, $T_3 = 25°C$, $V_3 = 0.01$ m^3.

Process 3→1, constant volume heating back to state 1.

Use the ideal gas equations with $C_p = 1.0$ kJ/(kg.K) to determine the entropy change, in each process.

Carnot Cycle

4.8 A Carnot engine operates between two reservoirs maintained at 400°C (673.15 K) and T_2, respectively. The heat transfer from the high-temperature reservoir is 25 kW. If the power developed by the engine is 13.74 kW, determine T_2. Calculate the heat loss at 30°C (303.15 K).

4.9 Two reservoirs maintained at 600°F (1059.67°R) and 50°F (409.67°R), are used to operate a Carnot engine. The rate of work done by the engine is 6.23 Btu/s. Calculate the heat transfer from the high-temperature reservoir. Hence, determine the heat loss at 50°F (409.67°R).

4.10 A refrigerator is cooling a space to –10°C (263.15 K) by transferring heat to the atmosphere at 20°C (293.15 K). By assuming a Carnot refrigerator, calculate the COP.

4.11 A Carnot engine operating between 60°C (333.15 K) and 400°C (673.15 K) is modified solely by raising the high temperature by 100°C and raising the low temperature by 100°C (383.15 K). Which of the following statements is false?

(a) More work is done during the isothermal expansion.

(b) More work is done during the isentropic compression.

(c) More work is done during the isentropic expansion.

(d) More work is done during the isothermal compression.

(e) Thermal efficiency is increased.

4.12 A Carnot refrigerator operates between –30°C (243.15 K) and 20°C (293.15 K). Find the change in coefficient of performance if the lower temperature is lowered by 10°C and the high temperature is increased by 5°C.

4.13 Find the coefficient of performance for a Carnot refrigerator (a) when it operates between –20°F (439.67°R) and 70°F (529.67°R), and (b) when the lower temperature is decreased by 20°F and the high temperature is raised by 10°F.

4.14 Determine the coefficient performance of a Carnot heat pump operating between –20°C (253.15 K) and 18°C (291.15 K).

4.15 Calculate the coefficient performance of a Carnot heat pump operating between two reservoirs at –5°F (454.67°R) and 60°F (519.67°R), respectively.

4.16 A Carnot cycle for water operates between 20 MPa and 0.2 MPa. At the higher pressure, water goes from the saturated liquid state to the saturated vapor state. Find the entropy change at this higher pressure. Thus, calculate the heat rejected to the low-temperature reservoir.

4.17 The load on a heat pump for a commercial building is 1×10^7 kJ/day to keep the building temperature at 22°C (295.15 K), when the external environment is at –5°C (268.15 K). Calculate the minimum theoretical operating cost of the pump, if electrical charges are 9 cents per kilowatthour.

4.18 The interior of a dwelling is to be maintained at 21°C (294.15 K) in the winter and 26°C (299.15 K) in the summer. A reversible heat pump is to be used to accomplish the heating and air-conditioning of the dwelling. Heat transfer through the walls, windows, and roof is about 3000 kJ/h per degree temperature difference between the interior dwelling temperature and the enviromental temperature. (a) If the outside summer temperature is 36°C (309.15 K), what is the minimum power required to run the heat pump in the reversed cycle? (b) To maintain the operating costs of the heat pump through the winter months the same as that through the summer months, the power input in the winter is the same as that calculated in (a). What is the minimum outside winter temperature for which the inside of the house can be kept at 21°C (294.15 K)?

Isentropic Processes

4.19 The outlet temperature of hydrogen gas from an adiabatic compressor is 125°C (398.15 K). Calculate its outlet pressure if the gas enters at 0.8 bar and 45°C (318.15 K), and undergoes a reversible process.

4.20 Determine the final pressure of a mass of hydrogen gas that is at 20 $lb_f/in.^2$ and 80°F (539.67°R), and is then compressed isentropically to a final temperature of 350°F (809.67°R).

4.21 Find the final pressure of helium gas that is compressed reversibly and adiabatically from an initial state of 1 bar and 15°C (288.15 K) to a final temperature of 200°C (473.15 K).

4.22 The outlet temperature of helium gas from a compressor is 420°F (879.67°R). Compute its outlet pressure if the helium gas enters at an initial state of 25 $lb_f/in.^2$ and 80°F (539.67°R), and undergoes a reversible adiabatic process.

4.23 Air is compressed isentropically, starting at 10°C (283.15 K) and 100 kPa. If the compression ratio V_1/V_2 is 7, determine the final temperature of the air.

4.24 An adiabatic piston-cylinder device compresses air from an initial state of 50°F (509.67°R) and 14.504 $lb_f/in.^2$ If the compression ratio V_1/V_2 is 12, find the final air temperature.

4.25 Compare the work required at steady state to compress water vapor isentropically to 5 MPa from the saturated vapor state at 0.1 MPa to the work required to pump liquid water isentropically to 5 MPa from the saturated liquid state at 0.1 MPa, each in terms of per kilogram of water flowing through the equipment. Kinetic and potential energy terms may be neglected.

Isentropic Efficiencies

4.26 Steam enters a turbine at $P_1 = 0.6$ MPa, $T_1 = 400$°C (673.15 K). The steam leaves at 0.1 MPa. If the isentropic efficiency of the turbine is 70%, determine the enthalpy of the steam at the outlet. Calculate the turbine work per unit mass of steam, in kilojoules per kilogram.

4.27 Consider a turbine operating at steady state. Steam enters at $P_1 = 90$ $lb_f/in.^2$, $T_1 = 750$°F (1209.67°R), and leaves at 15 $lb_f/in.^2$ If the isentropic efficiency of the turbine is 70%, compute the enthalpy of the exiting steam. Determine the work done by the turbine, in British thermal units per pound.

4.28 Water enters a pump at 0.1 MPa and 20°C (293.15 K). Water leaves the pump at 0.4 MPa. If the isentropic efficiency of the pump is 85%, determine the enthalpy of the water at the outlet. Calculate the pump work input per unit mass of water, in kilojoules per kilogram.

4.29 Consider a pump operating at steady state. Water enters at 14.7 $lb_f/in.^2$ and 70°F (529.67°R), and leaves at 60 $lb_f/in.^2$ If the isentropic efficiency of the pump is 75%, determine the enthalpy of the exiting water. Compute the work done on the pump, in British thermal units per pound.

4.30 Consider a nozzle operating at steady state. Steam enters at $P_1 = 1$ MPa, $T_1 = 370$°C (643.15 K) with a velocity of 50 m/s, and leaves at $P_2 = 2$ bars and $T_2 = 200$°C (473.15 K). There is no significant heat transfer. Calculate the nozzle efficiency.

4.31 Steam enters a nozzle at $P_1 = 140$ $lb_f/in.^2$, $T_1 = 700$°F (1159.67°R) with a velocity of 150 ft/s, and leaves at $P_2 = 20$ $lb_f/in.^2$ and $T_2 = 400$°F (859.67°R). The nozzle is operating under steady-state conditions, and heat transfer may be neglected. Compute the nozzle efficiency.

4.32 An electric compressor takes air from the environment and delivers it at a rate of 25 kg/s and a gage pressure of 4 bars, with no change in elevation. The isentropic efficiency of the compressor is 76%. Calculate the hourly cost of running the compressor, if electricity costs 9 cents per kilowatt hour.

Second Law in Entropy for a Control Volume

4.33 Steam at 8 MPa and 480°C (753.15 K) is throttled through a valve to a pressure of 2 MPa in a steady-flow process. Determine the entropy production for this process. Is the increase of entropy principle satisfied?

4.34 Steam at 300 $lb_f/in.^2$ and 1200°F (1659.67°R) is throttled through a valve to a pressure of 180 $lb_f/in.^2$ in a steady-flow process. Calculate the entropy of the steam before and after the valve. Is the increase of entropy principle satisfied?

4.35

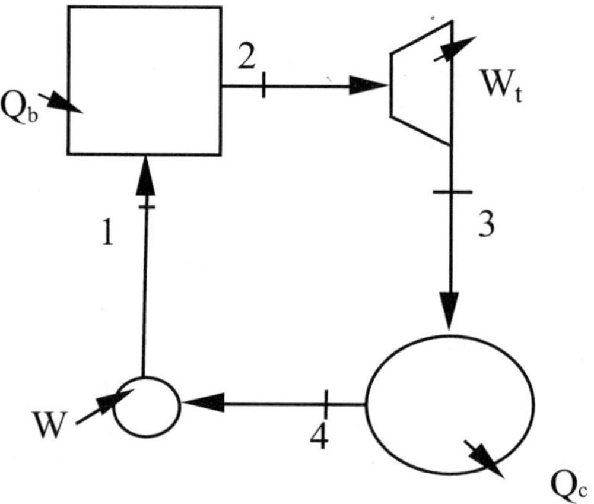

Schematic for Problem 4.35.

(a) In the ideal power cycle shown above, with water as the working fluid, the boiler pressure is 40 bars, the condenser pressure is 0.06 bar, and the highest temperature in the boiler is 440°C (713.15 K). Calculate the first law efficiency of the ideal power cycle.

(b) The isentropic efficiency of the turbine is actually 80%, and that of the pump is actually 90%. Calculate the entropy production in the turbine and in the pump. In addition, calculate the actual first law efficiency of the power cycle.

4.36 (a) Consider the ideal power cycle shown in Problem 4.35, with water as the working fluid. The boiler operates between the pressure limits of 900 $lb_f/in.^2$ and 5 $lb_f/in.^2$, with the steam leaving the boiler at 1000°F (1459.67°R). Compute the first law efficiency of the ideal power cycle.

(b) The isentropic efficiencies of the turbine and the pump are 78% and 85%, respectively. Compute the entropy production in the turbine and in the pump. Hence, compute the actual first law efficiency of the power cycle.

Second Law in Entropy

4.37 In any practical thermodynamic process, the overall entropy of an isolated system will

(a) Increase and then decrease

(b) Decrease and then increase

(c) Decrease only

(d) Increase only

(e) Remain the same.

4.38 Entropy is

(a) The change in enthalpy of a system

(b) A property of a system

(c) The internal energy of a substance

(d) The total heat content of a system

(e) The heat capacity of a substance.

4.39 For spontaneously occurring processes in an isolated system, which expression best evaluates dS?

(a) $dS = 0$

(b) $dS > 0$

(c) $dS < 0$

(d) $dS = C_P \dfrac{dT}{T} - R\dfrac{dP}{P}$

(e) $dS = dQ/T$

4.40 Which of the following statements about entropy is false?

(a) Entropy has the units of heat capacity.

(b) Entropy of a mixture is greater than that of its components under the same conditions.

(c) An irreversible process increases entropy of the universe.

(d) Entropy of a crystal at 0°C is zero.

(e) Net entropy change in any closed cycle is zero.

4.41 Consider a closed system and its surroundings as an isolated system. Answer true or false to each of the following statements, and explain.
(a) Processes in which the entropies of both the system and its surroundings remain unchanged are not allowed.
(b) The entropy of the system might increase, while the entropy of its surroundings decrease.
(c) Processes in which the entropies of both the system and its surroundings decrease are not allowed.
(d) All processes are such that the entropies of both the system and its surroundings increase.

4.42 A piston-cylinder device contains saturated liquid ammonia at 6°C (279.15 K). The ammonia adiabatically converts to the corresponding saturated vapor state. The stirrer causes the change of state, and the piston is free to move during the process. Compute the work per unit mass of ammonia, and the amount of entropy produced per unit mass.

4.43 In a piston-cylinder device, saturated liquid ammonia at 40°F (499.67°R) is adiabatically converted to the corresponding saturated vapor state. The piston is free to move during the process; the stirrer causes the change of state. Calculate the work per unit mass of ammonia, and the amount of entropy produced per unit mass.

4.44 A piston-cylinder device contains saturated liquid water at 30°C (303.15 K). A stirrer converts the water to the corresponding saturated vapor state in an adiabatic process, and the piston is free to move during the process. Determine the work per unit mass of water, and the amount of entropy produced per unit mass.

4.45 Saturated liquid water at 90°F (549.67°R) is adiabatically converted to the corresponding saturated vapor state, in a piston-cylinder device. The stirrer causes the change of state, and the piston is free to move during the process. Calculate the work per unit mass of water, and the amount of entropy produced per unit mass.

4.46 Refrigerant-134a is compressed adiabatically in a piston-cylinder device from saturated vapor at –4°C (269.15 K) to a final pressure of 0.4 MPa. Determine the minimum input work required per unit mass of refrigerant-134a.

4.47 A piston-cylinder device compresses refrigerant-134a from saturated vapor at 5°F (464.67°R) to a final pressure of 80 $lb_f/in.^2$ Find the minimum input work required per unit mass of refrigerant-134a.

4.48 Refrigerant-12 is compressed in a piston-cylinder device from saturated vapor at –15°C (258.15 K) to a final pressure of 0.8 MPa. There is no heat transfer. Calculate the minimum input work required per unit mass of refrigerant-12.

4.49 A piece of equipment that comprises a piston-cylinder, compresses refrigerant 12 adiabatically from saturated vapor at 0°F (459.67°R) to a pressure of 90 $lb_f/in.^2$ Compute the minimum input work required per unit mass of refrigerant 12.

4.50 A quantity of air is initially at 1 bar in a piston-cylinder device with a stirrer that performs work on the air. The air undergoes a process where its temperature increases from 0 to 250°C by work from the stirrer and also by heating from a reservoir maintained at 240°C. Recommend how this process may be accomplished with (a) minimum entropy production, and (b) maximum entropy production.

4.51 For each of the cases below, indicate whether the entropy change of the closed system is indeterminate, zero, positive, or negative:

(a) One kilogram of oxygen undergoing an isothermal process.

(b) Two pounds of ammonia undergoing an adiabatic process during which it is stirred.

(c) One kilogram of air undergoing an internally reversible process.

(d) Two pounds of water vapor undergoing an adiabatic process.

(e) Two kilograms of nitrogen modeled as an ideal gas undergoing an isobaric process to a lower temperature.

4.52 A closed system undergoes a process in which the work done on the system is 10 kJ and the heat transfer Q occurs only at temperature T_o. For each of the cases below, indicate whether the entropy change of the closed system is indeterminate, zero, positive, or negative:

(a) Internally reversible process, Q = 0.

(b) Internally reversible process, Q = +10 kJ.

(c) Internally reversible process, Q = –10 kJ.

(d) Internally irreversible, Q = 0.

(e) Internally irreversible, Q = +10 kJ.

(f) Internally irreversible, Q = –10 kJ.

Reversible Steady-Flow Processes

4.53 An air compressor at steady state, operates with inlet conditions of $P_1 = 1$ bar, $T_1 = 30°C$ (303.15 K), and outlet pressure of $P_2 = 8$ bars. The air undergoes a polytropic process with n = 1.1. Calculate the work and heat transfer per unit mass of air, in kilojoules per kilogram.

4.54 The air undergoes a polytropic process with n = 1.1 in a steady-state air compressor. The inlet state is defined by $P_1 = 14.7$ $lb_f/in.^2$, $T_1 = 90°F$ (549.67°R), and the outlet pressure $P_2 = 120$ $lb_f/in.^2$ Determine the work and heat transfer per unit mass of air, in British thermal units per pound.

Computer, Design, and General Problems

4.55 The hurricane has been likened to a natural Carnot engine. Discuss the possibility of harnessing the work output from such a natural phenomenon.

4.56 It has been proposed to develop a snow-skiing slope in Miami, Florida which has a semitropical climate. This winter sport facility is to cover an area of 10 acres, say. It has been proposed to use the steam produced for domestic and industrial applications. Discuss the feasibility of this proposal, and the energy consumption of such a facility.

4.57 The temperature of the ocean is not constant; it varies from locality to locality and with depth. This temperature difference has been used to produce work. Discuss the pros and cons, and the reasons that cause this technology to be less than popular.

4.58 An inventor has designed a ship that runs on the temperature difference of the water in the ocean. Discuss the practical feasibility of such a ship.

4.59 A perpetual motion machine of the third kind is one without friction. Design toys that may approach as nearly as possible such an ideal machine.

4.60 Discuss the pros and cons of the current use of the isentropic efficiency for equipment. Since isentropic processes are impossible to attain in practice, discuss the advantages and disadvantages of other efficiencies.

5

Availability (Exergy) Analysis

CONTENTS

Availability is the available energy that may be used. It is also called exergy. Some authors have a slightly different definition for exergy, but in this book exergy is used interchangeably with availability. Physically, availability may be viewed as the potential for use. The gaseous combustion gases in an internal combustion engine has availability. If it is not used, its availability is wasted. If the combustion gases are employed in a turbocharger, for instance, the availability has been used. Availability has the same units as energy. However, unlike energy, availability is not conserved.

5.1 Availability

The concept of availability comes from the fact that an opportunity arises for doing work whenever two systems at different states are engineered into communication. When one of the systems is idealized to represent the environment, and the other is some system of interest, availability is the maximum theoretical work obtainable as they interact to equilibrium.

5.1.1 Dead State

When a fixed mass of matter, a closed system, changes from the environment, an opportunity develops for doing work. Conversely, as the closed system changes to a state approaching the temperature and pressure of the environment, this opportunity decreases. When the closed system is in equilibrium with the environment, this opportunity does not exist. This state of the system is called the dead state. At the dead state, both the closed system and the environment possess energy, but the value of the availability of the system is zero because there is no chance of any interaction between the system and the environment.

5.1.2 Availability of a Substance

The availability of a mass with no boundary interactions is defined as

$$A = (E - U_o) + P_o(V - V_o) - T_o(S - S_o) \tag{5.1}$$

where E is the internal energy and kinetic energy and potential energy of the system. U_o, V_o, and S_o, respectively, are the internal energy, volume, and entropy of the mass when at dead state. The specific availability of the mass of matter is

$$a = (u - u_o) + P_o(v - v_o) - T_o(s - s_o) + V^2/2 + gz. \tag{5.2}$$

The units of specific availability are the same as those of specific energy. The change in availability between two states of a mass is the difference

$$A_2 - A_1 = (E_2 - E_1) + P_o(V_2 - V_1) - T_o(S_2 - S_1) \tag{5.3}$$

where P_o and T_o are the conditions of the environment.

Availability is thus an attribute of a system and its environment. If the conditions of the environment are fixed, then the availability may be treated like a property of the system.

Example 5.1

Problem: Five kilograms of water is initially saturated vapor at 140°C, the velocity is 25 m/s, and the elevation is 10 m. It undergoes a process which transforms it to a final state where it is saturated liquid at 20°C, the velocity is 10 m/s, and the elevation is 2 m. Determine in kilojoules (a) the availabilities corresponding to the initial and final states, and (b) the change in availability. The environmental temperature and pressure are $T_o = 25°C$ and $P_o = 100$ kPa. Take $g = 9.8\ m/s^2$.

Solution

Assumptions: (1) The velocities and elevations are given relative to the environment.

Analysis: The availability is given by

$$A = m[(u - u_o) + P_o(v - v_o) - T_o(s - s_o) + V^2/2 + gz].$$

The dead state for the water is characterized by

$$u_o = 104.86\ kJ/kg$$
$$v_o = 1.0029 \times 10^{-3}\ m^3/kg$$
$$s_o = 0.3673\ kJ/(kg.K).$$

The initial state is characterized by

$$u_1 = 2550\ kJ/kg$$
$$v_1 = 0.5089\ m^3/kg$$
$$s_1 = 6.93\ kJ/(kg.K).$$

The final state is characterized by

$$u_2 = 83.94\ kJ/kg$$
$$v_2 = 1.0018 \times 10^{-3}\ m^3/kg$$
$$s_2 = 0.2966\ kJ/(kg.K).$$

(a) The availability at the initial state is

$$\begin{aligned} A_1 = (5\,kg)\Big\{&(2550 - 104.86)\frac{kJ}{kg} \\ &+\left[\left(1.013\times10^3\frac{N}{m^2}\right)(0.5089 - 1.0029\times10^{-3})\frac{m^3}{kg}\right]\left(\frac{1\,kJ}{10^3\,N.m}\right) \\ &-(298.15\,K)(6.93 - 0.3673)\frac{kJ}{kg.K} \\ &+\left[\frac{(25\,m/s)^2}{2} + \left(9.8\frac{m}{s^2}\right)10\,m\right]\left(\frac{1\,N}{1\,kg.m/s^2}\right)\left(\frac{1\,kJ}{10^3\,N.m}\right)\Big\} \\ = {}&5(2445.14 + 51.45 - 1956.7 + 0.31 + 0.098) \\ = {}&2701\,kJ. \end{aligned}$$

The availability at the final state is

$$A_2 = 5\,(-20.93 - 0.000111 + 21.07 + 0.05 + 0.02) = 1.05 \text{ kJ}.$$

(b) The change in availability is

$$A_2 - A_1 = 1.05 - 2701 = -2700 \text{ kJ}.$$

The negative sign shows that the availability decreases in the process.

Example 5.2

Problem: An internal combustion engine cylinder contains gaseous combustion products at a pressure of 6 bars and a temperature of 717°C at the instant before it is exhausted. Calculate the specific availability of the gas in kilojoules per kilogram. The environmental temperature and pressure are $T_o = 25°C$ and $P_o = 1$ atm.

Solution

Assumptions: (1) We model the combustion products as air behaving like an ideal gas, and (2) neglect the effects of motion and gravity.

Analysis: With Assumption (2), the specific availability becomes

$$a = (u - u_o) + P_o(v - v_o) - T_o(s - s_o).$$

From the thermodynamic tables (either manual or computerized),

$$\begin{aligned} u - u_o &= 750.79 - 213.04 \\ &= 537.8 \text{ kJ/kg}. \end{aligned}$$

(The values of u, u_o may be obtained using the computerized tables by assuming that the pressure has very little effect on the values of the internal energy. For instance, u may be found for 1 mol of air by entering T = 717°C = 990.15 K, P = 0.1 MPa, even though the actual P is 0.6 MPa.)

$$\begin{aligned} s - s_o &= s^\circ(T) - s^\circ(T_o) - \frac{\overline{R}}{M}\ln\frac{P}{P_o} \\ &= 8.1237 - 6.8631 - \left(\frac{8.314}{28.97}\right)\ln\left(\frac{6}{1.013}\right) \\ &= 0.750 \text{ kJ}/(\text{kg.K}) \end{aligned}$$

$$T_o(s - s_o) = (298.15\text{ K})\{0.750 \text{ kJ}/(\text{kg.K})\} = 223.6 \text{ kJ/kg}.$$

The term $P_o(v - v_o)$ is calculated using the ideal gas equation of state.

Since

$$v = \left(\frac{\overline{R}}{M}\right)\frac{T}{P} \quad \text{and} \quad v_o = \left(\frac{\overline{R}}{M}\right)\frac{T_o}{P_o},$$

$$P_o(v - v_o) = \frac{\overline{R}}{M}\left(\frac{P_oT}{P} - T_o\right)$$

$$= \frac{8.314}{28.97}\left[\frac{(1.013)(990.15)}{6} - 298.15\right] = -37.6 \text{ kJ/kg}.$$

Hence, the specific availability of the gas is

$$a = 537.8 - 37.6 - 233.6 = 266.6 \text{ kJ/kg}.$$

5.2 Second Law in Availability for a Control Volume

Recall that an open system is one where there is flow of matter or mass in or out of the boundaries. The boundaries essentially define a control volume, that is, a volume defined as the space of interest in which we are concentrating our analysis. Figure 5.1 is identical to Figure 3.4. We consider the volume initially with a mass m_{ini}. Associated with this mass is its availability. The availability initially in the control volume is thus $m_{ini}\, a_{ini}$.

Consider a mass m_{in} entering the control volume [Figure 5.1, Part (b)]. The available energy associated with this mass is its availability and the energy required to push the mass into the control volume. The availability and the energy required to push the mass into the control volume *is* the flow availability of m_{in}. The flow availability associated with m_{in} is $m_{in}\, a_{f\,in}$, where the specific flow availability is defined as

$$a_f = (h - h_o) - T_o(s - s_o) + V^2/2 + gz. \tag{5.4}$$

Thus,

$$a_{f\,in} = (h_{in} - h_o) - T_o(s_{in} - s_o) + V_{in}^{\ 2}/2 + gz_{in}. \tag{5.5}$$

Consider a mass m_e exiting a control volume, as in Figure 5.1, Part (c). The available energy associated with this mass is its availability and the energy required to push the mass out of the control volume. The availability and the energy required to push the mass out of the control volume *is* the flow availability of m_e. The flow availability associated with m_e is $m_e a_{fe}$.

There is mass entering the control volume and mass exiting the control volume, so the mass within the control volume does not remain the same. Let us call the mass finally remaining in the control volume, m_{final}, as in

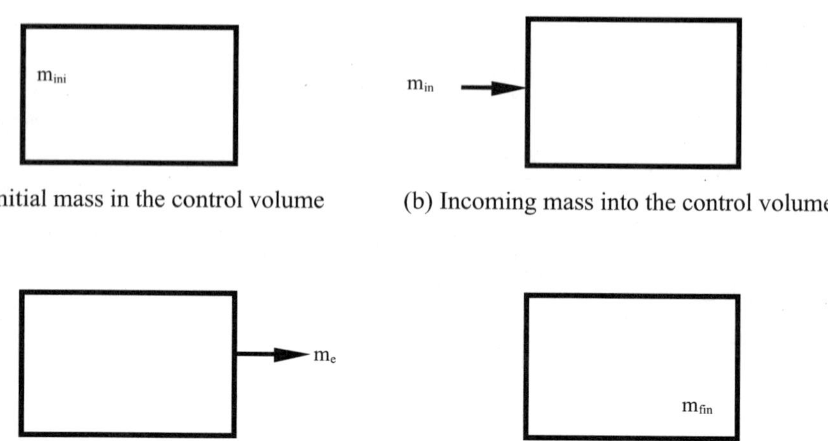

(a) Initial mass in the control volume

(b) Incoming mass into the control volume

(c) Mass exiting the control volume

(d) Final mass in the control volume

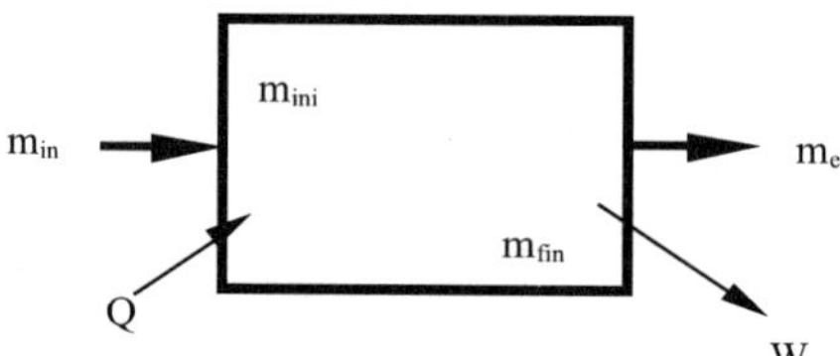

(e) Control volume with one inlet and one exit

FIGURE 5.1
Control volume.

Figure 5.1, Part (d). Associated with this mass is its availability. Hence, the availability left in the control volume is $m_{final}\, a_{final}$. In addition, there is heat entering the system, and work being done by the system. These are forms of available energy that have to be taken into account in the availability balance of the control volume. Heat Q_j brings with it an availability of

$$\left(1-\frac{T_o}{T_j}\right)Q_j\,,$$

where T_j is the temperature at which the heat is being transferred and T_o is the temperature of the dead state. The availability associated with the work done by the system is $W - P_o\,\Delta V$, where P_o is the pressure of the dead state and ΔV is the volume change of the system. Refer to Section 5.3 for a more detailed discussion on the availability transfer accompanying heat and work.

The final diagram in the series, Figure 5.1, Part (e), shows the various masses, and the heat and work interactions. The availability equation for the control volume in Figure 5.1, Part (e) is

$$\left(1-\frac{T_o}{T_j}\right)Q_j + m_{in}a_{f\,in} + m_{ini}a_{ini} = m_e a_{fe} + m_{final}a_{final} + W - P_o\Delta V + I_{cv}. \quad (5.6)$$

Note that the flow availability is used for masses that cross the system boundaries, and just the availability for masses that do not cross the system boundaries. Note that the sum of the availability due to incoming heat transfer, the initial availability of the system, and the incoming availability will be more than the sum of the final availability in the control volume, the outgoing availability, and the availability associated with the work done. The difference is made up in the availability balance equation by I_{cv}, which is the irreversibility term for the control volume. The irreversibility term represents the loss in availability. The second law equation, otherwise called the creation of entropy, is not a conservation equation. The irreversibility term I_{cv} is equal to $T_o\sigma_{cv}$. However, the form of the equation above is in a form similar to a conservation equation, with the irreversibility term included to make up the difference. Therefore, I_{cv} can be positive or zero; it can never be negative.

$I_{cv} > 0$ — actual or possible processes
$I_{cv} = 0$ — reversible processes
$I_{cv} < 0$ — impossible.

However, the change in availability can be positive, zero, or negative. Availability, unlike irreversibility, is a property of the system with respect to its environment.

Taking into consideration multiple inlet and exit streams and multiple heat transfer and work processes, the second law in availability equation becomes

$$\begin{aligned}\sum_k {}_k\left(1-\frac{T_o}{T_j}\right)Q_j + \sum_1 {}_1m_{in}\,{}_1a_{fin} + m_{ini}a_{ini} \\ = \sum_n {}_nm_e\,{}_na_{fe} + m_{final}a_{final} + \sum_m {}_m\left(W - P_o\Delta V\right) + I_{cv}.\end{aligned} \quad (5.7)$$

In a steady-state steady-flow (S.S.S.F.) process, the initial mass is equal to the final mass and the initial state is the same as the final state, i.e., $m_{ini} = m_{final}$ and $a_{ini} = a_{final}$. So the second law in availability equation for a S.S.S.F. process is

$$\left(1-\frac{T_o}{T_j}\right)Q_j + m_{in}a_{fin} = m_e a_{fe} + W + I_{cv} \quad (5.8)$$

since ΔV is zero for the control volume.

A closed system is one where no mass enters or leaves the system. Under these conditions, $m_{in} = m_e = 0$ and $m_{ini} = m_{final} = m$. So the second law in availability equation for a control mass or closed system is

$$\left(1-\frac{T_o}{T_j}\right)Q_j + ma_{ini} = ma_{final} + (W - P_o\Delta V) + I. \tag{5.9}$$

The above equations are similarly modified if there are more than one mass stream entering or more than one mass stream exiting the system for the S.S.S.F. process; if there is more than one heat transfer process, or more than one work process in either the S.S.S.F. process or the closed system, the equations are similarly modified to take into account the multiple processes.

The rate equation for the control volume for a single mass stream in and out of the system and single heat and work processes, is

$$\left(1-\frac{T_o}{T_j}\right)\dot{Q}_j + \dot{m}_{in}\, a_{fin} + \dot{m}_{ini}\, a_{ini} = \dot{m}_e a_{fe} + \dot{m}_{final}\, a_{final} + \dot{W} - P_o\Delta\dot{V} + \dot{I}_{cv}. \tag{5.10}$$

The corresponding equation for the multiple mass streams and boundary processes is

$$\begin{aligned}\sum_k \left(1-\frac{T_o}{T_j}\right)_k \dot{Q}_j + \sum_1 {}_1\dot{m}_{in} a_{fin} + \dot{m}_{ini}\, a_{ini} \\ = \sum_n {}_n\dot{m}a_{fe} + \dot{m}_{final}\, a_{final} + \sum_m {}_m\left(\dot{W} - P_o\Delta\dot{V}\right) + \dot{I}_{cv}.\end{aligned} \tag{5.11}$$

It is instructive to compare Equation (5.11) with Equation (3.14), which is repeated here for convenience.

$$\begin{aligned}\sum_k {}_k\dot{Q} + \sum_1 {}_1\dot{m}_{in}\left(h_{in} + \frac{V_{in}^2}{2} + gz_{in}\right)_1 + \dot{m}_{ini}\left(u_{ini} + \frac{V_{ini}^2}{2} + gz_{ini}\right) = \\ \sum_n {}_n\dot{m}_e\left(h_e + \frac{V_e^2}{2} + gz_e\right)_n + \dot{m}_{fin}\left(u_{fin} + \frac{V_{fin}^2}{2} + gz_{fin}\right) + \sum_p {}_p\dot{W}.\end{aligned} \tag{3.14}$$

The similarities between the two equations are emphasized here. Note the way in which the heat transfer and work terms have been appropriately modified. Associated with each mass is its availability. Since the availability is not conserved, the irreversibility term is introduced on the right-hand side to balance the availability equation.

Example 5.3

Problem: For a working fluid undergoing a S.S.S.F. process as shown, verify that the irreversibility I is equal to $T_o\sigma$. Heat Q_j is transferred at a temperature T_j.

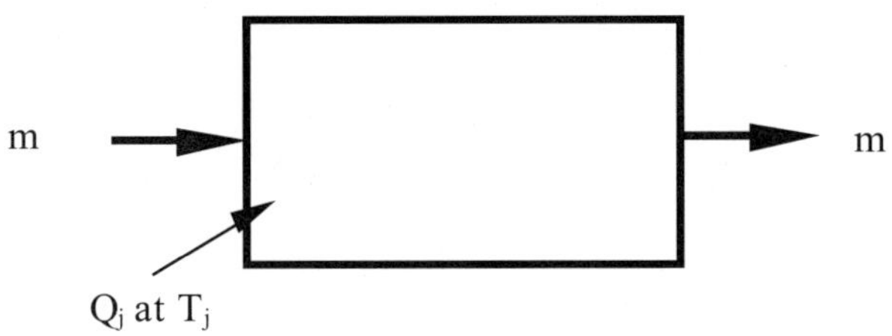

Schematic for Example 5.3.

Solution

Assumptions: (1) Changes in K.E. and P.E. are negligible, and (2) the dead state conditions are at T_o and P_o.

Analysis:

Applying the first law equation to the control volume,

$$Q_j = m(h_e - h_{in}). \tag{a}$$

Apply the second law in entropy equation to the control volume,

$$\sigma + \frac{Q_j}{T_j} + ms_{in} = ms_e. \tag{b}$$

Apply the second law in availability equation to the control volume,

$$\left(1 - \frac{T_o}{T_j}\right)Q_j + ma_{fin} = ma_{fe} + I. \tag{c}$$

From Equation (c),

$$\left(1 - \frac{T_o}{T_j}\right)Q_j + m\left(h_{in} - h_o - T_o\{s_{in} - s_o\}\right) = m\left(h_e - h_o - T_o\{s_e - s_o\}\right) + I.$$

Substituting for Q_j from Equation (a),

$$m(h_e - h_{in})\frac{T_o}{T_j}Q_j + m\left(h_{in} - h_o - T_o\{s_{in} - s_o\}\right) = m\left(h_e - h_o - T_o\{s_e - s_o\}\right) + I.$$

Simplifying,

$$m(s_{in} - s_e) = \frac{Q_j}{T_j} + \frac{I}{T_o}.$$

Comparing with Equation (b),

$$\frac{I}{T_o} = \sigma \quad \text{or} \quad I = T_o\sigma.$$

Example 5.4

Problem: Given the open system above from Examples 3.1 and 4.6, with three incoming masses and two outgoing masses, calculate the specific availabilities of the initial mass and the final mass, the specific flow availabilities of the incoming and outgoing masses, and the irreversibility. Take T_o = 25°C, P_o = 100 kPa.

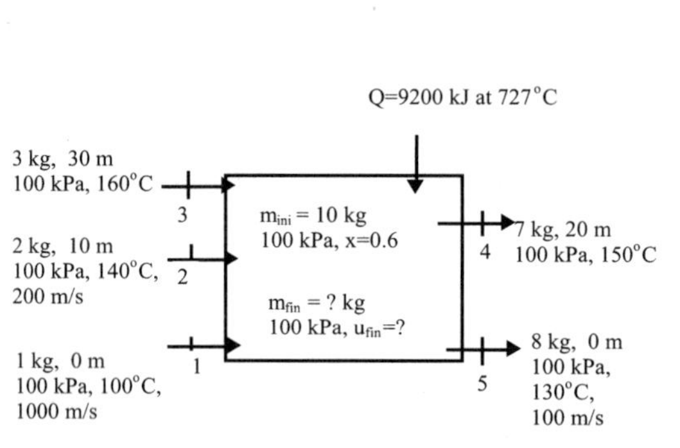

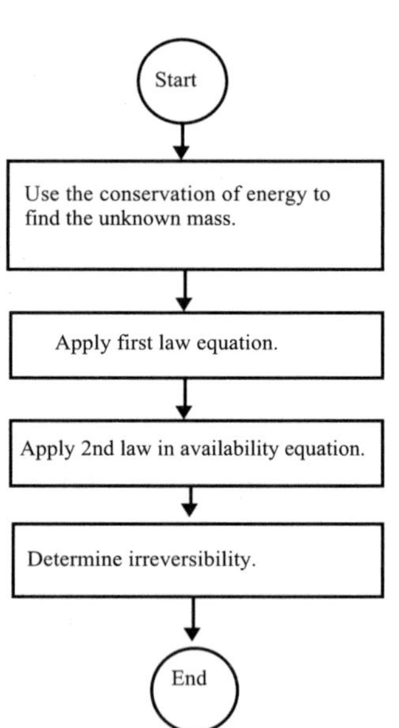

Schematic for Example 5.4.

Solution

Assumptions: (1) If the velocity of the mass is not given, its K.E. is neglected.

Analysis: The properties of water at the dead state are

$$h_o = 104.96 \text{ kJ/kg}, \qquad u_o = 104.86 \text{ kJ/kg}$$
$$s_o = 0.3673 \text{ kJ/(kg.K)}, \quad v_o = 0.0010029 \text{ m}^3\text{/kg}.$$

For the incoming mass at state 1,

$$a_{f1} = (2676 - 104.96)\text{kJ/kg} - (298.15\text{ K})(7.361 - 0.3673)\frac{\text{kJ}}{\text{kg.K}}$$
$$+\left[\frac{(1000\text{m/s})^2}{2} + 0\right]\left(\frac{1\text{ N}}{1\text{ kg.m/s}^2}\right)\left(\frac{1\text{kJ}}{10^3\text{ N.m}}\right)$$
$$= 2571 - 2085.17 + 500$$
$$= 985.83 \text{ kJ/kg}.$$

For the incoming mass at state 2,

$$a_{f2} = 2652 - 2145.4 + 20 = 526.6 \text{ kJ/kg}.$$

For the incoming mass at state 3,

$$a_{f3} = 2691.04 - 2174.3 + 0.294 = 517.03 \text{ kJ/kg}.$$

For the initial mass,

$$a_{ini} = (1670.6 - 104.86)\text{ kJ/kg}$$
$$+\left(1\times10^5 \frac{\text{N}}{\text{m}^2}\right)\left(1.0168 - 1.0029\times10^{-3}\right)\frac{\text{m}^3}{\text{kg}}\left(\frac{1\text{kJ}}{10^3\text{ N.m}}\right)$$
$$-(298.15\text{ K})(4.937 - 0.3673)\frac{\text{kJ}}{\text{kg.K}}$$
$$= 1565.74 + 101.58 - 1362.5$$
$$= 304.82 \text{ kJ/kg}.$$

For the final mass in the system and the solution given in Example 3.1,

$$
\begin{aligned}
s_{final} &= 4.892\,kJ/(kg.K),\\
v_{final} &= 1.0044\ m^3/kg\\
a_{final} &= (1656-104.86)kJ/kg\\
&\quad+\left(1\times10^5\frac{N}{m^2}\right)\left(1.0044-1.0029\times10^{-3}\right)\frac{m^3}{kg}\left(\frac{1\,kJ}{10^3\,N.m}\right)\\
&\quad-(298.15\,K)(4.892-0.3673)\frac{kJ}{kg.K}\\
&= 1551.14+100.34-1349.0\\
&= 302.48\ kJ/kg.
\end{aligned}
$$

For the outgoing mass at state 4,

$$a_{f4} = 2671.04 - 2160.3 + 0.196 = 510.94\ kJ/kg.$$

For the outgoing mass at state 5,

$$a_{f5} = 2632.04 - 2130.49 + 5 = 506.55\ kJ/kg.$$

The second law in availability equation for the control volume is

$$
\sum_k \left(1-\frac{T_o}{T_j}\right)_k Q_j + \sum_1 {}_1m_{in}\,{}_1a_{fin} + m_{ini}a_{ini} =
\sum_n {}_nm_e\,{}_na_{fe} + m_{final}a_{final} + I_{cv} - P_o\left(v_{final}-v_{ini}\right).
$$

Substituting the values for the masses, availabilities, heat transfer, and temperature of heat transfer,

$$
\begin{aligned}
&\{(1-0.298)9200+1.(985.83)+2.(526.6)+3.(517.03)+10.(304.82)\}kJ\\
&\quad=\{1.(302.48)+7.(510.94)+8.(506.55)\}kJ+I-(100)[(1\times1.0044)-(10\times1.0168)]\\
&(6458+6638.12)\,kJ = 7931.48\,kJ + I + 916.36\\
&I = 4250\ kJ.
\end{aligned}
$$

The irreversibility may also be calculated from the expression $I = T_o\sigma$. Using this relation, the entropy production σ is calculated to be 14.25 kJ/K. Compare this to the value of 14.3 kJ/K found in Example 4.6. The difference is due to round-off errors in computing the availabilities. If the objective is just to evaluate the irreversibility, the equation in entropy should be used because it would involve less round-off errors.

The logic diagram for solving the above problem is represented as shown. The mass conservation equation is used to find the unknown mass, u_{final}. The first law is used to determine the state of the water left finally in the control volume. The second law in availability equation is used to determine the irreversibility.

Example 5.5

Problem: Superheated steam is throttled via a valve from 30 bars, 1000°C, to 15 bars. Calculate in kilojoules the specific flow availability at the inlet and the outlet, as well as the irreversibility per unit mass of steam. Assume $T_o = 25°C$, $P_o = 1$ atm.

Solution

Assumptions: (1) There is no heat transfer and no work is done, and (2) changes in K.E. and P.E. are negligible.

Analysis: For a throttling process, the first law equation gives $h_2 = h_1$.

From the thermodynamic tables,

$$h_1 = 4632 \text{ kJ/kg}, \qquad s_1 = 8.401 \text{ kJ/(kg.K)}$$
$$h_2 = h_1, \qquad s_2 = 8.721 \text{ kJ/(kg.K)}.$$

With assumption (2), the specific flow availability is

$$a_f = (h - h_o) - T_o (s - s_o).$$

The availability of the steam at the inlet is

$$a_{f1} = (4632 - 104.96) - 298.15 (8.401 - 0.3673) = 2132 \text{ kJ/kg}.$$

The availability of the steam at the outlet is

$$a_{f2} = (4632 - 104.96) - 298.15 (8.721 - 0.3673) = 2036 \text{ kJ/kg}.$$

The steam undergoes a S.S.S.F. process as it passes through the valve. The availability rate equation for the control volume, the rate form of Equation (5.8), is

$$\left(1 - \frac{T_o}{T_j}\right)\cancel{\dot{Q}_j} + \dot{m}a_{f1} = \dot{m}a_{f2} + \cancel{\dot{W}} + \dot{I}_{cv}.$$

Since there is no heat transfer or work done, the equation gives

$$\dot{m}a_{f1} = \dot{m}a_{f2} + \dot{I}_{cv}.$$

Thus,

$$\frac{\dot{I}_{CV}}{\dot{m}} = (a_{f1} - a_{f2}) - 2132 - 2036 = 96 \text{ kJ/kg}.$$

Note that energy is conserved in the throttling process, but there is irreversibility. The irreversibility comes from the uncontrolled expansion.

The logic diagram for solving the above problem is represented as shown. The mass conservation equation is not used explicitly; a unit mass of the working fluid is considered as undergoing the process. The first law is used to determine state 2. The second law in availability equation is used to determine the irreversibility.

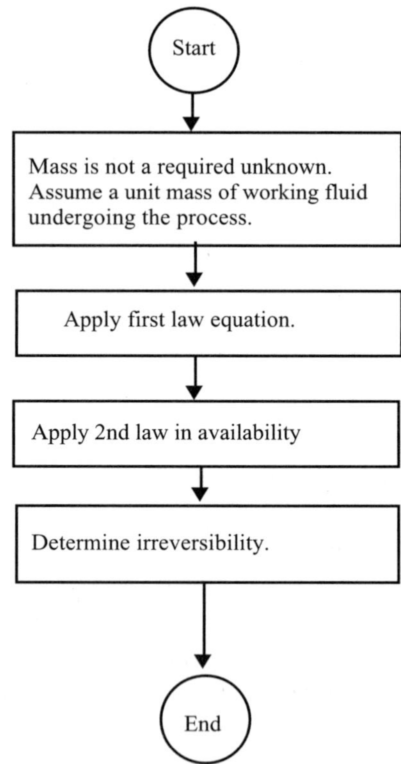

Logic diagram for Example 5.5.

Example 5.6

Problem: Steam at 5 MPa and 600°C enters a turbine at a steady rate of 10 kg/s and leaves at 0.2 MPa and 140°C. The steam loses 500 kW of heat to the environment, which is at 25°C and 1 atm. Determine (a) the availability of the

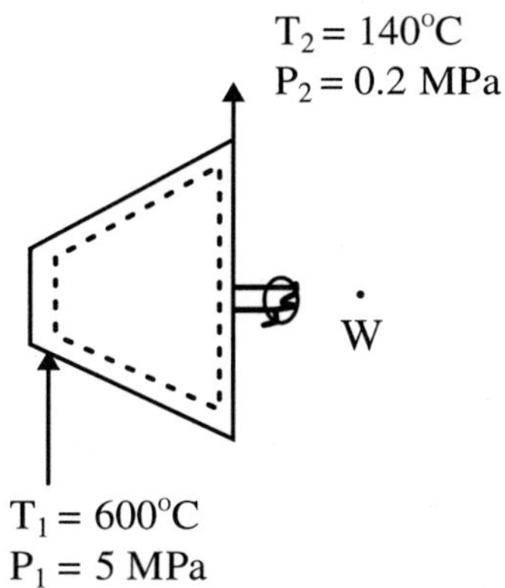

Schematic for Example 5.6.

steam at inlet, (b) the turbine power output, (c) the maximum possible turbine power output, and (d) the irreversibility.

Solution

Assumptions: (1) The changes in K.E. and P.E. are negligible.

Analysis:

Inlet state:	$P_1 = 5$ MPa,	$T_1 = 600°C$
	$h_1 = 3667$ kJ/kg,	$s_1 = 7.259$ kJ/(kg.K)
Outlet state:	$P_2 = 0.2$ MPa,	$T_2 = 140°C$
	$h_2 = 2748$ kJ/kg,	$s_2 = 7.228$ kJ/(kg.K)
Dead state:	$P_o = 101.3$ kPa,	$T_o = 25°C$
	$h_o = 104.96$ kJ/kg,	$s_o = 0.3673$ kJ/(kg.K).

(a) With assumption (1), the availability of the steam at inlet is

$$a_{f1} = (h_1 - h_o) - T_o\,(s_1 - s_o)$$
$$a_{f1} = [(3667 - 104.96) - 298.15\,(7.259 - 0.3673)]\ \text{kJ/kg}$$
$$= 1507\ \text{kJ/kg}.$$

(b) The first law rate equation for the turbine is

$$\dot{Q} + \dot{m}h_1 = \dot{m}h_2 + \dot{W}$$
$$\dot{W} = \dot{m}(h_1 - h_2) - \dot{Q}$$
$$\dot{W} = 10(3667 - 2748)\frac{\text{kJ}}{\text{s}} - 500\ \text{kW} = 8690\ \text{kW}.$$

The turbine power output is 8690 kW.

(c) The steam undergoes a S.S.S.F. process in the turbine. The second law in availability rate equation for the turbine, the rate form of Equation (5.8), is

$$\left(1-\frac{T_o}{T_j}\right)\dot{Q}_j + \dot{m}a_{f1} = \dot{m}a_{f2} + \dot{W} + \dot{I}_{cv}.$$

Since heat transfer is not significant for a turbine, the equation gives

$$\dot{m}a_{f1} = \dot{m}a_{f2} + \dot{W} + \dot{I}_{cv}.$$

The maximum possible turbine power is produced when $\dot{I}_{cv}$ is zero. This corresponds to a reversible turbine.

$$\begin{aligned}\dot{W}_{rev} &= \dot{m}(a_{f1} - a_{f2})\\ &= \dot{m}[(h_1 - h_2) - T_o(s_1 - s_2)]\\ \dot{W}_{rec} &= 10[(3667 - 2748) - 298.15(7.259 - 7.228)]\frac{kJ}{s} = 9098\text{ kW}.\end{aligned}$$

The maximum possible turbine power output is 9098 kW.

(d) The irreversibility is the difference between the turbine power output and the maximum possible turbine power output, which is

$$\dot{I}_{CV} = \dot{W}_{rev} - \dot{W} = 9098 - 8690 = 408\text{ kW}.$$

This quantity may also be calculated as $T_o\,\dot{\sigma}_{CV}$, where $\dot{\sigma}_{CV}$ is the rate of entropy generated in the process. It can be seen that considering the availability of inlet steam, only $(8690)/(1507 \times 10) = 58\%$ is converted to work.

The logic diagram for solving the above problem is very similar to that for Example 5.5. The second law in availability equation is used to determine the maximum possible turbine power output, before computing the irreversibility.

5.3 Second Law in Availability for a Control Mass

As shown in Section 5.2, the second law in availability for a closed system or control mass is

$$\left(1-\frac{T_o}{T_j}\right)Q_j + ma_{ini} = ma_{final} + (W - P_o\Delta V) + I. \tag{5.9}$$

The time-dependent second law in availability equation for the control mass is

$$\left(1-\frac{T_o}{T_j}\right)\dot{Q}_j + \dot{m}a_{ini} = \dot{m}a_{final} + (\dot{W} - P_o\Delta\dot{V}) + \dot{I}. \tag{5.12}$$

The corresponding equation for multiple boundary processes is

$$\sum_k \left(1-\frac{T_o}{T_j}\right)_k \dot{Q}_j + \dot{m}a_{ini} = \dot{m}a_{final} + \sum_m \left(\dot{W} - P_o\Delta\dot{V}\right)_m + \dot{I}. \tag{5.13}$$

It is easy to see that Equation (5.13) is a special case or subset of Equation (5.11). Since there is no mass moving in or out of a control mass system, there is no need to consider incoming availability with incoming mass or outgoing entropy with outgoing mass. Care has to be taken to consider all boundary interactions.

Example 5.7

Problem: An adiabatic piston-cylinder assembly contains 0.2 kg of steam at 2 MPa and 400°C. The steam expands to a final state of 400 kPa and 200°C, while doing work. The environment is at $P_o = 100$ kPa and $T_o = 25$°C. Determine (a) the availability of the steam at the initial and final states, (b) the availability change of the steam, and (c) irreversibility of the process.

Solution

Assumptions: (1) Changes in K.E. and P.E. are negligible.

Analysis:

Initial state: $P_1 = 2$ MPa, $T_1 = 400$°C
$u_1 = 2945$ kJ/kg, $v_1 = 0.1512$ m³/kg
$s_1 = 7.127$ kJ/(kg.K)

Final state: $P_2 = 400$ kPa, $T_2 = 200$°C
$u_2 = 2647$ kJ/kg, $v_2 = 0.5342$ m³/kg
$s_2 = 7.171$ kJ/(kg.K)

Dead state: $P_o = 100$ kPa, $T_o = 25$°C
$u_o = 104.86$ kJ/kg, $v_o = 0.0010$ m³/kg
$s_o = 0.3673$ kJ/(kg.K).

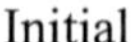

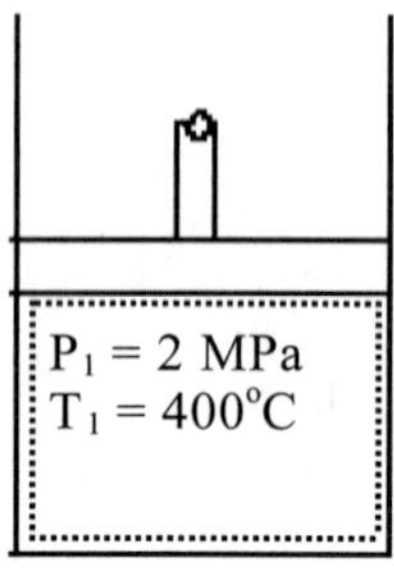

$P_o = 100$ kPa
$T_o = 25°C$

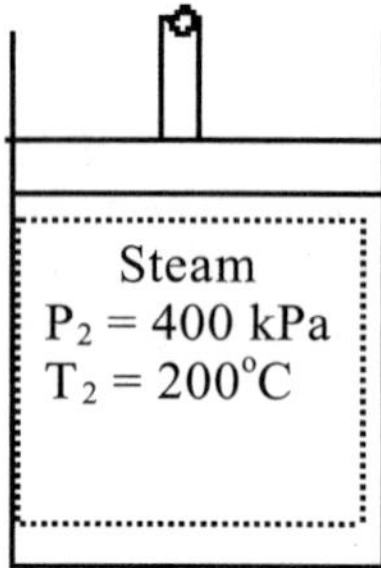

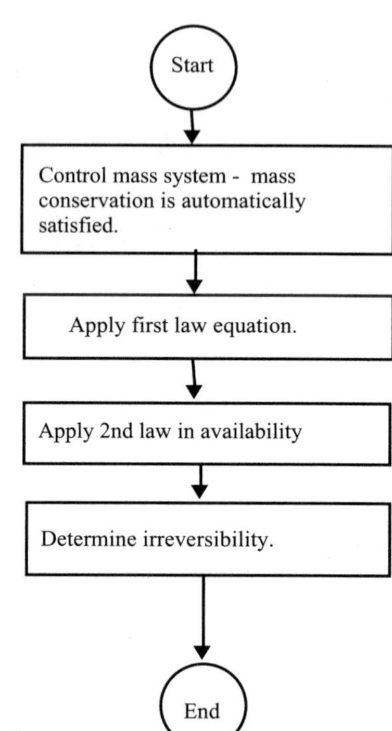

Final

Schematic and Logic Diagram for Example 5.7.

(a) Availability at the initial state is

$$\begin{aligned} A_1 &= m\left[(u_1 - u_o)\right] - T_o(s_1 - s_o) + P_o(v_1 - v_o) \\ &= (0.2\text{ kg})\{(2945 - 104.86)\text{ kJ/kg} \\ &\qquad -(298.15\text{ K})\left[(7.127 - 0.3673)\text{kJ}/(\text{kg.K})\right] \\ &\qquad +(100\text{ kPa})\left[(0.1512 - 0.0010)\text{ m}^3/\text{kg} \times \text{kJ}/(\text{kPa.m}^3)\right]\} \\ &= 168\text{ kJ.} \end{aligned}$$

Availability at the final state is

$$A_2 = (0.2)\{(2647 - 104.86) \\ -298.15(7.171 - 03673) \\ +100\,(0.5342 - 0.0010)\} = 113.4 \text{ kJ}.$$

(b) The availability change of the steam is

$$A_2 - A_1 = (113.4 - 168) \text{ kJ} = -54.6 \text{ kJ}.$$

(c) The first law equation for the control mass is

$$U_1 = U_2 + W.$$

Thus, $W = m\,(u_1 - u_2) = (0.2)(2945 - 2647) = 59.6$ kJ.
The availability equation for the control mass is

$$A_1 = A_2 + W - P_o\Delta V + I$$
$$I = (A_1 - A_2) - W + P_o\Delta V$$
$$I = (54.6 - 59.6)\text{ kJ}$$
$$+(100 \text{ kPa})(0.2 \text{ kg})\left[(0.5342 - 0.1512)\text{m}^3/\text{kg}\left(\frac{1 \text{ kJ}}{1 \text{ kPa.m}^3}\right)\right]$$
$$= 2.66 \text{ kJ}.$$

The irreversibility can also be determined from

$$I = T_o\sigma$$
$$I = T_o[m(s_2 - s_1)]$$
$$= (298.15 \text{ K})[(0.2 \text{ kg})(7.171 - 7.127)\text{kJ}/(\text{kg.K})]$$
$$= 2.6 \text{ kJ}$$

which is about the same as before.

The logic diagram to solve the above problem may be represented as shown. Since the system is a control mass, mass conservation is automatically satisfied. The first law equation is used to find the work done. Then the second law in availability equation is used to determine the irreversibility.

Example 5.8

Problem: The wall of a house is 7 m × 8 m and 25 cm thick. Its thermal conductivity, k, is 0.7 W/(m.°C). The house is heated to 25°C on a day when the outside environment is at 7°C. The temperatures of the outer and inner surfaces of the wall are determined to be 10°C and 18°C, respectively. Calculate (a) the rate of heat transfer through the wall, (b) the irreversibility rate in the wall, and (c) the rate of total irreversibility associated with this heat conduction process.

Solution

Assumptions: (1) Availability change of wall is zero, and (2) heat conduction is one-dimensional.

Analysis: The wall is taken to be a control mass. Fourier's law of heat conduction gives

$$\dot{Q} = kA\left(\frac{\Delta T}{L}\right)_{wall} = [0.7\,W/(m.^{\circ}C)][(7 \times 8)m^2]\frac{(18-10)^{\circ}C}{0.25\,m} = 1254\,W$$

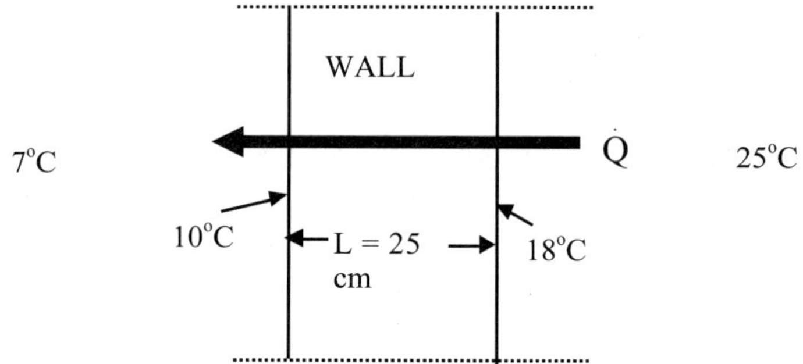

Schematic for Example 5.8.

where

A = cross-sectional area,
L = thickness of the wall.

The availability rate equation of the wall is

$$\dot{Q}\left(1-\frac{T_o}{T}\right)_{in} = \dot{Q}\left(1-\frac{T_o}{T}\right)_{out} + \dot{I}$$

$$(1254W)\left(1-\frac{280.15\,K}{291.15\,K}\right) - (1254\,W)\left(1-\frac{280.15\,K}{283.15\,K}\right) = \dot{I}.$$

Thus, $\dot{I} = 34.1$ W.

The irreversibility rate in the wall is 34.1 W.

To compute the rate of total irreversibility, the boundaries are expanded to include regions on both sides of the wall that undergo a temperature change. The boundaries of the expanded system (system + immediate surroundings) are at 280.15 K and 298.15 K instead of 283.15 and 291.15 K, respectively. The availability rate equation of the expanded system is

$$\dot{I} = (1254\,\text{W})\left(1 - \frac{280.15\ \text{K}}{298.15\ \text{K}}\right) - (1254\,\text{W})\left(1 - \frac{280.15\ \text{K}}{280.15\ \text{K}}\right).$$

Thus, $\dot{I} = 75.7$ W.

The total irreversibility rate associated with the heat conduction process is 75.7 W. The difference between the two irreversibilities is 41.6 W, and is the availability destroyed in the air layers adjacent to the wall on both sides. This availability destruction is caused by irreversible heat transfer through a finite temperature difference.

5.4 Availability Transfer

Consider a closed system where heat transfer Q_j takes place across a system boundary where the temperature T_j is constant at $T_j > T_o$. The accompanying availability transfer is given by

$$(\text{availability transfer accompanying heat}) = \left(1 - \frac{T_o}{T_j}\right)Q_j. \qquad (5.14)$$

The right-hand side of this equation is seen to be the work that could be developed by a reversible power cycle receiving Q_j at temperature T_j and supplying energy by heat transfer to the environment at T_o. When T_j is less than T_o, the sign of the availability transfer would be opposite to the sign of the heat transfer, so that the heat transfer and the accompanying availability transfer would be oppositely directed.

If there is no change in the system volume during the work process, the availability accompanying work will be equal to the work W of the system. Now, consider the system when the volume increases, $V_2 > V_1$, as it does work and there is no heat transfer. Since the system would do work on the surroundings equal to $P_o(V_2 - V_1)$, the maximum amount of work that could be derived from the system is

$$W_{max} = W - P_o(V_2 - V_1), \qquad (5.15)$$

which is the availability transfer associated with the work W.

5.5 Second Law (Exergetic) Efficiency

For a closed system undergoing a cycle with no work, the first law equation and second law equation in availability equation, respectively, are

$$\dot{Q}_s = \dot{Q}_j + \dot{Q}_l. \tag{5.16}$$

$$\left(1 - \frac{T_o}{T_s}\right)\dot{Q}_s = \left(1 - \frac{T_o}{T_j}\right)\dot{Q}_j + \left(1 - \frac{T_o}{T_l}\right)\dot{Q}_l + \dot{I}. \tag{5.17}$$

The energy balance equation shows that of the heat energy transferred in, part is used, $\dot{Q}_j$, and part is lost, $\dot{Q}_l$. This may be described by an efficiency in terms of a ratio (desired output)/input as

$$\eta = \frac{\dot{Q}_j}{\dot{Q}_s}. \tag{5.18}$$

Theoretically, the value of η may reach 100% if the heat energy loss $\dot{Q}_l$ is reduced to zero.

The availability equation shows that the availabilities corresponding to $\dot{Q}_s$ and $\dot{Q}_j$ account for the availability corresponding to the heat energy transferred in, and the shortfall is due to destruction by irreversibilities in the system. This can also be described by an efficiency in terms of a ratio (desired output)/input as

$$\eta_{II} = \frac{(1 - T_o / T_j)\dot{Q}_j}{(1 - T_o / T_s)\dot{Q}_s}. \tag{5.19}$$

This may be called a second law or exergetic efficiency.

Introducing Equation (5.18) into Equation (5.19), we obtain

$$\eta_{II} = \eta \frac{(1 - T_o / T_j)}{(1 - T_o / T_s)}. \tag{5.20}$$

To increase the value of η_{II}, the use temperature T_j should be made to approach the source temperature T_s.

5.5.1 Second Law Ratio to Measure Thermal Environmental Impact

From Equation (5.17), it can be seen that the availability lost may be reduced if T_l is brought as close as possible to T_o. Consider the internal combustion engine. If the combustion gases are exhausted to the environment at T_o, the availability of the gases is wasted. Say, this occurs at a temperature T_{l1}. However, if a turbocharger is employed to extract part of this availability of the combustion gases, the new lost temperature T_{l2} is brought closer to T_o. Since $T_{l1} > T_{l2} > T_o$, the term corresponding to $\dot{Q}_l$ is thus reduced in the availability balance equation, and the ratio η_{II} gives appropriately a higher value.

In addition, the exhausted gases may be viewed as thermal environmental contaminants, as well as being chemical contaminants. If the lost temperature of the exhausted gases is lowered, they become less of a thermal problem. For example, consider an environmental temperature T_o of 25°C. If the combustion gases are exhausted at 50°C, they are more of a thermal contaminant than if they were exhausted at 35°C. The environmental aspect of thermal systems may be measured by another ratio (losses)/input, where the ratio is

$$r_{II} = \frac{(1 - T_o / T_l)\dot{Q}_l}{(1 - T_o / T_s)\dot{Q}_s}. \tag{5.21}$$

The objective is to reduce this ratio to as close to zero as possible for minimum environmental impact. Such a ratio is useful as a measure of the thermal environmental impact of thermal systems.

If a similar ratio (losses)/input is used in the energy balance equation, we get

$$r_I = \frac{\dot{Q}_l}{\dot{Q}_s} = 1 - \eta. \tag{5.22}$$

Consider the example above of the internal combustion engine exhausting into an environment at 20°C. If the magnitude of $\dot{Q}_l$ remains unchanged, r_I remains unchanged for the cases of $T_l = 50°C$ or $T_l = 35°C$. However, r_{II} decreases when T_l decreases from 50°C to 35°C. This is a good illustration of the usefulness of r_{II} employed as a measure of the thermal environmental impact of thermal systems.

The foregoing discussion provides the background for the statement that the *second law* deals with the *quality* of energy, whereas the *first law* deals with the *quantity* of energy. The availability of a working fluid is a measure of the quality of energy it possesses. As good engineers, we have to try to reduce the quality as well as the quantity of energy that is wasted. In so doing, we are also reducing the adverse environmental impact caused by the waste energy.

Example 5.9

Problem: The gaseous combustion products in an internal combustion engine are at a pressure of 6 bars and a temperature of 717°C. Without a turbocharger, the gases are exhausted at this state into the environment. With a turbocharger, the gases are reduced to a temperature of 197°C at the exhaust.

Assume that the heat loss ratio $\dot{Q}_l / \dot{Q}_s$ is the same, 30% in both cases, and that $T_s = 907°C$. Calculate the value of r_{II} for the two cases. The environmental temperature and pressure are $T_o = 15°C$ and $P_o = 1$ atm.

Solution

From Equation (5.19),

$$r_{II} = \frac{(1 - T_o / T_l)\dot{Q}_l}{(1 - T_o / T_s)\dot{Q}_s}.$$

Without a turbocharger,

$$r_{II} = \left(\frac{1 - \dfrac{288.15}{990.15}}{1 - \dfrac{288.15}{1180.15}} \right) 30\% = 0.938(30\%) = 0.28.$$

With a turbocharger,

$$r_{II} = \left(\frac{1 - \dfrac{288.15}{470.15}}{1 - \dfrac{288.15}{1180.15}} \right) 30\% = 0.512(30\%) = 0.15$$

the value of r_{II} decreases. The adverse thermal environmental impact of the case with a turbocharger is less than that without a turbocharger.

Example 5.10

Problem: Two textile dryers use steam heating, with the steam at 212°F. The energy supplied is 1178 Btu / lb_m of textile in the first dryer and 1071 Btu / lb_m of textile in the second dryer. The heat loss is 589 Btu / lb_m of textile in the first dryer. If both machines are to have the same r_{II} value, calculate the heat energy lost in the second dryer. Assume that the lost temperature is the same in both dryers. The environmental temperature and pressure are $T_o = 77°F$ and $P_o = 1$ atm.

Solution

From Equation (5.19),

$$r_{II} = \frac{(1 - T_o / T_l)\dot{Q}_l}{(1 - T_o / T_s)\dot{Q}_s}.$$

If both r_{II} values are the same, and T_I's and T_s's are the same,

$$\left(\frac{\dot{Q}_l}{\dot{Q}_s}\right)_1 = \left(\frac{\dot{Q}_l}{\dot{Q}_s}\right)_2$$

$$0.5 = \frac{\dot{Q}_l}{1071}.$$

Thus, the heat loss from the second dryer is 535.5 Btu/lb_m of textile.

5.5.2 Second Law (Exergetic) Efficiencies of Systems

In general, the second law efficiency measures the degree to which a reversible operation has been accomplished; its value should vary from zero in the worst situation (100% availability destruction) to unity for the best case (no availability destruction). Accordingly, the second law efficiency of a system during a process is

$$\begin{aligned}\eta_{II} &= \frac{\text{Desired availability output}}{\text{Availability input}} \\ &= 1 - \frac{\text{Availability destroyed}}{\text{Availability input}} \\ &= 1 - \frac{\text{Irreversibility}}{\text{Availability input}}.\end{aligned} \tag{5.23}$$

We can also define a second law efficiency to measure the performance of a system undergoing a cycle. For a heat engine, we can define a second law efficiency η_{II} as the ratio of the actual thermal efficiency to the maximum possible thermal efficiency under identical conditions

$$\eta_{II} = \frac{\eta_I}{\eta_{I,rev}} \quad \text{(heat engines)}. \tag{5.24}$$

For other cyclic devices such as refrigerators and heat pumps, the second law efficiency can be defined in terms of the coefficients of performance as

$$\eta_{II} = \frac{COP}{COP_{rev}} \quad \text{(refrigerators and heat pumps)}. \tag{5.25}$$

The second law efficiency of any thermal system may also be defined as the ratio (desired output)/input, and the terms in the ratio being obtained from the availability equation. Some more common systems are analyzed in this section.

Example 5.11

Problem: In Example 5.7, calculate the second law efficiency of the process.

Solution

The second law efficiency of the process is

$$\eta_{II} = \frac{\text{Desired availability output}}{\text{Availability input}} = 1 - \frac{\text{Irreversibility}}{\text{Availability input}}$$

$$\eta_{II} = 1 - \frac{2.6}{54.6} = 0.95.$$

Example 5.12

Problem: In a refrigeration cycle, the $Q_{in} = 120$ kJ/s and $Q_{out} = 180$ kJ/s. The corresponding ideal cycle would have $Q_{out} = 160$ kJ/s with the same Q_{in}. Determine the second law efficiency of this cycle.

Solution

Since

$$W_{cycle} = Q_{out} - Q_{in},$$

$$W_{cycle}\big|_{actual} = 60 \text{ kJ/s}$$

$$W_{cycle}\big|_{ideal} = 40 \text{ kJ/s}.$$

Thus,

$$\beta_{actual} = \left|\frac{Q_{in}}{W_{cycle}}\right|_{actual} = \frac{120}{60} = 2$$

$$\beta_{ideal} = \left|\frac{Q_{in}}{W_{cycle}}\right|_{ideal} = \frac{120}{40} = 3$$

$$\eta_{II} = \frac{\beta_{actual}}{\beta_{ideal}} = \frac{2}{3} = 0.67.$$

The second law efficiency of the cycle is 0.67.

5.5.2.1 Turbines, Compressors, and Pumps

For an adiabatic turbine operating at steady state, the second law in availability equation is

$$\dot{m}a_{f1} = \dot{m}a_{f2} + \dot{W}_{cv} + \dot{I}_{cv}$$
$$a_{f1} - a_{f2} = \frac{\dot{W}_{cv}}{\dot{m}} + \frac{\dot{I}_{cv}}{\dot{m}}. \tag{5.26}$$

This equation indicates that the flow availability drops because the turbine does work, $\dot{W}_{cv} / \dot{m}$, and the availability is destroyed, $\dot{I}_{CV} / \dot{m}$. A parameter that measures the ratio (desired output)/input is

$$\eta_{II} = \frac{\dot{W}_{CV} / \dot{m}}{a_{f1} - a_{f2}}. \tag{5.27}$$

This second law efficiency of the turbine measures how well the flow availability has been converted into work output. It has been sometimes called the turbine effectiveness. Note that this ratio is different from the isentropic turbine efficiency defined in Chapter 4, Section 4.8.1.

For adiabatic compressors and pumps operating at steady state, the second law in availability equation may be written as

$$\left(-\frac{\dot{W}_{cv}}{\dot{m}}\right) = a_{f2} - a_{f1} + \frac{\dot{I}_{cv}}{\dot{m}}.$$

This equation shows that the availability input to the device, $(-\dot{W}_{CV} / \dot{m})$, is either used to increase the flow availability of the working fluid, or is destroyed. A parameter that measures the ratio (desired output)/input is

$$\eta_{II} = \frac{a_{f2} - a_{f1}}{\left(-\frac{\dot{W}_{CV}}{\dot{m}}\right)}. \tag{5.28}$$

Note that this second law efficiency is different from the isentropic compressor efficiency or isentropic pump efficiency defined in Chapter 4, Section 4.8.2.

Example 5.13

Problem: In Example 5.6, calculate the second law efficiency of the turbine.

Solution

The second law efficiency of the turbine is

$$\eta_{II} = \frac{\dot{W}_{CV} / \dot{m}}{a_{f1} - a_{f2}}$$

$$= \frac{\dot{W}}{\dot{W}_{rev}} = \frac{8690}{9098} = 0.96.$$

5.5.2.2 Non-Contact Heat Exchangers

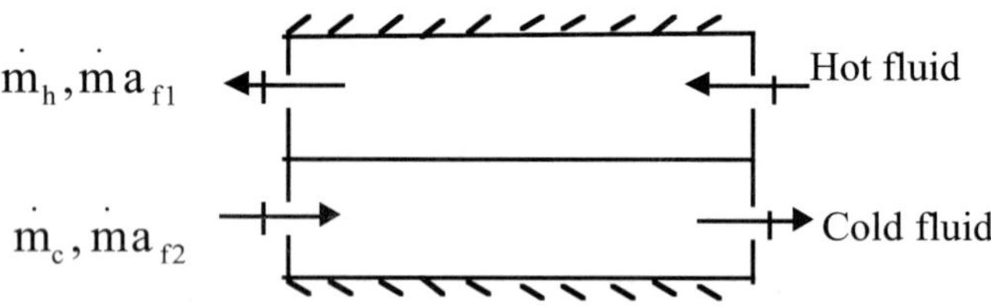

FIGURE 5.2
Counterflow non-contact heat exchanger.

A non-contact heat exchanger, as shown in Figure 5.2, operates at steady state. There is no heat exchange with its surroundings. Heat is transferred from the hot fluid to the cold fluid. The availability rate equation is

$$\dot{m}_h a_{f1} + \dot{m}_c a_{f3} = \dot{m}_h a_{f2} + \dot{m}_c a_{f4} + \dot{I}_{cv}. \tag{5.29}$$

The equation indicates that the availability input via the hot fluid is converted to the desired output, availability increase of the cold fluid, plus the amount destroyed. A parameter that measures the ratio (desired output)/input is

$$\eta_{II} = \frac{\dot{m}_c (a_{f4} - a_{f3})}{\dot{m}_h (a_{f1} - a_{f2})}. \tag{5.30}$$

The value of this second law efficiency is typically less than one.

5.5.2.3 Direct Contact Heat Exchangers

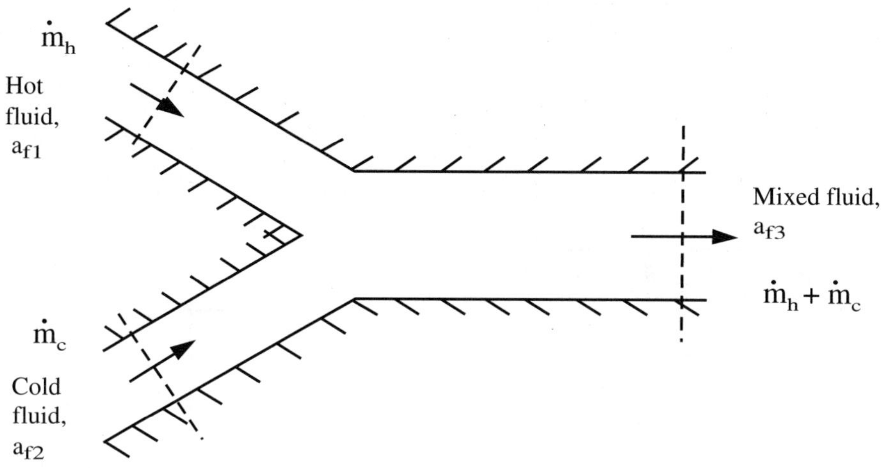

FIGURE 5.3
Direct contact heat exchanger.

A direct contact heat exchanger, as shown in Figure 5.3, operates at steady state. There is no heat exchange with its surroundings. The hot fluid mixes with the cold fluid to produce a warm mixture. The availability rate equation is

$$\dot{m}_h a_{f1} + \dot{m}_c a_{f2} = (\dot{m}_h + \dot{m}_c) a_{f3} + \dot{I}_{cv}.$$

This may be written as

$$\dot{m}_h (a_{f3} - a_{f1}) = \dot{m}_c (a_{f3} - a_{f2}) + \dot{I}_{cv}. \qquad (5.31)$$

The equation indicates that the availability input via the hot fluid is converted to the desired output, availability increase of the cold fluid, plus the amount destroyed. A parameter that measures the ratio (desired output)/input is

$$\eta_{II} = \frac{\dot{m}_c (a_{f3} - a_{f2})}{\dot{m}_h (a_{f3} - a_{f1})}. \qquad (5.32)$$

5.5.3 Application to the Power Cycle

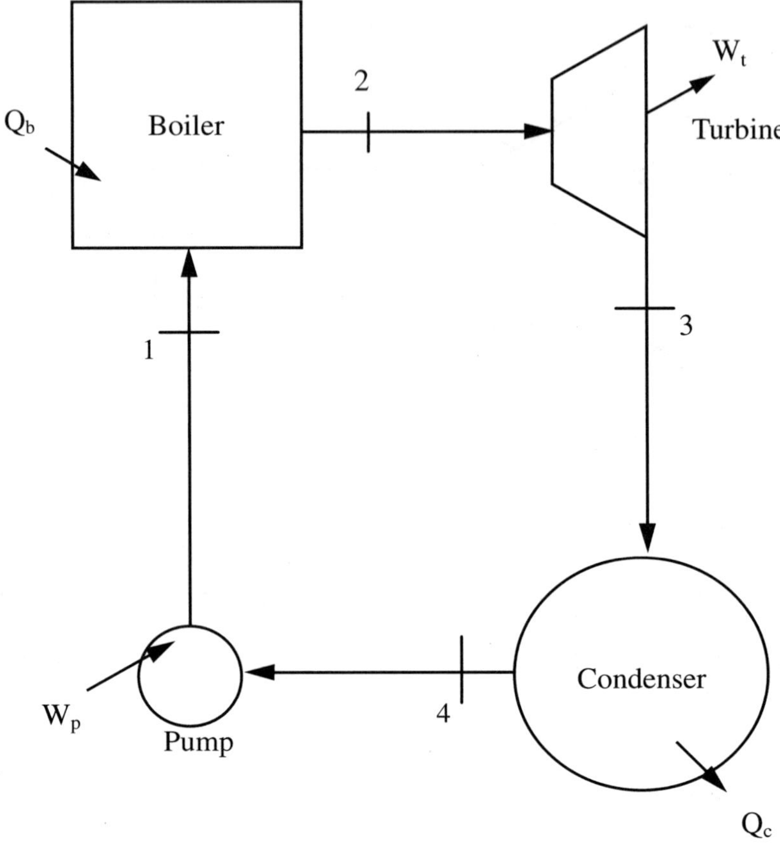

FIGURE 5.4
Power cycle schematic.

In the power cycle above, the second law in availability for the whole system which is a closed system, is

$$\left(1-\frac{T_o}{T_b}\right)Q_b-\left(1-\frac{T_o}{T_c}\right)Q_c=W_t-W_p+I_{cy}. \tag{5.33}$$

If a boundary is drawn around the boiler, it will define an open system undergoing a S.S.S.F. process. The states of the water are designated 1, 2, 3, and 4, respectively. The second law in availability equation for the boiler, Equation (5.8), is

$$\left(1-\frac{T_o}{T_b}\right)Q_b + m_1 a_{f1} = m_2 a_{f2} + \cancel{W_b} + I_b. \tag{5.34}$$

Since $m_1 = m_2 = m$, and there is no work done in the boiler,

$$\left(1-\frac{T_o}{T_b}\right)Q_b + ma_{f1} = ma_{f2} + I_b. \tag{5.35}$$

Similarly, each of the other equipment, the turbine, the condenser, and the pump, may be defined as an open system undergoing a S.S.S.F. system. The second law in availability equation for the turbine, Equation (5.8), is

$$\left(1-\frac{T_o}{T_t}\right)\cancel{Q_t} + ma_{f2} = ma_{f3} + W_t + I_t.$$

Since there is no heat transfer in the turbine, the equation becomes

$$ma_{f2} = ma_{f3} + W_t + I_t. \tag{5.36}$$

The second law in availability equation for the condenser, Equation (5.8), is

$$\left(1-\frac{T_o}{T_c}\right)Q_c + ma_{f3} = ma_{f4} + \cancel{W_c} + I_c.$$

There is no work done in a condenser, so the equation becomes

$$\left(1-\frac{T_o}{T_c}\right)Q_c + ma_{f3} = ma_{f4} + I_c. \tag{5.37}$$

The second law in availability equation for the pump, Equation (5.8), is

$$\left(1-\frac{T_o}{T_p}\right)\cancel{Q_p} + ma_{f4} = ma_{f1} + W_p + I_p.$$

Since there is no heat transfer in the pump, availability equation for the pump becomes

$$ma_{f4} = ma_{f1} - W_p + I_p. \tag{5.38}$$

Second law efficiencies may be defined to evaluate the performance of each piece of equipment. The function of the boiler is to convert liquid water into vapor by heating. An evaluation parameter could be the change in availability of the water divided by the availability transfer with the heat. For the boiler,

$$\eta_{II} = \frac{ma_{f2} - ma_{f1}}{\left(1 - \frac{T_o}{T_b}\right)Q_b}. \tag{5.39}$$

Similarly, the second law efficiency of the condenser may be defined as

$$\eta_{II} = \frac{ma_{f4} - ma_{f3}}{\left(1 - \frac{T_o}{T_c}\right)Q_c}. \tag{5.40}$$

The efficiency compares the change in availability of the water in the condenser with the availability transfer with heat. The second law efficiency of the turbine may be defined as

$$\eta_{II} = \frac{W_t}{ma_{f2} - ma_{f3}}. \tag{5.41}$$

This compares the work done by the turbine with the change in availability of the working fluid. In the case where the process is isentropic, the second law efficiency of the turbine is 100%. This corresponds to an isentropic efficiency of 100%; even though the two efficiencies have different definitions. The second law efficiency of the pump may be defined as

$$\eta_{II} = \frac{ma_{f1} - ma_{f4}}{W_p}. \tag{5.42}$$

This compares the change in availability of the water in the pump with the work done on the pump. In the case where the process is isentropic, the second law efficiency of the pump is 100%. This corresponds to an isentropic efficiency of 100%, even though the two efficiencies have different definitions.

Note that the addition of the four availability equations for the four pieces of equipment gives us the availability equation for the whole system. In addition, it can be seen that

$$I_{cy} = I_b + I_t + I_c + I_p. \tag{5.43}$$

As discussed in Section 5.5.2, the second law efficiency of the actual cycle may also be expressed as

$$\eta_{II} = \frac{\eta_I}{\eta_{I,rev}}.$$

Example 5.14

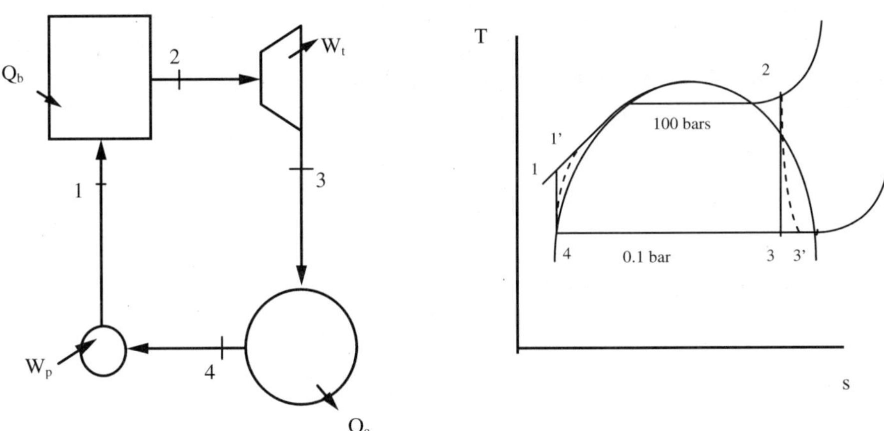

Schematics for Example 5.14.

Problem: The operating parameters of the ideal power cycle shown above, with water as the working fluid, are 100 bars for the boiler, 0.1 bar for the condenser, and the highest temperature in the boiler is 600°C. Calculate the first law efficiency of the ideal power cycle.

The isentropic efficiency of the turbine is actually 90%, and that of the pump is actually 90%. The environment is at $T_o = 25°C$ and $P_o = 1$ bar. Calculate the second law efficiencies of the boiler, turbine, condenser, and pump. In addition, calculate the actual first law efficiency of the power cycle.

Solution

Assumptions: (1) Changes in kinetic and potential energy can be neglected.

Analysis: First, identify the state points on the temperature-entropy T-s diagram. Find the properties for each state, starting at state 2 since both T_2 and P_2 are known.

$$h_2 = 3625 \text{ kJ/kg} \qquad s_2 = 6.903 \text{ kJ/(kg.K)}$$
$$h_3 = 2187 \text{ kJ/kg} \qquad s_3 = 6.903 \text{ kJ/(kg.K)}$$
$$h_4 = 191.81 \text{ kJ/kg} \qquad s_4 = 0.6492 \text{ kJ/(kg.K)}$$
$$h_1 = 201.9 \text{ kJ/kg} \qquad s_1 = 0.6492 \text{ kJ/(kg.K)}.$$

The thermal efficiency of the ideal power cycle, or first law efficiency is

$$\eta_I = \frac{W_t - W_P}{Q_b} = \frac{(h_2 - h_3) - (h_1 - h_4)}{(h_2 - h_1)} = 41.7\%.$$

Since the isentropic efficiency of the turbine is 90%, $(h_2 - h_{3'}) = 0.9\ (h_2 - h_3)$, $h_{3'} = 2330.8$ kJ/kg, and $s_{3'} = 7.354$ kJ/(kg.K). The increase in entropy in the turbine = 0.451 kJ/(kg.K), and the irreversibility = 134.5 kJ/kg.

Since the isentropic efficiency of the pump is 90%, $0.9\ (h_{1'} - h_4) = (h_1 - h_4)$, $h_{1'} = 203.0$ kJ/kg, and $s_{1'} = 0.6528$ kJ/(kg.K) The increase in entropy in the pump is 0.0036 kJ/(kg.K), and the irreversibility is 1.07 kJ/kg.

The actual first law efficiency of the cycle is

$$\eta_I = \frac{W_{t'} - W_{p'}}{Q_{b'}} = \frac{(h_2 - h_{3'}) - (h_{1'} - h_4)}{(h_2 - h_{1'})} = 37.5\%.$$

The second law efficiency of the cycle may be expressed as

$$\eta_{II} = \frac{\eta_I}{\eta_{I,rev}} = \frac{37.5\%}{41.7\%} = 0.90.$$

We expect this value for η_{II} since the isentropic efficiency of the turbine and of the pump are both 90%.

For the boiler,

$$\eta_{II} = \frac{ma_{f2} - ma_{f1'}}{\left(1 - \frac{T_o}{T_{b'}}\right) Q_{b'}} = \frac{1559.5 \text{ kJ/kg}}{(0.6585)\ 3422 \text{ kJ/kg}} = 0.69.$$

Since $T_{b'}$ is not given explicitly, an assumption is made in the choice of $T_{b'}$. It may be any temperature at or higher than 600°C. This will give rise to different values of η_{II}; however, the choice has to be consistent throughout the analysis. It is taken to be 600°C in this example.

Similarly, the second law efficiency of the condenser may be defined as

$$\eta_{II} = \frac{ma_{f4} - ma_{f3'}}{\left(1 - \frac{T_o}{T_{c'}}\right) Q_{c'}} = 0.997.$$

The condenser operates only at one temperature, so the selection of T_c is the saturation temperature at 0.1 bar, which is 45.81°C.

The second law efficiency of the turbine may be defined as

$$\eta_{II} = \frac{W_{t'}}{ma_{f2} - ma_{f3'}} = 0.906.$$

The second law efficiency of the pump may be defined as

$$\eta_{II} = \frac{ma_{f1'} - ma_{f4}}{W_{p'}} = 0.905.$$

5.6 Practical Considerations

Second law efficiencies are useful in comparing the performances of energy systems. These efficiencies may also be used as a tool to evaluate the effectiveness of improvements performed or to be performed on existing energy systems. The limit of 100% second law efficiency is not necessarily the practical objective in all cases of interests. Decisions on engineering matters are based in general on the optimal use of man (manpower), money, materials, and time. For instance, the attainment of no availability losses and destruction and, hence, 100% second law efficiency, may require too much time for an operation to be profitable.

Thermoeconomics is a field combining energy systems and engineering economics. Sometimes, the term thermoeconomics is used for the smaller field that marries availability (exergy) analysis and engineering economics for the optimal design and operation of energy systems. Thermoeconomics is a natural offshoot of the current study of the fundamentals of thermodynamics.

Another natural offshoot is the study of exergy as applied to environmental assessment. Exergy as a direct measure of environmental impact has been used in Extended Exergy Accounting or EEA. In the spirit of improving productivity and competitiveness, the optimal design and use of energy systems could spur the study of EEA and related fields.

Problems

Availability

5.1 A total of 3 kg of water is initially saturated vapor at 110°C (383.15 K), the velocity is 40 m/s, and the elevation is 8 m. It undergoes a process which transforms it to a final state where it is saturated liquid at 25°C (298.15 K), the velocity is 35 m/s, and the elevation is 1 m. Determine in kilojoules (a) the availabilities corresponding to the initial and final states, and (b) the change in availability. The environmental temperature and pressure are $T_o = 25°C$ (298.15 K) and $P_o = 0.1$ MPa. Take $g = 9.8\ m/s^2$.

5.2 A total of 6 lb of water is initially saturated vapor at 230°F (689.67°R), the velocity is 120 ft/s, and the elevation is 24 ft. It undergoes a process which transforms it to a final state where it is saturated liquid at 77°F (536.67°R), the velocity is 105 ft/s, and the elevation is 3 ft. Calculate in British thermal units (a) the availabilities corresponding to the initial and final states, and (b) the change in availability. The environmental temperature and pressure are $T_o = 77°F$ (536.67°R) and $P_o = 14.504\ lb_f/in.^2$ Take $g = 32.2\ ft/s^2$.

5.3 A total of 2 kg of ammonia is initially saturated vapor at 110°C (383.15 K), the velocity is 40 m/s, and the elevation is 10 m. It undergoes a process which transforms it to a final state where it is saturated liquid at 25°C (298.15 K), the velocity is 35 m/s, and the elevation is 1 m. Find in kilojoules (a) the availabilities corresponding to the initial and final states, and (b) the change in availability. The environmental temperature and pressure are $T_o = 25°C$ (298.15 K) and $P_o = 0.1$ MPa. Take $g = 9.8\ m/s^2$.

5.4 A total of 6 lb of ammonia is initially saturated vapor at 230°F (689.67°R), the velocity is 120 ft/s, and the elevation is 24 ft. It undergoes a process which transforms it to a final state where it is saturated liquid at 77°F (536.67°R), the velocity is 105 ft/s, and the elevation is 3 ft. Compute in British thermal units (a) the availabilities corresponding to the initial and final states, and (b) the change in availability. The environmental temperature and pressure are $T_o = 77°F$ (536.67°R) and $P_o = 14.7\ lb_f/in.^2$ Take $g = 32.2\ ft/s^2$.

5.5 An internal combustion engine cylinder contains gaseous combustion products at a pressure of 8 bars and a temperature of 900°C (1173.15 K) at the instant before it is exhausted. Calculate in kilojoules per kilogram the specific availability of the gas. The environmental temperature and pressure are $T_o = 25°C$ (298.15 K) and $P_o = 1$ atm.

5.6 Consider saturated liquid ammonia at –10°C (263.15 K). Determine the specific availability in kilojoules per kilogram, if $T_o = 25°C$ (298.15 K), $P_o = 0.1$ MPa, and there is negligible kinetic energy or potential energy.

5.7 Assume that the kinetic energy and potential energy of a quantity of saturated liquid ammonia at 10°F (469.67°R) are negligible. Calculate the specific availability in British thermal units per pound if $T_o = 77°F$ (536.67°R), $P_o = 1$ atm.

Second Law in Availability for a Control Volume

5.8 Superheated steam is throttled via a valve from 30 bars, 300°C (573.15 K) to 5 bars. Calculate in kilojoules per kilogram the change in specific flow availability between the inlet and the outlet, as well as the irreversibility per unit mass of steam. Assume $T_o = 25°C$ (298.15 K), $P_o = 0.1$ MPa.

5.9 Superheated steam is throttled via a valve from 440 $lb_f/in.^2$, 570°F (1029.67°R) to 70 $lb_f/in.^2$ Compute in British thermal units per pound the change in specific flow availability between the inlet and the outlet, as well as the irreversibility per unit mass of steam. Assume $T_o = 77°F$ (536.67°R), $P_o = 1$ atm.

5.10 Superheated steam is throttled via a valve from 32 bars, 1100°C (1373.15 K) to 16 bars. Find in kilojoules per kilogram the specific flow availability at the inlet, as well as the irreversibility per unit mass of steam. Assume $T_o = 25°C$ (298.15 K), $P_o = 0.1$ MPa.

5.11 Superheated steam is throttled via a valve from 470 $lb_f/in.^2$, 2000°F (2459.67°R) to 250 $lb_f/in.^2$ Determine in British thermal units per pound the specific flow availability at the outlet, as well as the irreversibility per unit mass of steam. Assume $T_o = 77°F$ (536.67°R), $P_o = 1$ atm.

5.12 Ammonia is throttled via a valve from 10 bars, 280°C (553.15 K) to 6 bars. Calculate in kilojoules per kilogram the specific flow availability at the inlet and the outlet, as well as the irreversibility per unit mass of steam. Assume $T_o = 25°C$ (298.15 K), $P_o = 0.1$ MPa.

5.13 Ammonia is throttled via a valve from 150 $lb_f/in.^2$, 360°F (819.67°R) to 90 $lb_f/in.^2$ Find in British thermal units per pound the specific flow availability at the inlet and the outlet, as well as the irreversibility per unit mass of steam. Assume $T_o = 77°F$ (536.67°R), $P_o = 1$ atm.

5.14 At steady state, 7 kg/s of steam enters a turbine at 5 MPa and 700°C (973.15 K) and leaves at 0.1 MPa and 150°C (423.15 K). If the turbine power output is 7466 kW, determine (a) the heat loss by the steam to the environment, which is at 25°C (298.15 K) and 0.1 MPa. In addition, determine (b) the availability of the steam at the outlet.

5.15 At steady state, 6 lb_m/s of steam enters a turbine at 600 $lb_f/in.^2$ and 500°F (959.67°R) and leaves at 25 $lb_f/in.^2$ and 250°F (709.67°R). The steam loses 50 Btu/s of heat to the environment, which is at 77°F (536.67°R) and 1 atm. Determine (a) the availability of the steam at inlet, (b) the turbine power output, (c) the maximum possible turbine power output, and (d) the irreversibility.

5.16 A technical salesperson claims that he/she has a steam turbine that produces 3000 kW. The inlet steam conditions are 600 kPa, 300°C (573.15 K) and the outlet pressure is 14 kPa, and the rate of steam flow is 3.5 kg/s. Is the claim valid? If he/she changes the rate of steam flow to 4.5 kg/s, what is your conclusion?

5.17 Ammonia at state 1 is a saturated liquid at 0.7 MPa. It is throttled to state 2 at a pressure of 0.1 MPa. Calculate the entropy production. Hence, deduce the irreversibility for the process. Assume $T_o = 293$ K, $P_o = 0.1$ MPa. The valve is then replaced by a turbine; find the maximum work that can be produced. Neglect heat transfer to the surroundings and changes in K.E. and P.E.

5.18 The inlet state of air entering a steady-state compressor is at 40°C (313.15 K) and 0.5 MPa. The air is compressed at constant temperature and without internal irreversibilities to 2 MPa. The mass flow rate of the air is 0.1 kg/s. Assume $T_o = 25$°C (298.15 K), $P_o = 0.1$ MPa. Neglect changes in K.E. and P.E.

(a) Consider the control volume comprising the compressor alone. Calculate the availability transfers accompanying heat and work, and the irreversibility.

(b) Consider the control volume comprising the compressor and its immediate surroundings such that heat transfer takes place at T_o. Do the calculations in Part (a).

Second Law in Availability for a Control Mass

5.19 An adiabatic piston-cylinder assembly contains 0.22 kg of steam at 2.2 MPa and 450°C (723.15 K). The steam expands to a final state of 450 kPa and 250°C (523.15 K) while doing work. The environment is at $P_o = 100$ kPa and $T_o = 25$°C (298.15 K). Determine (a) the availability of the steam at the initial and final states, (b) the availability change of the steam, and (c) irreversibility of the process.

5.20 A piston-cylinder assembly contains 0.5 lb_m of steam at 300 $lb_f/in.^2$ and 850°F (1309.67°R). The steam expands adiabatically to a final state of 60 $lb_f/in.^2$ and 480°F (939.67°R), doing work in the process. The environment is at $P_o = 1$ atm and $T_o = 77$°F (536.67°R). Calculate (a) the availability of the steam at the initial and final states, and (b) irreversibility of the process.

5.21 The wall of a house is 7.5 m × 10 m and 40 cm thick. Its thermal conductivity is 0.8 W/(m.°C). The house is heated to 20°C (293.15 K) on a day when the outside environment is at 5°C (278.15 K). The temperatures of the outer and inner surfaces of the wall are 8°C (281.15 K) and T_a, respectively. If the rate of heat transfer through the wall is 1200 W, calculate (a) the temperature of the wall, T_a, (b) the irreversibility rate in the wall, and (c) the rate of total irreversibility associated with this heat conduction process.

5.22 The wall of a house is 20 ft × 24 ft and 12 in. thick. Its thermal conductivity is 5×10^{-4} Btu/(ft.s.°F). The house is heated to 80°F (539.67°R) on a day when the outside environment is at 40°F (499.67°R). The temperatures of the outer and inner surfaces of the wall are T_b and 66°F (525.67°R), respectively. If the rate of heat transfer through the wall is 2.4 Btu/s, calculate (a) the temperature of the wall, T_b, (b) the irreversibility rate in the wall, and (c) the rate of total irreversibility associated with this heat conduction process.

5.23 A well-insulated, rigid tank is made up of two compartments, one having a volume of 4 m^3, and the other twice the volume of the first. Initially, one compartment is evacuated, that of 4 m^3 contains steam at 0.3 MPa and 200°C (473.15 K), and the valve between the compartments is shut. The valve is opened and the steam expands to an eventual equilibrium state, filling the total volume. Determine (a) the final temperature in degrees Celsius and the final pressure in atmospheres, and (b) the irreversibility.

5.24 The initial state of 4 kg of water is at 250°C (523.15 K) and 0.12 MPa. It expands reversibly and isothermally to 0.1 MPa while obtaining heat from a source at 550°C (823.15 K). For the water, determine the work, the heat transfer, the availability transfer accompanying work, and the availability transfer accompanying heat. Assume T_o = 25°C (298.15 K), P_o = 0.1 MPa.

Second Law Ratio to Measure Thermal Environmental Impact

5.25 The gaseous combustion products in an internal combustion engine are at a pressure of 8 bars and a temperature of 820°C (1093.15 K). Without a turbocharger, the gases are exhausted at this state into the environment. With a turbocharger, the gases are reduced to a temperature of 187°C (460.15 K) at the exhaust. Assume that the heat loss ratio $\dot{Q}_l/\dot{Q}_s$ is 25% without a turbocharger and only 20% with a turbocharger, and that T_s = 900°C (1173.15 K). Calculate the value of r_{II} for the two cases. The environmental temperature and pressure are T_o = 15°C (288.15 K) and P_o = 1 atm.

5.26 The gaseous combustion products in an internal combustion engine are at a pressure of 120 $lb_f/in.^2$ and a temperature of 1500°F (1959.67°R). Without a turbocharger, the gases are exhausted at this state into the environment. With a turbocharger, the gases are reduced to a temperature of 370°F (829.67°R) at the exhaust. Assume that the heat loss ratio $\dot{Q}_l/\dot{Q}_s$ is 20% without a turbocharger and only 15% with a turbocharger, and that T_s = 1650°F (2109.67°R). Calculate the value of r_{II} for the two cases. The environmental temperature and pressure are T_o = 60°F (519.67°R) and P_o = 1 atm.

5.27 The combustion gases in a propulsion gas turbine reach a maximum pressure and temperature of 10 bars and a temperature of 1400 K in the combustor. Compare the thermal enviornmental impact of a turbine that expands these combustion gases to 1.15 bars at 700 K to one that expands them to the same pressure at 600 K. Assume that the heat loss ratio $\dot{Q}_l / \dot{Q}_s$ is 35% in both cases. The environmental temperature and pressure are $T_o = 20°C$ (293.15 K) and $P_o = 1$ bar.

5.28 The combustion gases in a propulsion gas turbine reach a maximum pressure and temperature of 145 $lb_f/in.^2$ and a temperature of 2520°R in the combustor. Compare the thermal enviornmental impact of a turbine that expands these combustion gases to 17 $lb_f/in.^2$ at 1260°R to one that expands them to the same pressure at 1080°R. Assume that the heat loss ratio $\dot{Q}_l / \dot{Q}_s$ is 25% in both cases. The environmental temperature and pressure are $T_o = 68°F$ (527.67°R) and $P_o = 1$ atm.

5.29 Two textile dryers use steam heating, with the steam at 100°C (373.15 K). The energy supplied is 2710 kJ/kg of textile in the first dryer and 2463 kJ/kg of textile in the second dryer. The heat loss is 1355 kJ/kg of textile in the first dryer. If the r_{II} value of the first dryer is 90% that of the second dryer, calculate the heat energy loss in the second dryer. Assume that the loss temperature is the same in both dryers. The environmental temperature and pressure are $T_o = 20°C$ (293.15 K) and $P_o = 1$ atm.

5.30 Two textile dryers use steam heating, with the steam at 212°F (671.67°R). The energy supplied is 1200 Btu/lb_m of textile in the first dryer and 1100 Btu/lb_m of textile in the second dryer. The heat loss is 600 Btu/lb_m of textile in the first dryer. If the r_{II} value of the second dryer is 79% that of the first dryer, calculate the heat energy loss in the second dryer. Assume that the loss temperature is the same in both dryers. The environmental temperature and pressure are $T_o = 68°F$ (527.67°R) and $P_o = 1$ atm.

5.31 The gaseous combustion products in a boiler at pressure of 1.0 bar and a temperature of 700°C (973.15 K). Without an economizer and an air preheater, the gases are exhausted at this state into the environment. With an economizer and an air preheater, the gases are reduced to a temperature of 300°C (573.15 K) at the exhaust. Assume that the heat loss ratio $\dot{Q}_l / \dot{Q}_s$ is 30% without the economizer and air preheater, and is only 15% with an economizer and air preheater, and that $T_s = 950°C$ (1223.15 K). Determime the value of the second law ratio to measure thermal environmental impact, r_{II}, for the two cases. The environmental temperature and pressure are $T_o = 20°C$ (293.15 K) and $P_o = 1$ bar.

Second Law Efficiencies of Systems

5.32 The first law or thermal efficiency of a simple power plant is 30%. The maximum possible thermal efficiency of the plant is 63%. Determine its second law efficiency.

5.33 A non-contact counterflow heat exchanger operates at steady state with oil as the hot fluid and water as the cold fluid. Oil and water may be considered as incompressible with specific heats of 0.5 and

1.0 kJ/(kg.K), respectively. Water is heated from 10°C (283.15 K) to 90°C (363.15 K), while oil is cooled from 150°C (423.15 K) to T. The mass flow rates of oil and water are 4.8 kg/s and 1 kg/s, respectively. There is no significant pressure change in either fluid. Changes in K.E. and P.E. can be neglected. The heat exchanger may be considered adiabatic. Calculate (a) the temperature T, and (b) the second law efficiency.

5.34 In a refrigeration cycle, the Q_{in} = 100 kJ/s and Q_{out} =200 kJ/s. The corresponding ideal cycle would have Q_{out} =180 kJ/s with the same Q_{in}. Determine the second law efficiency of this cycle.

5.35 In a refrigeration cycle, the Q_{in} = 200 Btu/s and Q_{out} = 400 Btu/s. The corresponding ideal cycle would have Q_{out} =300 Btu/s with the same Q_{in}. Determine the second law efficiency of this cycle.

5.36 In a heat pump cycle, the Q_{in} = 200 kJ/s and Q_{out} = 400 kJ/s. The corresponding ideal cycle would have Q_{out} = 360 kJ/s with the same Q_{in}. Determine the second law efficiency of this cycle.

5.37 In a heat pump cycle, the Q_{in} = 400 Btu/s and Q_{out} = 650 Btu/s. The corresponding ideal cycle would have Q_{out} = 600 Btu/s with the same Q_{in}. Determine the second law efficiency of this cycle.

5.38 Evaluate whether the following equipment can operate at steady state as described. Assume T_o = 25°C (298.15 K).

(a) Heat is transferred to the equipment at its surface where the temperature is 125°C (398.15 K). The equipment provides electricity to its environment at the rate of 10 kW. There are no other energy transfers.

(b) Electricity is supplied to an equipment at the rate of 10 kW. Heat leaves the equipment at its surface where the temperature is 125°C (398.15 K). There are no other energy transfers.

Application to the Power Cycle

5.39 For the ideal power cycle using water shown above, the high pressure is 60 bars and the low pressure is 0.04 bar, and the highest temperature in the boiler is 500°C (773.15 K).

The isentropic efficiency of the turbine is actually 85%, and that of the pump is actually 70%. The environment is at T_o = 25°C (298.15 K) and P_o = 1 bar. Calculate the second law efficiencies of the boiler, turbine, condenser, and pump. Calculate the difference between the actual first law efficiency of the cycle with that of the ideal cycle.

5.40 In the ideal power cycle utilizing water, the high pressure is 900 $lb_f/in.^2$, the low pressure is 1 $lb_f/in.^2$, and the highest temperature in the cycle is 1000°F (1459.67°R).

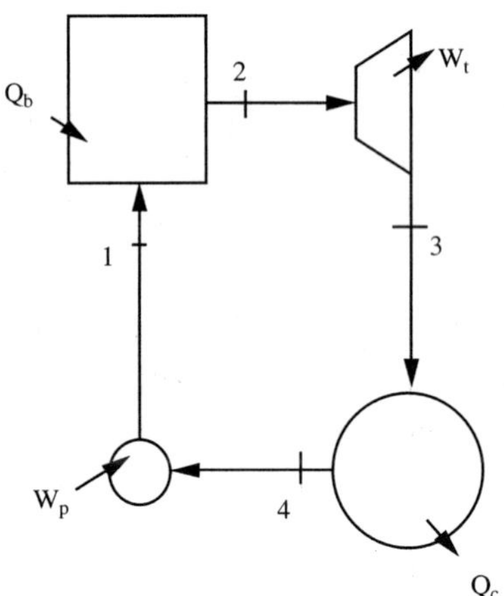

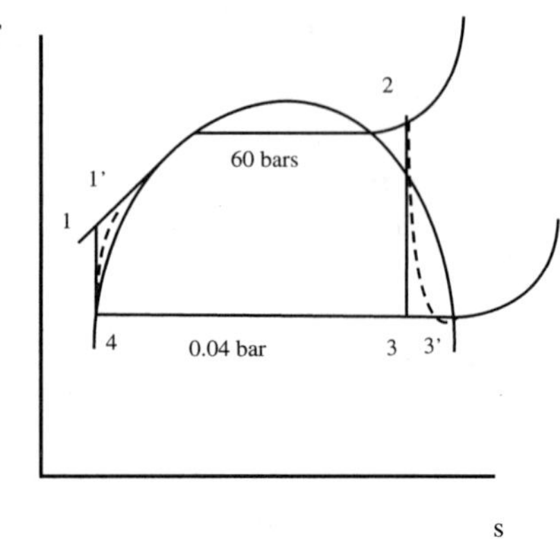

Schematics for Problem 5.39.

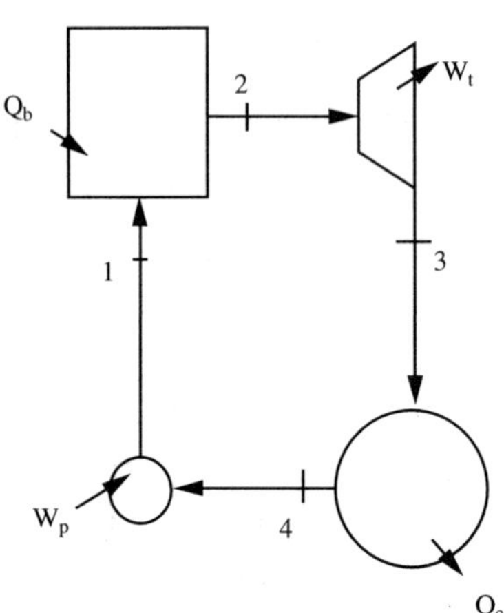

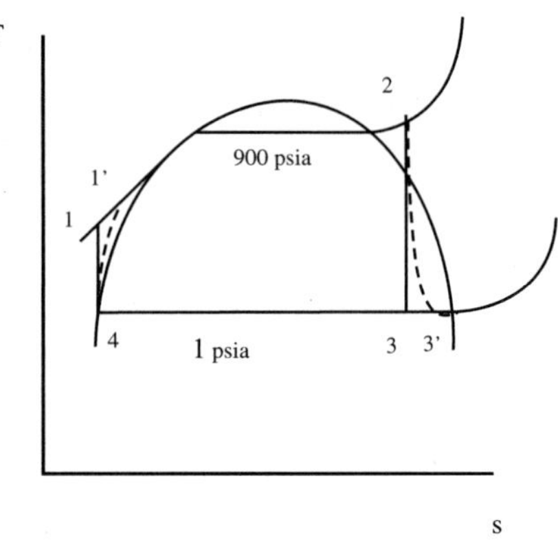

Schematics for Problem 5.40.

The isentropic efficiencies of the turbine and the pump are 85 and 70%, respectively. The environment is at $T_o = 77°F$ (536.67°R) and $P_o = 14.7$ $lb_f/in.^2$ Determine the second law efficiencies of the boiler, turbine, condenser, and pump. Determine the difference between the actual first law efficiency of the cycle and that of the ideal power cycle.

Computer, Design, and General Problems

5.41 Perform a literature survey of recent articles regarding second law efficiency and exergetic efficiency. Discuss the differences, if any, between the definitions used.

5.42 Heat may be supplied to a building either via steam, a heat pump, or electrical resistance. The price charged per kilowatt hour is the same regardless of the form in which heat is supplied by the utility company. Discuss this pricing policy.

5.43 Obtain operating data of several thermal systems. Evaluate the second law ratio that measures the thermal environmental impact of each system.

5.44 Write a computer program that allows the calculation of the second law ratio to measure the thermal environmental impact of any thermal system. Keeping the use temperature and source temperature constant, determine the behavior of this ratio as the environmental temperature T_o varies.

5.45 Three operating parameters define an ideal steam power cycle: the boiler pressure, the highest temperature in the boiler, and the condenser pressure. The isentropic efficiency of the turbine and of the pump range in value from 70 to 90%. The environmental pressure is 1 bar, and the environmental temperature can range from –10 to 30°C. Write a computer program to calculate the second law efficiencies of the boiler, turbine, condenser, and pump, and the actual first law efficiency of the power cycle. The program should accept different values of the operating parameters in practical ranges, different values of the isentropic efficiencies, and different environmental temperatures.

5.46 Perform a literature survey of recent articles regarding thermoeconomics.

5.47 Select a power plant in your locality. Obtain general information about the operating states before and after various components of the plant, and the production figures. Set up an availability costing system for that plant based on the information gathered.

5.48 Select an energy-intensive plant (other than a power plant) in your locality, e.g., an oil refinery. Repeat Problem 5.47 above.

5.49 Research about the contributions of the Odum brothers, and "emergy" or "embedded energy." Write an essay on this topic.

5.50 Perform a literature survey of recent articles regarding "emergy."

5.51 Research and write an essay on Extended Exergy Accounting.

6

Vapor Power Systems

CONTENTS

6.1 The Carnot Vapor Cycle

The Carnot cycle is the most efficient cycle operating between two specified temperature levels. The Carnot cycle operating within the two-phase region of a pure substance is shown in Figure 6.1. The working fluid is heated reversibly and isothermally in a boiler (process 1–2), expanded isentropically in a turbine (process 2–3), condensed reversibly and isothermally in a condenser (process 3–4), and compressed isentropically by a compressor to the initial state (process 4–1). The Carnot cycle is represented by a rectangle on a T-s diagram.

There are some impracticalities in the cycle as described:

1. Operating the cycle within the saturation dome of a pure substance severely limits the maximum temperature that can be used in the cycle. This limits the maximum temperature in the cycle, which

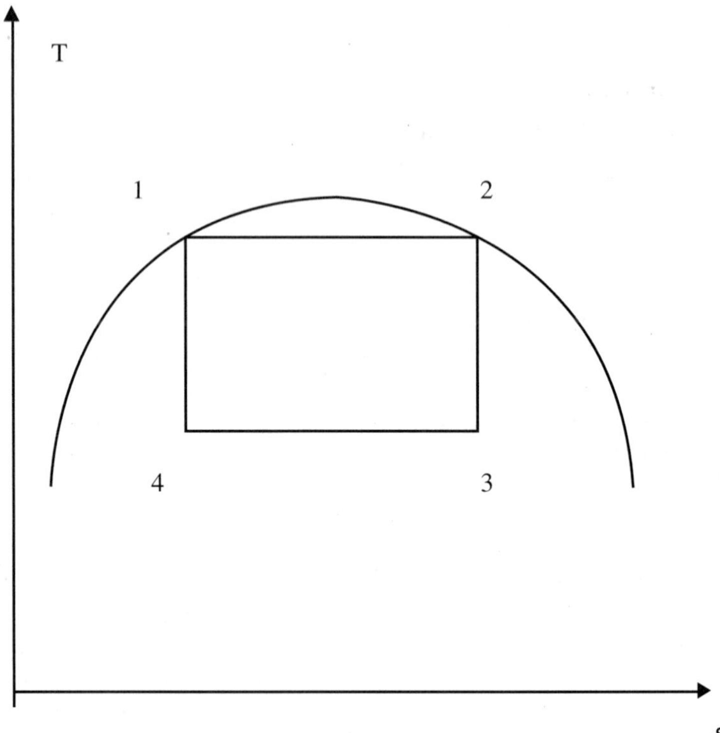

FIGURE 6.1
Carnot cycle.

thus limits the thermal efficiency. Raising the maximum temperature in the cycle will require heat transfer to the working fluid in a single phase, which is not easily done isothermally.

2. The isentropic expansion can be approached practically by a well-designed turbine. However, the turbine will have to handle steam with low quality, that is, steam containing a large portion of moisture. The impingement of liquid droplets on the turbine blades causes erosion and the turbine blades cannot last very long under such conditions. Steam with less than 90% quality is not used practically in steam turbines.
3. The isentropic compression process in the pump causes two problems. It is difficult to control the condensation process so as to end up with the desired quality at state 4. It is not practical to engineer a compressor that will handle two phases.

6.2 The Rankine Cycle: Ideal Cycle for Vapor Power Cycles

The impracticalities of the Carnot cycle can be eliminated by superheating the steam in the boiler and completely liquefying it in the condenser. The ideal Rankine cycle, shown in Figure 6.2, does not have any internal irreversibilities and comprises the following processes:

Process 1–2: Constant pressure heat addition in a boiler.
Process 2–3: Isentropic expansion in a turbine.
Process 3–4: Constant pressure heat rejection in a condenser.
Process 4–1: Isentropic compression in a pump.

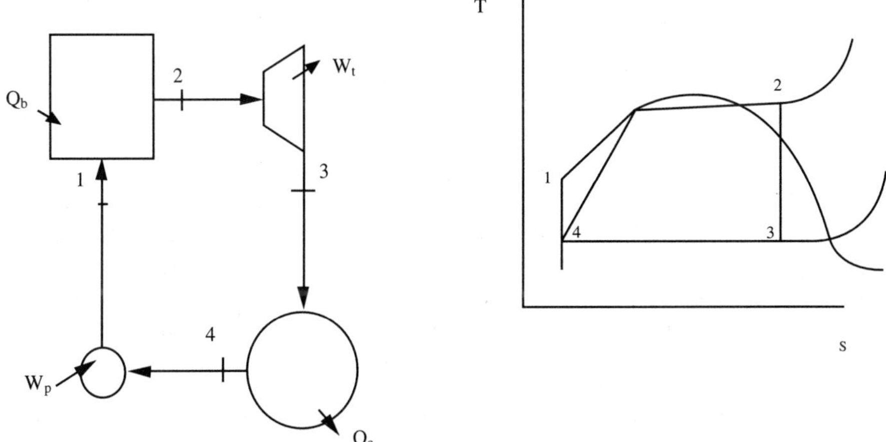

FIGURE 6.2
The ideal Rankine cycle.

As first indicated in Example 3.1, the first law analysis of the Rankine cycle is discussed in the following paragraphs.

For the Rankine cycle, the first law for the whole system, which is a closed one, is $Q_b - Q_c = W_t - W_p$.

If a boundary is drawn around the boiler, it will define an open system undergoing a S.S.S.F. process. The states of the water are designated 1, 2, 3, and 4, respectively. Neglecting the K.E. and P.E. terms, the first law equation for the boiler is $Q_b + m_1 h_1 = m_2 h_2$.

Since $m_1 = m_2 = m$, then $Q_b + m h_1 = m h_2$.

Similarly, each of the other pieces of equipment, i.e., the turbine, the condenser, and the pump, may be defined as an open system undergoing a S.S.S.F. system.

Neglecting the K.E. and P.E. terms, the first law equation for the turbine is $mh_2 = mh_3 + W_t$.

Neglecting the K.E. and P.E. terms, the first law equation for the condenser is $-Q_c + mh_3 = mh_4$.

Neglecting the K.E. and P.E. terms, the first law equation for the pump is $mh_4 = mh_1 - W_p$.

The thermal efficiency of the Rankine cycle is determined from

$$\eta_{th} = \frac{W_{net}}{Q_b} = 1 - \frac{Q_c}{Q_b} \tag{6.1}$$

where $W_{net} = Q_b - Q_c = W_t - W_p$.

In the U.S. and elsewhere, the conversion efficiency of power plants is often expressed in terms of heat rate, which is the amount of heat supplied in British thermal units to generate 1 kWh of electricity. The smaller the heat rate, the greater the efficiency. Since 1 kWh = 3412 Btu, the relation between the heat rate and the thermal efficiency can be expressed as

$$\eta_{th} = \frac{3412\ (\text{Btu}/\text{kWh})}{\text{Heat rate}\ (\text{Btu}/\text{kWh})}. \tag{6.2}$$

It can be seen from the T-s diagram of the Rankine cycle, that the thermal efficiency of the cycle is represented by the area within the cycle divided by the area below the boiler line and the entropy axis. The efficiency can be increased by increasing the area within the cycle. Therefore, the following ways have been recognized as useful in increasing the thermal efficiency.

1. The condenser pressure can be lowered; this increases the area within the cycle. The condensers of steam power plants usually operate well below the atmospheric pressure. However, dropping the condenser pressure increases the moisture content of the steam at the final stages of the turbine. This increased moisture content erodes the turbine blades and decreases the efficiency of the turbine. With a fixed boiler exit state, lowering the condenser pressure increases the pump work input, the turbine work output, the heat supplied, and the thermal efficiency of the cycle since the heat rejected is decreased.

   Referring to Figure 6.3, dropping the condenser pressure will change the cycle 1-2-3-4-5 to 1-2-3′-4′-5′. The net work will be increased by the amount represented by the shaded area.
2. The steam can be superheated to high temperatures without increasing the boiler pressure; the area within the cycle is increased. With the boiler and condenser pressures fixed, superheating the steam causes the pump work input to remain the same while the turbine output increases. The heat supplied and the heat rejected increase, and the thermal efficiency increases.

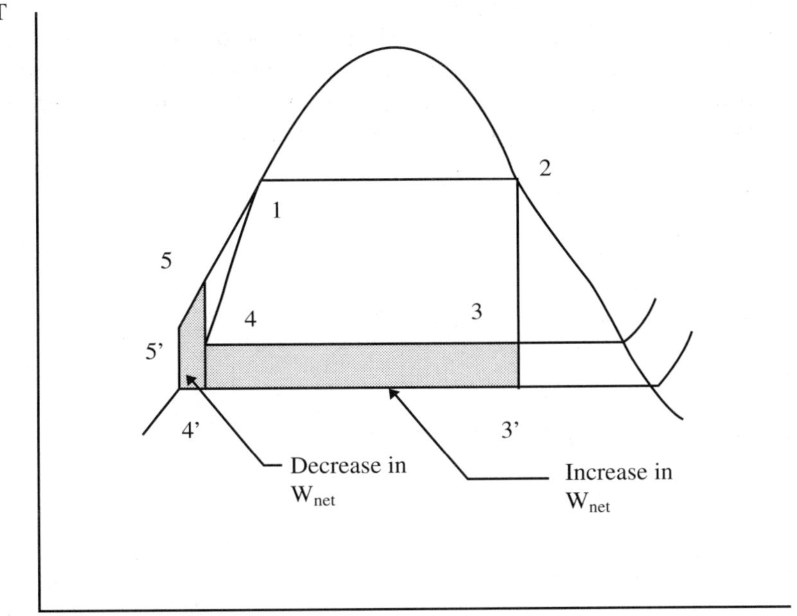

FIGURE 6.3
Effect of condenser pressure on net work.

Superheating the steam to higher temperatures has the desirable effect of decreasing the moisture content of the steam at the turbine exit. The maximum temperature to which steam can be superheated is limited by metallurgical factors. The highest practical steam temperature allowed at the turbine inlet is about 650°C at the present time.

3. Another way of increasing the area within the cycle is to increase the boiler pressure. The effect of increasing the boiler pressure at a fixed boiler outlet temperature is that the cycle shifts to the left and the moisture content of the steam at the turbine exit increases. This unwanted effect can be corrected by reheating the steam, as discussed in a later section. With a fixed boiler outlet temperature and condenser pressure, increasing the boiler pressure increases both the turbine work output and the pump work input. The heat supplied increases while the heat rejected decreases, so the thermal efficiency increases.

Over the years the operating pressures of boilers have increased. Nowadays, many modern steam power plants operate at supercritical pressures

(P > 22.09 MPa) and have thermal efficiencies of about 40% for fossil-fuel plants and 34% for nuclear ones.

Referring to Figure 6.4, increasing the boiler pressure will change the cycle 1-2-3-4-5 to 1″-2″-3″-4-5″. The net work will be increased by the amount indicated, and decreased by the amount represented by the shaded area. Since the increase is greater than the decrease, the net effect is to increase the work done.

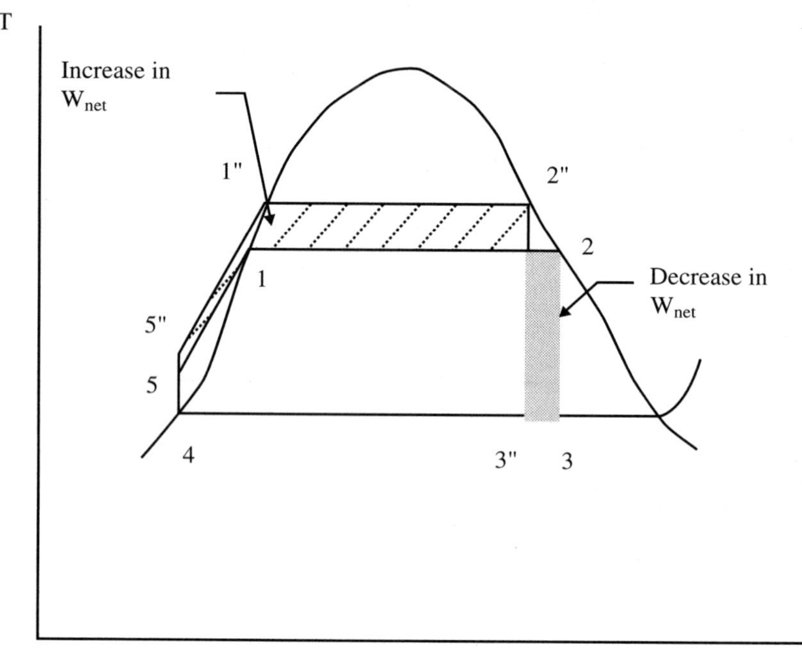

FIGURE 6.4
Effect of boiler pressure on net work.

Example 6.1

Problem: Consider an ideal Rankine cycle using water. Steam leaves the boiler in a superheated condition at 650°C and 9 MPa, and the condenser pressure is 0.005 MPa. The net power output of the cycle is 100 MW. Determine:

(a) Mass flow rate of the steam, in kg/s
(b) The rate of heat transfer in the boiler, $\dot{Q}_b$, in megawatts
(c) The rate of heat transfer in the condenser, $\dot{Q}_c$, in megawatts

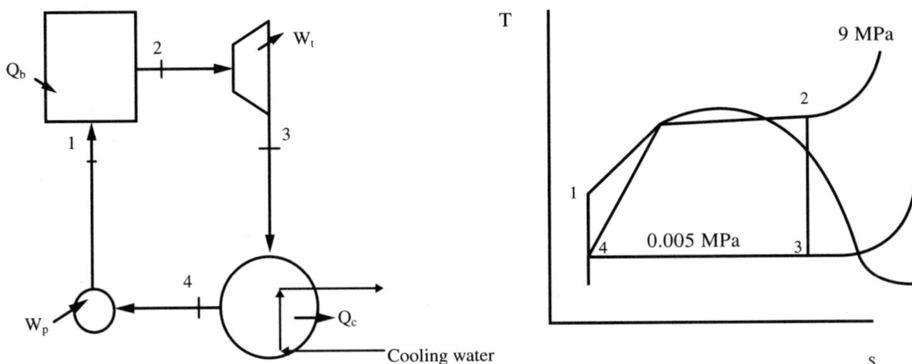

Schematic for Example 6.1. **T-s Diagram for Example 6.1.**

(d) The mass flow rate of the condenser cooling water, in kilograms per second, if cooling water enters the condenser at 16°C and leaves at 30°C

(e) The thermal efficiency of the cycle

Solution

Assumptions:

(1) The turbine and pump work adiabatically, and

(2) K.E. and P.E. changes are negligible.

Analysis:

From the thermodynamic tables,

$h_2 = 3755$ kJ/kg, $s_2 = 7.094$ kJ/(kg.K)

$h_3 = 2163$ kJ/kg, $s_3 = s_2 = 7.094$ kJ/(kg.K)

$h_4 = 137.8$ kJ/kg, $s_4 = 0.4763$ kJ/(kg.K), $v_4 = 0.001005$ m^3/kg

$h_1 \approx h_4 + v_4(P_4 - P_1) = 146.8$ kJ/kg, $s_1 = s_4 = 0.4763$ kJ/(kg.K)

(a) The turbine power is

$$\dot{w}_T = \dot{m}\left(h_2 - h_3\right).$$

The pump power is

$$\dot{w}_P = \dot{m}\left(h_4 - h_1\right).$$

Net power of the cycle

$$= \dot{W}_T + \dot{W}_P$$
$$= \dot{m}\left[(h_2 - h_3) + (h_4 - h_1)\right]$$
$$= 100\,\text{MW} = 100 \times 10^3\ \text{kJ/s}.$$

Hence,

$$\dot{m} = 63.17\ \text{kg/s}.$$

(b) The boiler heat transfer is

$$\dot{Q}_b = \dot{m}(h_2 - h_1) = 227.9\ \text{MW}.$$

(c) The condenser heat transfer is

$$\dot{Q}_c = \dot{m}(h_4 - h_3) = -127.9\ \text{MW}.$$

(d) Select as control volume the condenser, which includes both the condensing steam side and the cooling water side. The first law for the control volume gives

$$\dot{m}_{cw}\, h_{cw,in} + \dot{m}h_3 = \dot{m}_{cw}\, h_{cw,out} + \dot{m}h_4.$$

Rearranging,

$$\dot{m}_{cw}\left(h_{cw,in} - h_{cw,out}\right) = \dot{m}(h_3 - h_4)$$
$$\dot{m}_{cw} = \frac{\dot{m}(h_3 - h_4)}{\left(h_{cw,out} - h_{cw,in}\right)}.$$

For the cooling water, $h \approx h_f(T)$, so the enthalpy values may be obtained from the saturation tables. Thus, the mass flow rate of the condenser cooling water is

$$\dot{m}_{cw} = \frac{127{,}900}{(125.77 - 67.18)} = 2183\ \text{kg/s}.$$

(e) The thermal efficiency of the cycle is

$$\eta = \frac{\text{net power}}{\dot{Q}_b}$$
$$= 43.9\%.$$

6.3 The Reheat Rankine Cycle

The reheat cycle has been conceived to solve the problem of excessive moisture at the final stages of the turbine. This will allow the temperature in a constant-pressure boiler to be increased, or the boiler pressure to be increased, in order to improve boiler efficiency.

The reheat Rankine cycle is different from the simple Rankine cycle in that the expansion process in the turbine takes place in two stages and the steam is reheated between the stages.

In the high-pressure turbine (first stage), steam is expanded isentropically to an intermediate pressure and returned to the boiler where it is reheated at constant pressure, usually to the inlet temperature of the first turbine stage [Figure 6.5, Parts (a) and (b)]. Steam is then expanded in the low-pressure turbine (second stage) to the condenser pressure. The total heat input and the total turbine work output for a reheat cycle become

$$q_{in} = q_{boiler} + q_{reheat} = (h_2 - h_1) + (h_4 - h_3) \tag{6.3}$$

and

$$w_{turbine} = w_{turbine,I} + w_{turbine,II} = (h_2 - h_3) + (h_4 - h_5). \tag{6.4}$$

The use of more than two reheat stages is not practical since the theoretical improvement in efficiency from the second reheat is about half that which results from a single reheat, which is about 4 to 5%. Double reheat would result in superheated exhaust steam if the turbine inlet pressure is not high enough. This will not be desirable because cycle efficiency will drop as the average temperature of heat rejection has increased. The double reheat is used usually with supercritical-pressure ($P > 22.09$ MPa) power plants.

Example 6.2

Problem: A steam power plant operates on an ideal Rankine cycle with reheat. The steam leaves the boiler at 10 MPa and 620°C and the condenser pressure is 0.005 MPa. If the moisture content of the steam at the outlet of the low-pressure (LP) turbine is not to exceed 15%, find (a) the pressure at which the steam has to be reheated, and (b) the thermal efficiency of the cycle. The steam is reheated to 620°C.

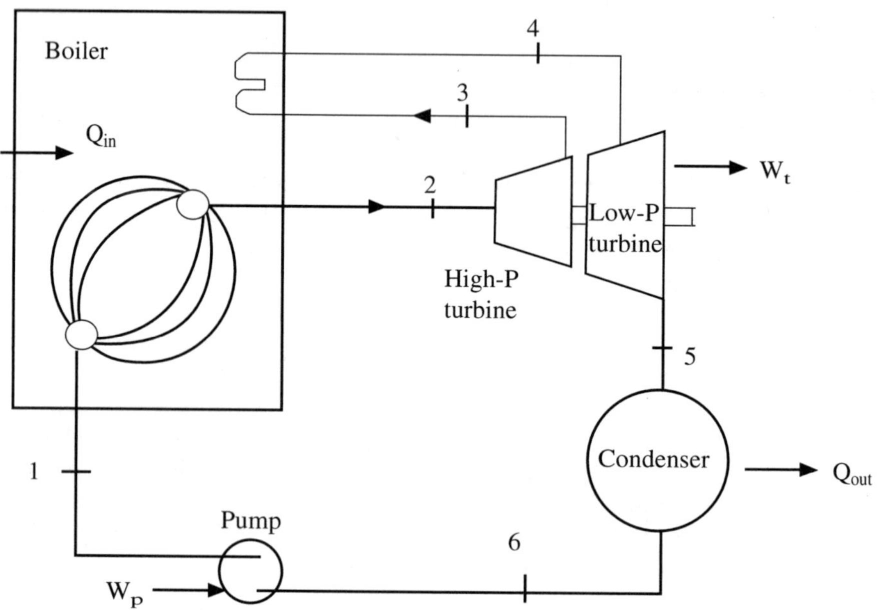

FIGURE 6.5(a)
Reheat Rankine cycle.

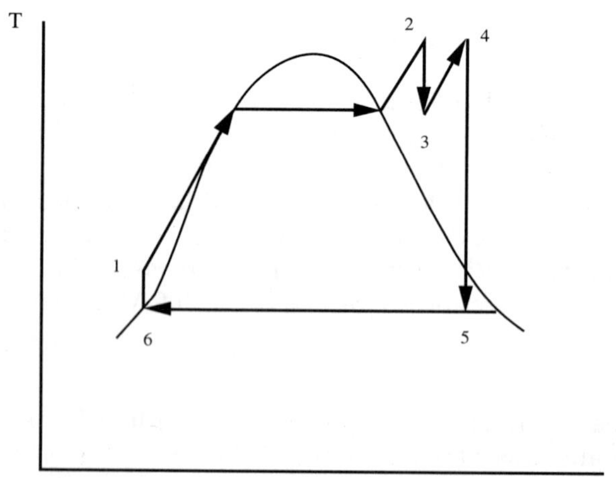

FIGURE 6.5(b)
Reheat Rankine cycle.

Solution

Assumptions:

(1) Changes in K.E. and P.E. are negligible, and

(2) Assume that both the high-pressure (HP) turbine and the LP turbine are isentropic.

Analysis:

(a) State 5 is obtained from its quality $x_5 = 0.85$, and its pressure, $P_5 = 0.005$ MPa.

Hence,

$$h_5 = 2198 \text{ kJ/kg, and } s_5 = 7.207 \text{ kJ/(kg.K)}.$$

Since

$$s_4 = s_5, \text{ and } T_4 = 620°C,\ P_4 = 6.2 \text{ MPa, and } h_4 = 3704 \text{ kJ/kg}.$$

Therefore, the steam should be reheated at a pressure of 6.2 MPa or lower to prevent the moisture content from exceeding 15% at the LP turbine outlet.

(b) From the thermodynamic tables,

$$h_2 = 3675 \text{ kJ/kg}, \qquad s_2 = 6.959 \text{ kJ/(kg.K)}$$
$$h_3 = 3494 \text{ kJ/kg}, \qquad s_3 = s_2$$
$$h_6 = 137.8 \text{ kJ/kg}, \qquad s_6 = 0.4763 \text{ kJ/(kg.K)}$$
$$h_1 = 147.8 \text{ kJ/kg}, \qquad s_1 = s_6.$$

The work output comes from both the HP and the LP turbines, and the work input is for the pump. Hence,

$$\begin{aligned} w_{net} &= (h_2 - h_3) + (h_4 - h_5) + (h_6 - h_1) \\ &= 1677 \text{ kJ/kg}. \end{aligned}$$

The heat input is the heat added to the water in the boiler (including reheating), thus

$$\begin{aligned} q_{in} &= (h_2 - h_1) + (h_4 - h_3) \\ &= 3737 \text{ kJ/kg}. \end{aligned}$$

The thermal efficiency of the cycle is

$$\eta_{th} = \frac{w_{net}}{q_{in}} = 44.9\%.$$

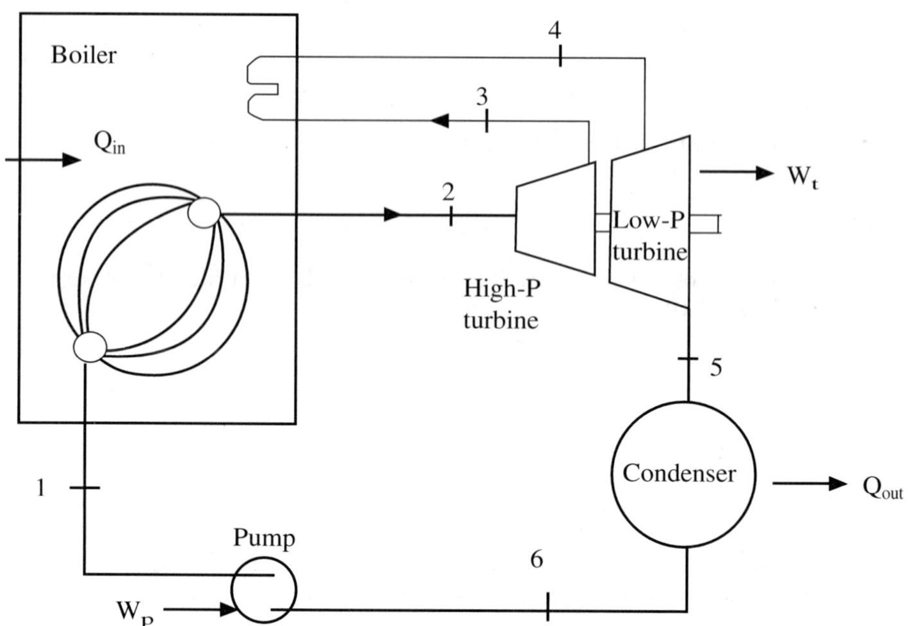

Schematic for Example 6.2.

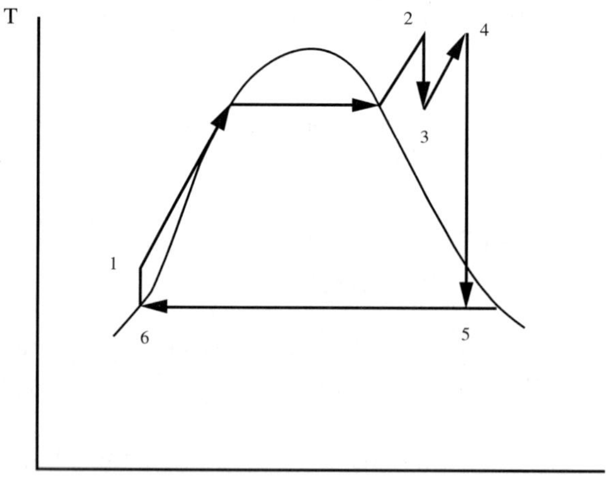

T-s Diagram for Example 6.2.

Example 6.3

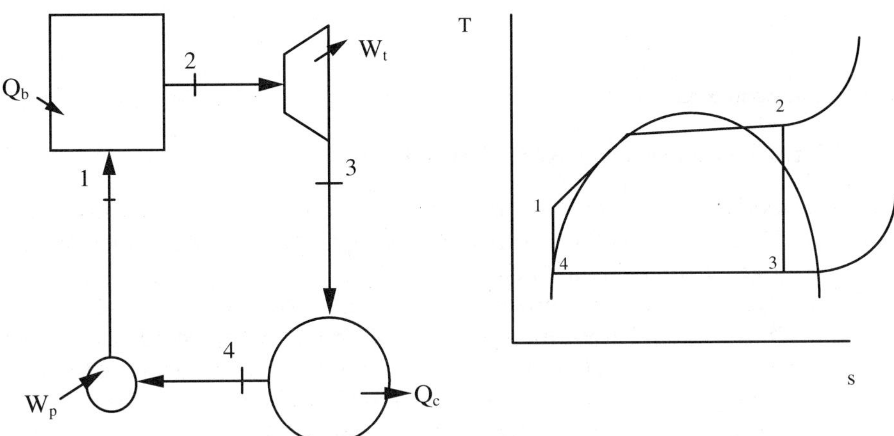

Schematic for Example 6.3. **T-s Diagram for Example 6.3.**

Problem: In Example 6.2, compute the thermal efficiency of the corresponding ideal Rankine cycle without reheat. Determine (a) the moisture content at the turbine outlet, and (b) the thermal efficiency of the cycle. Comment on the effects of reheating.

Solution

Assumptions: (1) Changes in K.E. and P.E. are negligible.

Analysis:

(a) Without reheat,

State 2 is the same as before in Example 6.2.

State 4 is the same as state 6 in Example 6.2.

State 1 is the same as before in Example 6.2.

Since

$$s_3 = s_2 \text{ and } P_3 = 0.005 \text{ MPa},$$
$$h_3 = 2122 \text{ kJ/kg}, \; x_3 = 0.8187.$$

Hence, the moisture content at the turbine outlet is 18.13%.

(b) The thermal efficiency of the ideal Rankine cycle is

$$\eta_{th} = \frac{w_{net}}{q_b} = \frac{(h_2 - h_3) + (h_4 - h_1)}{(h_2 - h_1)} = 43.8\%.$$

Thus, reheating reduces the moisture content at the turbine outlet, and increases the thermal efficiency from 43.8 to 44.9%.

6.4 The Regenerative Rankine Cycle

Heat is added at a relatively low average temperature when water is at a liquid state. This lowers the cycle efficiency. Regeneration seeks to increase the average temperature at which heat is added.

The practical regeneration process involves the extraction or "bleeding" of steam from the turbine at various points. This steam, which could have been used to produce more work in the turbine, is used instead to heat the feedwater (water entering the boiler). The regenerator or feedwater heater is the equipment used to heat the feedwater by regeneration.

A feedwater heater is a heat exchanger where heat is transferred from the steam to the feedwater either by mixing the two streams (open feedwater heaters) or without mixing them (closed feedwater heaters).

6.4.1 Open Feedwater Heaters

An open feedwater heater (FWH) is a mixing chamber where the steam bled from the turbine is mixed with the feedwater exiting the first pump. The schematic of a steam power plant with one open feedwater heater and the T-s diagram of the cycle are shown in Figure 6.6, Parts (a) and (b).

In the regenerative Rankine cycle which is ideal, steam enters the turbine at the boiler pressure (state 2) and expands isentropically to an intermediate pressure (state 3). A portion of the steam is bled at this state and routed to the feedwater heater, while the remaining steam continues to expand isentropically to the condenser pressure (state 4). This steam leaves the condenser as a saturated liquid at the condenser pressure (state 5). The condensed water or feedwater then enters an isentropic pump (pump I), where it is compressed to the feedwater heater pressure (state 6) and is routed to the feedwater heater, where it mixes with the steam bled from the turbine. The portion of steam bled is such that the mixture leaves the heater as a saturated liquid at the heater pressure (state 7). The second pump raises the pressure of the water to the boiler pressure (state 1). The cycle is completed by heating the water in the boiler to the turbine inlet state (state 2).

In the analysis of steam power plants, it is often convenient to draw a control volume around each individual equipment and consider each separately. Then, each equipment can be considered as undergoing a steady-state steady-flow (S.S.S.F.) process. Then it is convenient to consider quantities expressed per unit mass of steam flowing through the boiler. For each 1 kg of steam leaving the boiler, y kg expands partially in the turbine and is

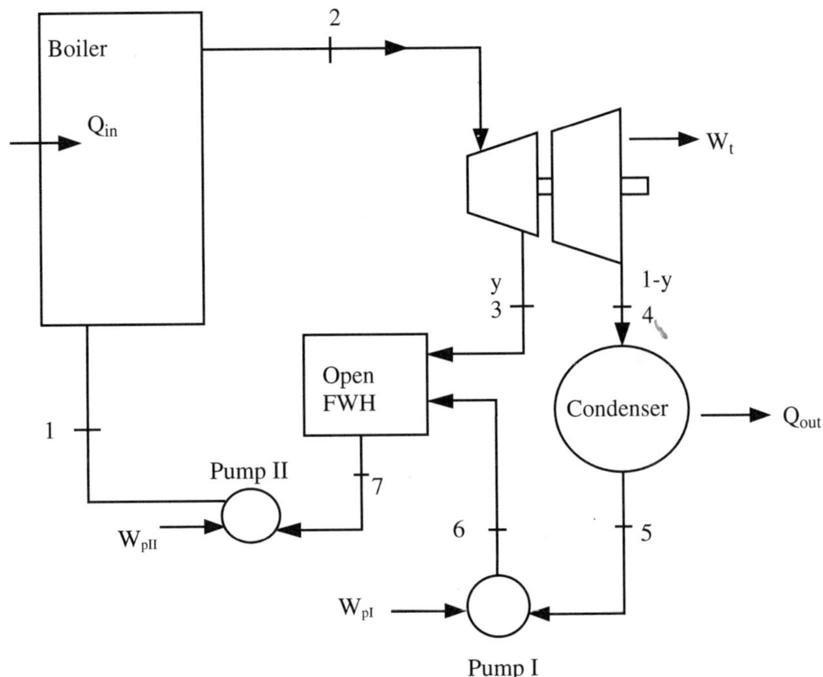

FIGURE 6.6(a)
Regenerative Rankine cycle with open feedwater heater.

extracted at state 3. The remaining (1–y) kg expands completely to the condenser pressure. Thus, the mass flow rates are different from the turbine to the open feedwater heater, and from the turbine through the condenser and pump I to the open feedwater heater. For example, if the mass flow rate through the boiler is $\dot{m}$, then it will be (1–y) $\dot{m}$ through the condenser and pump I. The heat and work interactions of a regenerative Rankine cycle with one feedwater heater can be expressed per unit mass of steam flowing through the boiler as follows:

$$q_{in} = h_2 - h_1 \tag{6.5}$$

$$q_{out} = (1 - y)(h_4 - h_5) \tag{6.6}$$

$$w_{t,out} = (h_2 - h_3) + (1 - y)(h_3 - h_4) \tag{6.7}$$

$$w_{p,in} = (1 - y)\, w_{p\,I,in} + (1 - y)\, w_{p\,II,in} \tag{6.8}$$

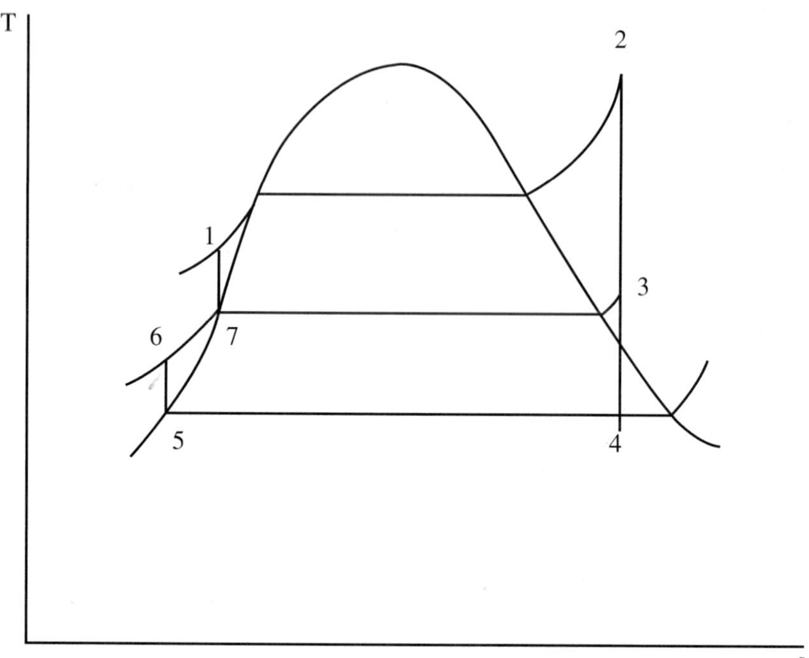

FIGURE 6.6(b)
Regenerative Rankine cycle with open feedwater heater.

where

$$y = \dot{m}_3 / \dot{m}_2 \qquad \text{(fraction of steam extracted)}$$

$$w_{pI,in} = v_5(P_6 - P_5)$$

$$w_{pII,in} = v_7(P_1 - P_7).$$

Regeneration increases the efficiency because it raises the average temperature at which heat is added to the steam in the boiler by raising the temperature of the water before it enters the boiler. The cycle efficiency can be increased further by increasing the number of feedwater heaters. The optimum number of feedwater heaters is determined by economical factors. An additional feedwater heater is justified if it saves more in fuel costs than its own cost.

6.4.2 Closed Feedwater Heaters

A closed feedwater heater is an equipment where the heat is transferred from the bled steam to the feedwater without any mixing. The schematic and the

T-s diagram of a steam power plant with one closed feedwater heater are shown in Figure 6.7, Parts (a) and (b). In an ideal heater, the feedwater is heated to the exit temperature of the bled steam, which ideally leaves the heater as a saturated liquid at the bleed pressure. In practical power plants, the feedwater exits the heater at a temperature below the exit temperature of the bled steam because a temperature difference has to exist for heat transfer to occur.

The condensed steam is then pumped to the mixing chamber on the feedwater line as shown in the figure, or routed to the condenser through an equipment called a trap. A trap allows the liquid to be throttled to a lower pressure but traps the vapor. The fluid undergoes an isenthalpic process during the throttling process.

The closed feedwater heater is more expensive than the open one because it is more complex owing to its inner tubing. The closed feedwater heater is a less efficient heat exchanger than the open feedwater heater because the two streams are not allowed to be in direct contact. However, a pump is not required for the closed heater if used in conjunction with a trap, because the two streams are at different pressures. Most practical power plants use both open and closed feedwater heaters.

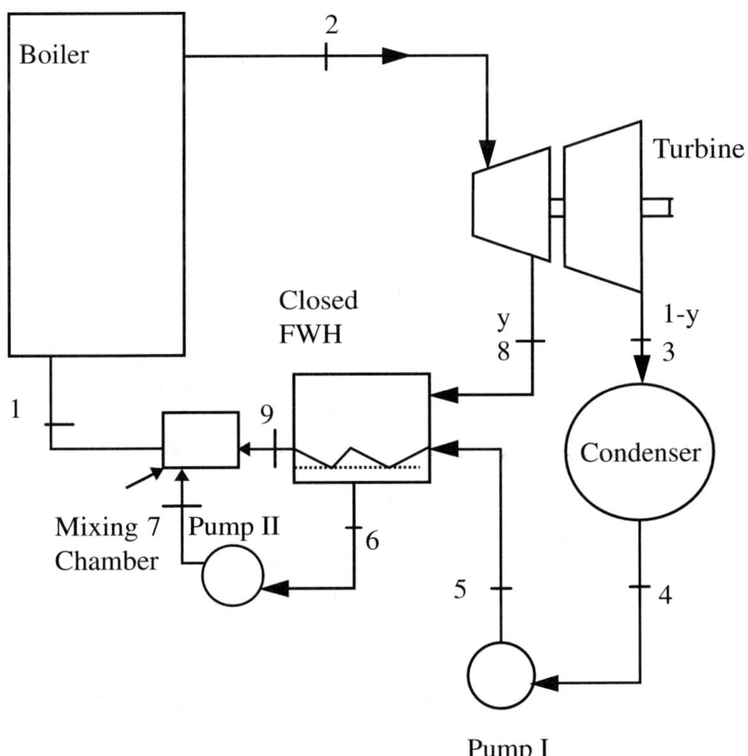

FIGURE 6.7(a)
Regenerative Rankine cycle with closed feedwater heater.

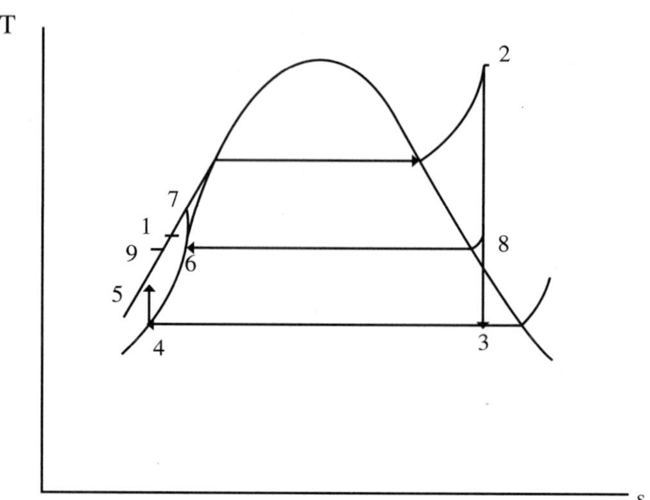

FIGURE 6.7(b)
Regenerative Rankine cycle with closed feedwater heater.

Example 6.4

Problem: It is interesting to evaluate the effect of the number of open feedwater heaters on the thermal efficiency of an ideal cycle where steam enters the turbine at 18 MPa, 620°C and the pressure in the condenser is 8 kPa. Compute the thermal efficiency for the cases described below.

(a) No feedwater heater.
(b) One feedwater heater, working at 800 kPa.
(c) Two feedwater heaters, one at 4 MPa and the other at 400 kPa.

Solution

(a) No feedwater heater — see Schematic 6.4(a) and T-s diagram 6.4(a).

Assumptions: (1) Consider unit mass of water going through the whole system.

From the thermodynamic tables,

$$h_2 = 3610 \text{ kJ/kg}, \qquad s_2 = 6.631 \text{ kJ/(kg.K)}$$
$$h_3 = 2074 \text{ kJ/kg}, \qquad s_3 = s_2 = 6.631 \text{ kJ/(kg.K)}, \; x_3 = 0.7908$$
$$h_4 = 173.9 \text{ kJ/kg}, \qquad s_4 = 0.5925 \text{ kJ/(kg.K)}$$
$$h_1 = 191.9 \text{ kJ/kg}, \qquad s_1 = s_4 = 0.5925 \text{ kJ/(kg.K)}.$$

From the quality, we know state 3 is in the two-phase region.

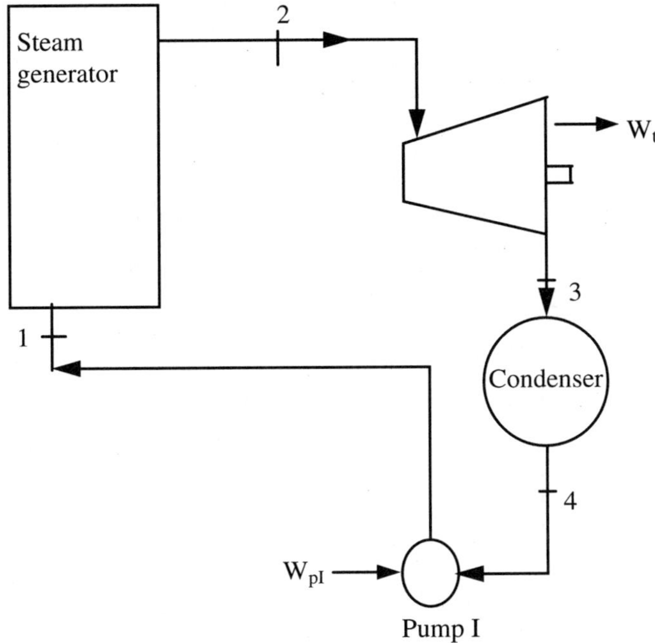

Schematic of a Rankine cycle with no feedwater heater for Example 6.4(a).

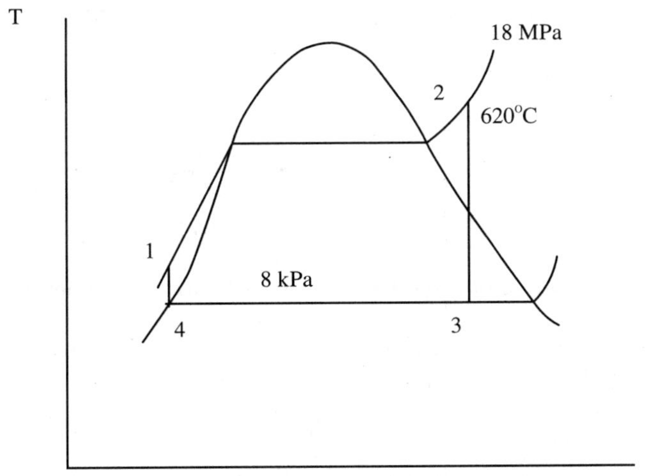

T-s diagram of a Rankine cycle with no open feedwater heater for Example 6.4(a).

The thermal efficiency of the cycle with no feedwater heater is

$$\eta_{th} \frac{w_T - w_P}{q_b} = \frac{(h_2 - h_3) - (h_1 - h_4)}{(h_2 - h_1)} = 44.4\%.$$

(b) Regenerative Rankine cycle with one open feedwater heater (FWH) — see Schematic 6.4(b) and T-s diagram 6.4(b).

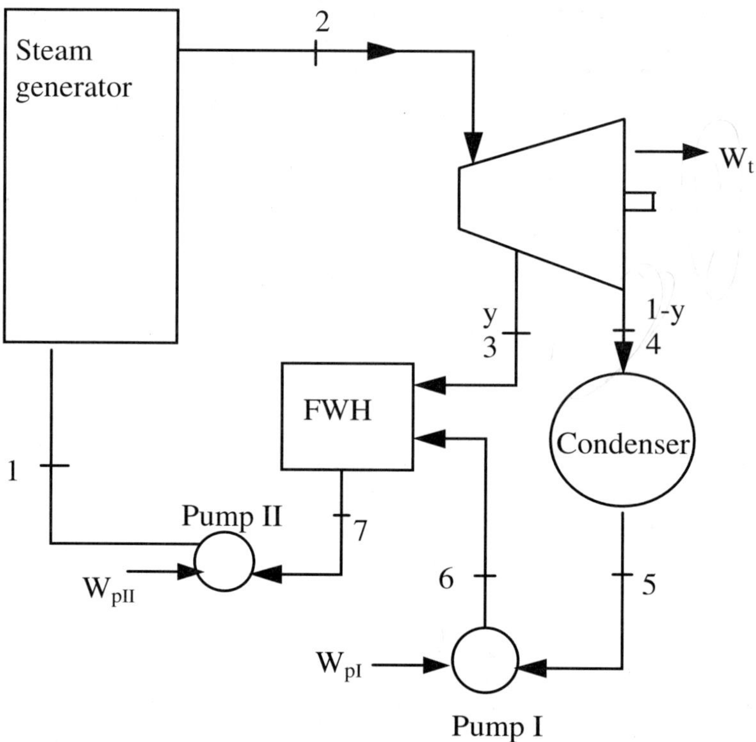

Schematic of a regenerative Rankine cycle with one open feedwater heater for Example 6.4(b).

Assumptions: (1) Consider a unit mass of water going through the whole system. Let y be the fractional mass bled off at 800 kPa.

From (a), $h_2 = 3610$ kJ/kg, $s_2 = 6.631$ kJ/(kg.K)

From tables, $h_3 = 2755$ kJ/kg, $s_3 = s_2$, $x_3 = 0.9931$

From (a), $h_4 = 2074$ kJ/kg, $s_4 = s_3 = s_2$, $x_4 = 0.7908$

From (a), $h_5 = 173.9$ kJ/kg, $s_5 = 0.5925$ kJ/(kg.K), $v_5 = 0.0010084$ m^3/kg

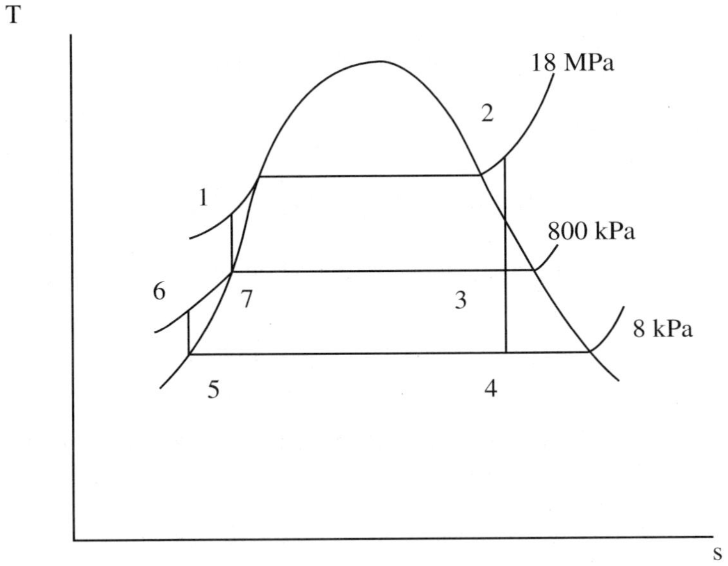

T-s diagram of a regenerative Rankine cycle with one open feedwater heater for Example 6.4(b).

Since $s_6 = s_5 = 0.5925$ kJ/(kg.K), $h_6 \approx h_5 + (P_6 - P_5)v_5 = 174.7$ kJ/kg

From tables, $h_7 = 721.1$ kJ/kg, $s_7 = 2.046$ kJ/(kg.K), $v_7 = 0.0011148$ m³/kg

Since $s_1 = s_7 = 2.046$ kJ/(kg.K), $h_1 \approx h_7 + (P_1 - P_7)v_7 = 740.3$ kJ/kg.

From the qualities, we know states 3 and 4 are in the two-phase region.
Select the FWH as a control volume. The first law gives

$$yh_3 + (1-y)h_6 = 1.h_7$$

$$y = \frac{h_7 - h_6}{h_3 - h_6} = 0.2118 \text{ kg}.$$

Select the turbine as a control volume. The first law gives

$$w_T = (h_2 - h_3) + (1-y)(h_3 - h_4)$$
$$= 1391.8 \text{ kJ/kg}.$$

Select pump I as a control volume. The first law gives

$$w_{pI} = (1-y)(h_5 - h_6) = -0.552 \text{ kJ/kg}.$$

Similarly, the work associated with pump II is

$$w_{pII} = y(h_7 - h_1) = -4.07 \text{ kJ/kg}.$$

Net work done in the cycle is

$$w_T + w_{pI} + w_{pII} = 1387.2 \text{ kJ/kg}.$$

The thermal efficiency of the cycle with one open feedwater heater is

$$\eta_{th} = \frac{w_{net}}{q_b} = \frac{w_{net}}{(h_2 - h_1)} = 48.3\%.$$

(c) Regenerative Rankine cycle with two open feedwater heaters — see Schematic 6.4(c) and T-s diagram 6.4 (c).

Assumptions: (1) Consider unit mass of water going through the whole system. Let y be the fractional mass bled off at 4 MPa, and y′ be the fractional mass bled off at 400 kPa.

From (a), $h_2 = 3610$ kJ/kg, $s_2 = 6.631$ kJ/(kg.K)

From tables, $h_3 = 3123$ kJ/kg, $s_3 = s_2$, $v_3 > v_{sat\ vap}$

From tables, $h_4 = 2628$ kJ/kg, $s_4 = s_3 = s_2$, $x_4 = 0.9483$

From (a), $h_5 = 2074$ kJ/kg, $s_5 = s_4 = s_3 = s_2$, $x_5 = 0.7908$

From (a), $h_6 = 173.9$ kJ/kg, $s_6 = 0.5925$ kJ/(kg.K), $v_6 = 0.0010084$ m^3/kg

Since $s_7 = s_6 = 0.5925$ kJ/(kg.K), $h_7 \approx h_6 + (P_7 - P_6)v_6 = 174.3$ kJ/kg

From tables, $h_8 = 604.7$ kJ/kg, $s_8 = 1.7766$ kJ/(kg.K), $v_8 = 0.0010836$ m^3/kg

Since $s_9 = s_8 = 1.7766$ kJ/(kg.K), $h_9 \approx h_8 + (P_9 - P_8)v_8 = 608.6$ kJ/kg

From tables, $h_{10} = 1087.3$ kJ/kg, $s_{10} = 2.796$ kJ/(kg.K), $v_{10} = 0.0012522$ m^3/kg

Since $s_1 = s_{10} = 2.796$ kJ/(kg.K), $h_1 \approx h_{10} + (P_1 - P_{10})v_{10} = 1104.8$ kJ/kg

From the qualities, we know states 4 and 5 are in the two-phase region.

Select the high-pressure (HP) FWH as a control volume. The first law gives

$$yh_3 + (1 - y)h_9 = 1.h_{10}$$

$$y = \frac{h_{10} - h_9}{h_3 - h_9} = 0.1904 \text{ kg}.$$

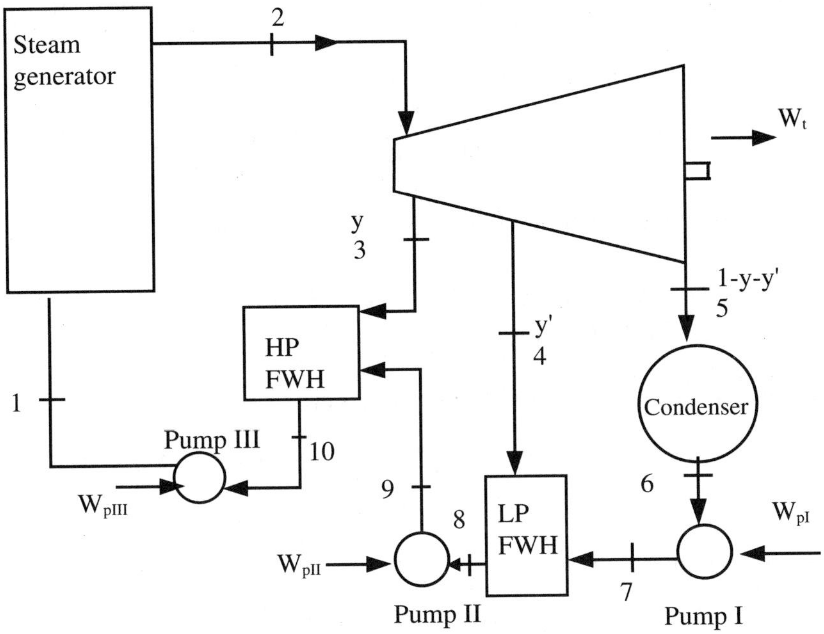

Schematic of a regenerative Rankine cycle with two open feedwater heaters for Example 6.4(c).

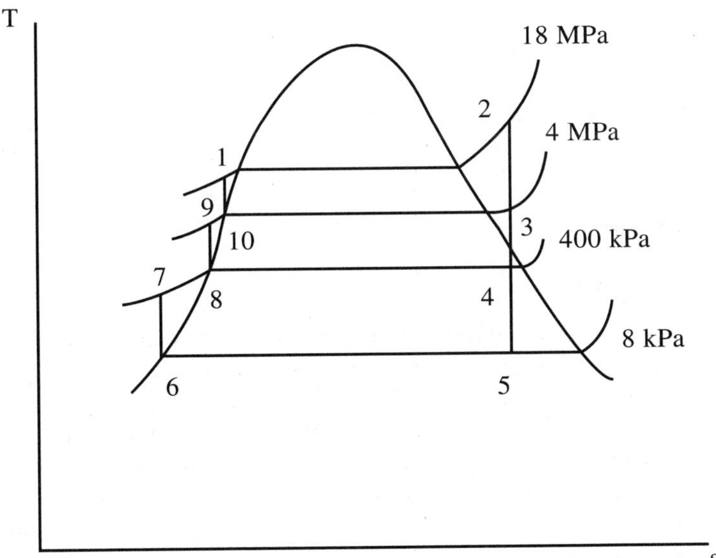

T-s diagram of a regenerative Rankine cycle with two open feedwater heaters for Example 6.4(c).

Select the low-pressure (LP) FWH as a control volume. The first law gives

$$y'h_4 + (1 - y - y')h_7 = (1 - y).h_8$$

$$y' = (1 - y)\frac{h_8 - h_7}{h_4 - h_7} = 0.1420 \text{ kg}.$$

Select the turbine as a control volume. The first law gives

$$w_T = (h_2 - h_3) + (1 - y)(h_3 - h_4) + (1 - y - y')(h_4 - h_5)$$
$$= 1257.6 \text{ kJ/kg}.$$

The work associated with the three pumps are

$$w_{pI} + w_{pII} + w_{pIII} = (1 - y - y')(h_6 - h_7) + (1 - y)(h_8 - h_9) + 1(h_{10} - h_1)$$
$$= -20.9 \text{ kJ/kg}.$$

Net work done in the cycle is

$$w_T + w_{pI} + w_{pII} + w_{pIII} = 1237 \text{ kJ/kg}.$$

The thermal efficiency of the cycle with two open feedwater heaters is

$$\eta_{th} = \frac{w_{net}}{q_b} = \frac{w_{net}}{(h_2 - h_1)} = 49.4\%.$$

Hence, an increase in the number of feedwater heaters increases the thermal efficiency of the cycle.

6.5 Air Preheater

In our discussions so far, we have concentrated on the water side of the Rankine cycle, and not as much on the heat production side. Typically, a fuel is burned in the boiler to generate the heat. If the exhaust (flue) gases are simply vented after the boiler, the gases would still be at a relatively high temperature. This high-temperature gas can be used to preheat the air entering the boiler. By increasing the average temperature of the incoming air, the efficiency of the combustion is increased. This decreases the heat rate of the boiler, and thus improves its efficiency.

Example 6.5

Problem: The exhaust gases in a boiler, before venting, are at 1000°C and at a pressure of about 1 atm. Compute the flow availability of the gases in kilojoules per kilogram. Assume the gases to have the properties of air. The environment is at 25°C and 1 atm. If the gases are used in an air preheater, they are vented at 500°C instead of 1000°C. Compute the percentage of availability used in this case. In Chapter 5, Section 5.5.1, a second law ratio to measure thermal environmental impact was introduced, where

$$r_{II} = \frac{\left(1 - \frac{T_o}{T_j}\right)\dot{Q}_l}{\left(1 - \frac{T_o}{T_s}\right)\dot{Q}_s}.$$

Determine the percentage change of this ratio owing to the use of the preheater.

Solution

Assumptions: (1) Changes in K.E. and P.E. are negligible.

Analysis:

At the dead state,

$$h_o = 298.6\ \text{kJ/kg, and } s_o = 6.86\ \text{kJ/(kg.K)}.$$

At 1000°C and 1 atm,

$$h_1 = 1364.1\ \text{kJ/kg, and } s_1 = 8.41\ \text{kJ/(kg.K)}.$$

The flow availability of the exhaust gases is

$$a_{f1} = (h_1 - h_o) - T_o\,(s_1 - s_o) = 603\ \text{kJ/kg}.$$

At 500°C and 1 atm,

$$h_2 = 792.9\ \text{kJ/kg, and } s_2 = 7.85\ \text{kJ/(kg.K)}.$$

The flow availability of the exhaust gases is then

$$a_{f2} = (h_2 - h_o) - T_o(s_2 - s_o) = 199\ \text{kJ/kg}.$$

Hence, the percentage of availability used by the air preheater is

$$= \frac{603 - 199}{603} = 67\%.$$

Without the preheater, the value of the second law ratio is

$$r_{II} = \frac{\left(1 - \frac{298.15}{1273.15}\right)\dot{Q}_1}{\left(1 - \frac{T_o}{T_s}\right)\dot{Q}_s}.$$

With the preheater, the value of the second law ratio is

$$r_{II} = \frac{\left(1 - \frac{298.15}{773.15}\right)\dot{Q}_1}{\left(1 - \frac{T_o}{T_s}\right)\dot{Q}_s}.$$

Hence,

$$\frac{r_{II}\big|_{preheater}}{r_{II}\big|_{original}} = \frac{\left(1 - \frac{298.15}{773.15}\right)}{\left(1 - \frac{298.15}{1273.15}\right)} = 0.802.$$

The second law ratio to measure thermal environmental impact is reduced to 80.2% of its original value. The percentage change is thus a reduction of 19.8%.

6.6 Economizer

The flue gases from the boiler may also be used to heat the feedwater. This is done in a heat exchanger called the economizer. Typically, the economizer is placed just before the air preheater, if one is present. Heat is recovered from the flue gases by the water, so the heat rate is reduced. Consequently, the boiler efficiency is increased.

Example 6.6

Problem: In Example 6.6, the exhaust gases are used in an economizer instead of the air preheater. The water entering the economizer is at 10 MPa and 39.29°C. It leaves at 10 MPa and 50°C. Calculate the ratio $\dot{m}_{water} / \dot{m}_{gases}$ if the gases are vented at 500°C, after the economizer.

Solution

Assumptions:

(1) Changes in K.E. and P.E. are negligible, and
(2) The economizer is an adiabatic, ideal heat exchanger.

Analysis:

For the water, the incoming properties are

$$h_{w1} = 173.42 \text{ kJ/kg, and } s_{w1} = 0.559 \text{ kJ/(kg.K)}.$$

The outgoing properties of the water are

$$h_{w2} = 217.9 \text{ kJ/kg, and } s_{w2} = 0.699 \text{ kJ/(kg.K)}.$$

For the ideal economizer, the first law equation gives

$$\dot{m}_{water}\left(h_{w2} - h_{w1}\right) = \dot{m}_{gases}\left(h_1 - h_2\right).$$

Thus, the ratio

$$\frac{\dot{m}_{water}}{\dot{m}_{gases}} = \frac{\left(h_1 - h_2\right)}{\left(h_{w2} - h_{w1}\right)} = 12.84.$$

6.7 Availability Analysis of Vapor Power Cycles

This topic was partly discussed in Chapter 5, Section 5.5.3. The current section completes the discussion. The ideal Rankine cycles (simple, reheat, or regenerative) are internally reversible, that is, they do not have irreversibilities that are inside the system where water is the working fluid. However, there are irreversibilities associated with heat transferred to the water in the boiler, and with the heat lost by the water in the condenser.

For the boiler, the irreversibility rate may be found from the entropy generation, according to

$$\dot{I}_b = T_o \dot{\sigma} = T_o\left(\sum \dot{m}_e s_e - \sum \dot{m}_i s_i - \frac{Q_b}{T_b}\right). \tag{6.9}$$

On a unit-mass basis for a one-inlet, one-outlet, steady-flow boiler, the equation may be written as

$$i = T_o \frac{\dot{\sigma}}{\dot{m}} = T_o\left(s_e - s_i - \frac{q_b}{T_b}\right). \quad (6.10)$$

For the condenser, the irreversibility rate may be found from the entropy generation, according to

$$\dot{I}_c = T_o \dot{\sigma} = T_o\left(\sum \dot{m}_e s_e - \sum \dot{m}_i s_i - \frac{Q_c}{T_c}\right). \quad (6.11)$$

On a unit-mass basis for a one-inlet, one-outlet, steady-flow condenser, the equation may be written as

$$i = T_o \frac{\dot{\sigma}}{\dot{m}} = T_o\left(s_e - s_i - \frac{q_c}{T_c}\right). \quad (6.12)$$

Example 6.7

Problem: Compute the irreversibility of each of the four major components of the Rankine cycle described in Example 6.1. Heat is transferred to the working fluid in the furnace of the boiler at 1250°C and heat is transferred to the cooling water in the condenser at 16°C and 100 kPa. Calculate the availability of the steam before and after the turbine.

Solution

Assumptions: (1) Changes in K.E. and P.E. are negligible, and (2) the environment is at 16°C and 100 kPa.

Analysis: The processes through the turbine and the pump are isentropic, hence the irreversibility in the turbine as well as in the pump is zero.

The irreversibility of the boiler is given by

$$i = T_o\left(s_2 - s_1 - \frac{q_b}{T_b}\right)$$

$$i = 289.15\left\{7.094 - 0.4763 - \frac{227.9 \times 10^3}{63.17(1523.15)}\right\} \text{kJ/kg} = 1229\,\text{kJ/kg}.$$

The irreversibility of the condenser is given by

$$i = T_o\left(s_4 - s_3 - \frac{q_c}{T_c}\right)$$

$$i = 289.15\left\{0.4763 - 7.094 + \frac{127.9 \times 10^3}{63.17(289.15)}\right\} kJ/kg = 111.2 \text{ kJ/kg}.$$

The conditions at dead state (16°C, 0.1 MPa) are given by

$$h_o = 67.27 \text{ kJ/kg}, \; s_o = 0.239 \text{ kJ/kg}.$$

Availability of the steam before the turbine is

$$a_{f2} = (h_2 - h_o) - T_o\,(s_2 - s_o)$$
$$a_{f2} = (3755 - 67.27) - 289.15\,(7.094 - 0.239) = 1705.6 \text{ kJ/kg}.$$

Availability of the steam after the turbine is

$$a_{f3} = (h_3 - h_o) - T_o\,(s_3 - s_o)$$
$$a_{f3} = (2163 - 67.27) - 289.15\,(7.094 - 0.239) = 113.6 \text{ kJ/kg}.$$

6.8 Cogeneration

Cogeneration is the production of more than one useful form of energy (such as electric power and process heat) from one energy source.

Since the heat required for processes is usually at a relatively low temperature, good use is not made of the relatively high-temperature products of combustion obtained by burning fuel. This inefficiency can be removed by a cogeneration arrangement in which fuel is used to produce both electricity and steam (or process heat). Schematics of such systems are shown in Figures 6.8 and 6.9. Industries such as food processing and chemical production, which require steam for different processes and electricity for operating machines, lighting, etc. are well suited for the use of cogeneration.

Another cogeneration arrangement that is increasingly popular in the U.S. is district heating. In this arrangement the power plant not only supplies electricity for industrial, commercial, and domestic use but also steam for process needs, space heating, and domestic water heating. District heating is popular in northern Europe.

A schematic of an ideal cogeneration plant is shown in Figure 6.8. The ideal cogeneration plant is not practical because it cannot handle variations in power demand or process heat load. The power demand has to be such that the heat rejected has to match exactly the process heat requirement.

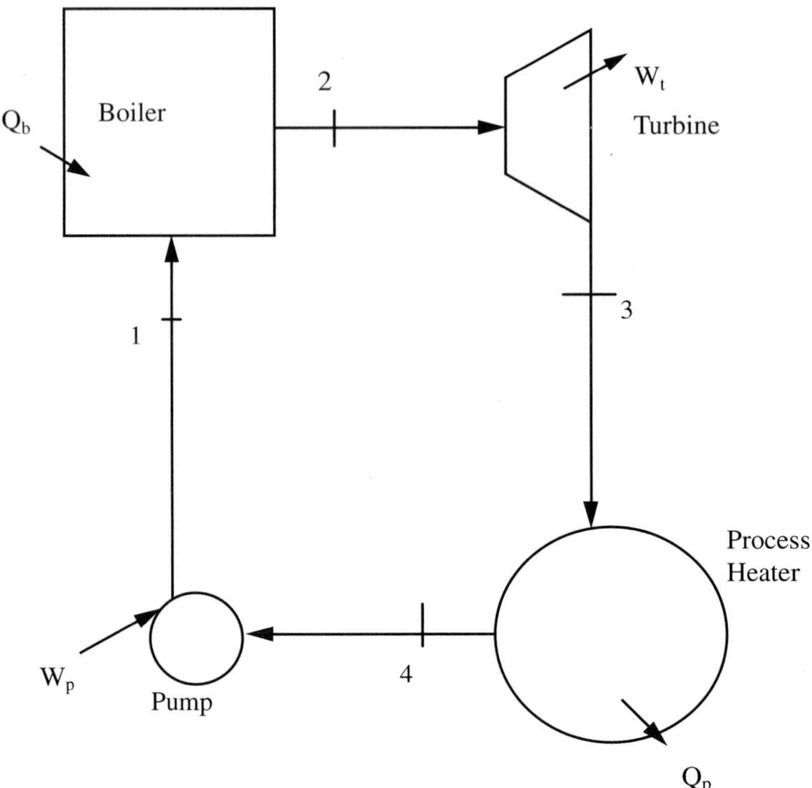

FIGURE 6.8
Ideal cogeneration plant.

The schematic of a practical cogeneration plant is shown in Figure 6.9. Most of the time, a portion of the steam is bled from the turbine at some intermediate pressure P_4. This bled steam is used in the process heater. The major portion of the steam expands to the condenser pressure P_5, where it is cooled to the saturated liquid state. Both pumps, one after the condenser, and the other after the process heater, raise the pressure of the water to the boiler pressure at state 1. Under these conditions, the first law equations give

$$\dot{Q}_b = \dot{m}_1(h_2 - h_1) \tag{6.13}$$

$$\dot{Q}_c = \dot{m}_1(h_5 - h_6) \tag{6.14}$$

$$\dot{Q}_{Pr} = \dot{m}_3 h_3 + \dot{m}_4 h_4 - \dot{m}_7 h_7 \tag{6.15}$$

$$\dot{W}_t = (\dot{m}_2 - \dot{m}_3)(h_2 - h_4) + \dot{m}_5(h_4 - h_5) \tag{6.16}$$

$$\dot{W}_p = \dot{m}_1(h_6 - h_1) + \dot{m}_7(h_7 - h_1). \tag{6.17}$$

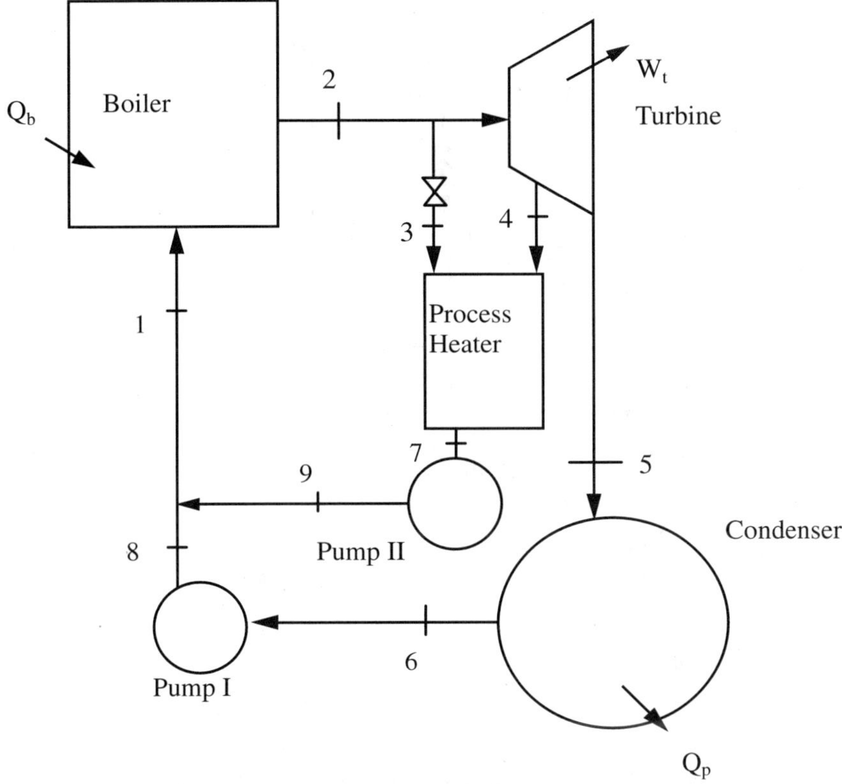

FIGURE 6.9
Schematic of a practical cogeneration plant.

During periods when there is no requirement for process heat, all the steam is expanded through the turbine and the condenser $(\dot{m}_3 = \dot{m}_4 = 0)$. This is the case when the cogeneration plant operates like an ordinary steam power plant. During periods when there is a high demand for process heat, all the steam is directed to the process heater; none goes through the condenser $(\dot{m}_5 = 0)$. No heat is rejected via the condenser. If this still does not produce enough process heat, some steam is expanded via a pressure-reducing valve (PRV) to a pressure P_4 and directed to the process heater. When all the steam generated in the boiler is passed through the PRV $(\dot{m}_3 = \dot{m}_2,\ \dot{m}_5 = \dot{m}_4 = 0)$, no power is produced by the turbine and maximum process heating is produced.

Example 6.8

Problem:

A cogeneration plant as shown in the figure, operates with a boiler pressure of 8 MPa and a condenser pressure of 7 kPa. The superheated steam leaves the boiler at 550°C, and at a rate of 10 kg/s. The steam is bled from the turbine

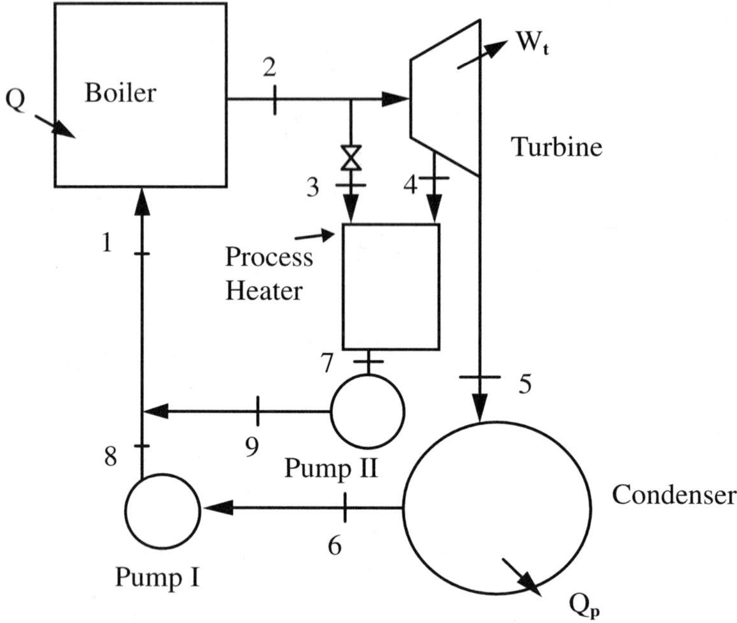

Schematic for Example 6.8.

at 1 MPa for process heating. When there is a large demand for process heat, some steam is expanded through the PRV to 1 MPa and directed to the process heater. The bled fractions are adjusted so that the working fluid leaves the process heater as a saturated liquid at 1 MPa. Calculate:

(a) The maximum power produced;
(b) The maximum rate at which process heat is supplied;
(c) The rate of process heat supply when 20% of the steam is directed through the PRV, and 50% of the steam is bled from the turbine at 1 MPa for process heating.

Solution

Assumptions: (1) Changes in K.E. and P.E. are negligible, and (2) neglect any heat losses and pressure drops in the piping.

Analysis:

From the thermodynamic tables,

$h_2 = 3521$ kJ/kg, $s_2 = 6.878$ kJ/(kg.K)
$h_3 = h_2 = 3521$ kJ/kg, $P_3 = P_4$, $s_3 = 7.816$ kJ/(kg.K)
$h_4 = 2918$ kJ/kg, $s_4 = s_2$
$h_5 = 2136$ kJ/kg, $s_5 = s_4 = s_2$

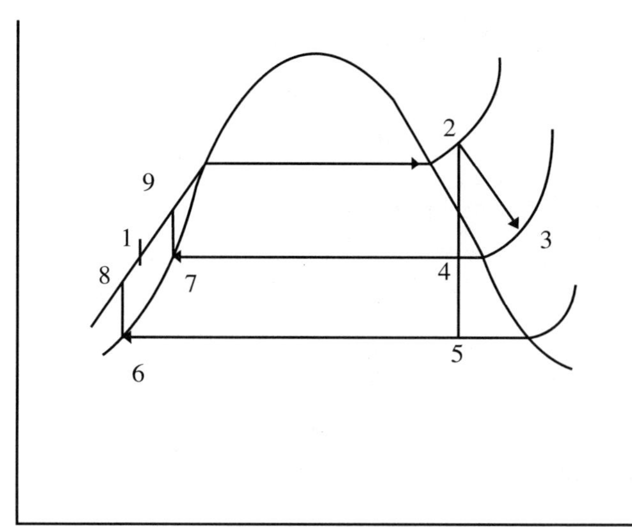

T-s Diagram for Example 6.8.

$h_6 = 163.4$ kJ/kg, $s_6 = 0.5591$ kJ/(kg.K) $v_6 = 0.001007$ m^3/kg
$h_7 = 762.8$ kJ/kg, $s_7 = 2.139$ kJ/(kg.K) $v_7 = 0.001127$ m^3/kg
$h_8 \approx h_6 + v_6 (P_8 - P_6) = 171.4$ kJ/kg, $s_8 = s_6$
$h_9 \approx h_7 + v_7 (P_9 - P_7) = 770.7$ kJ/kg, $s_9 = s_7$.

(a) Maximum power is produced when $\dot{m}_3 = \dot{m}_4 = 0$.

$$\dot{m}_2 = \dot{m}_5 = 10 \text{ kg/s}.$$

Maximum power is thus

$$\dot{W}_t = \dot{m}_2(h_2 - h_4) + \dot{m}_5(h_4 - h_5)$$
$$= \dot{m}_2(h_2 - h_5) = 13,850 \frac{\text{kJ}}{\text{s}} = 13.85 \text{ MW}.$$

(b) Maximum rate at which process heat is supplied is when $\dot{m}_4 = 0$ and $\dot{m}_3 = \dot{m}_7 = 10$ kg/s.
Maximum process heat rate is thus

$$\dot{Q}_{Pr} = \dot{m}_3 h_3 - \dot{m}_7 h_7$$
$$= \dot{m}_3(h_3 - h_7) = 27,582 \frac{\text{kJ}}{\text{s}} = 27.58 \text{ MW}.$$

(c) When

$$\dot{m}_3 = 2\ \text{kg/s},$$
$$\dot{m}_4 = 5\ \text{kg/s},$$
$$\dot{m}_7 = 7\ \text{kg/s},$$

rate of process heat supply is

$$\dot{Q}_{Pr} = \dot{m}_3 h_3 + \dot{m}_4 h_4 - \dot{m}_7 h_7$$
$$= 16{,}292\ \frac{\text{kJ}}{\text{s}} = 16.29\ \text{MW}.$$

6.9 Binary Vapor Cycles

The binary cycle came about because there is not one single fluid that has the perfect characteristics for a vapor power cycle. Let us examine the characteristics that are best suited for a vapor power cycle.

1. It should be plentiful and low in cost, nontoxic, chemically stable, and relatively noncorrosive.
2. It should have a large enthalpy of vaporization so that heat transfer will approach being isothermal and large mass flow rates need not be required.
3. A critical temperature that is above the metallurgically allowed maximum temperature (about 620°C) allows a large portion of the heat transfer to take place isothermally as the fluid changes phase in the boiler. This factor allows the cycle to approach the Carnot cycle, which requires isothermal heat transfer at the high temperature.
4. A condenser pressure that is not too low; pressures much lower than atmospheric create air-leakage problems.
5. A saturation dome that is an inverted U. This reduces the moisture in the last stages of the turbine.

Although water has its shortcomings in meeting factors 3 and 5, it has been found more satisfactory overall for large power plants for the utility companies. However, other cycles may use working fluids that are best suited for their purposes. Cycles working at lower temperatures may use refrigerants.

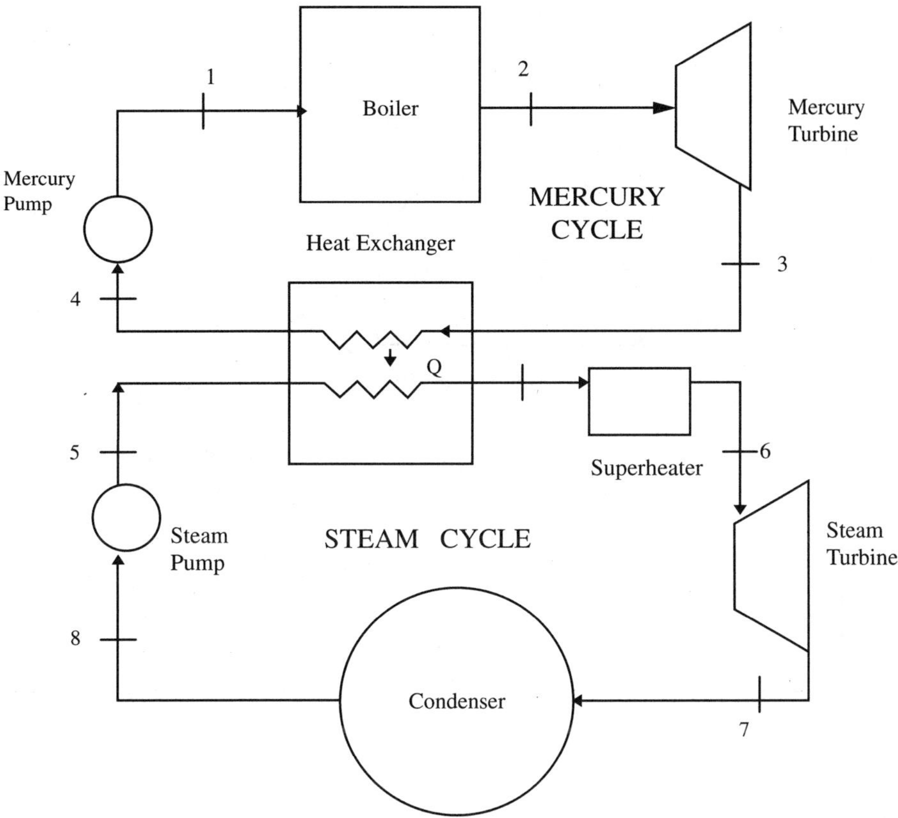

FIGURE 6.10(a)
Mercury-water binary vapor cycle.

Power systems for space applications may use mercury because of its desirable performance characteristics at the relatively high temperatures of such systems. In addition, water may be used in conjunction with another working fluid in a binary vapor cycle to achieve better overall performance than a single vapor cycle.

In a binary vapor cycle, two working fluids are used: one with good operating characteristics at the high temperature and the other with good characteristics at the low temperature. A schematic diagram and the T-s diagram of a binary vapor cycle using mercury and water are shown in Figure 6.10, Parts (a) and (b). The heat rejection of the high-temperature cycle (the topping cycle) is used as the heat addition of the low-temperature cycle (the bottoming cycle). Since the decrease of specific enthalpy of mercury is much less

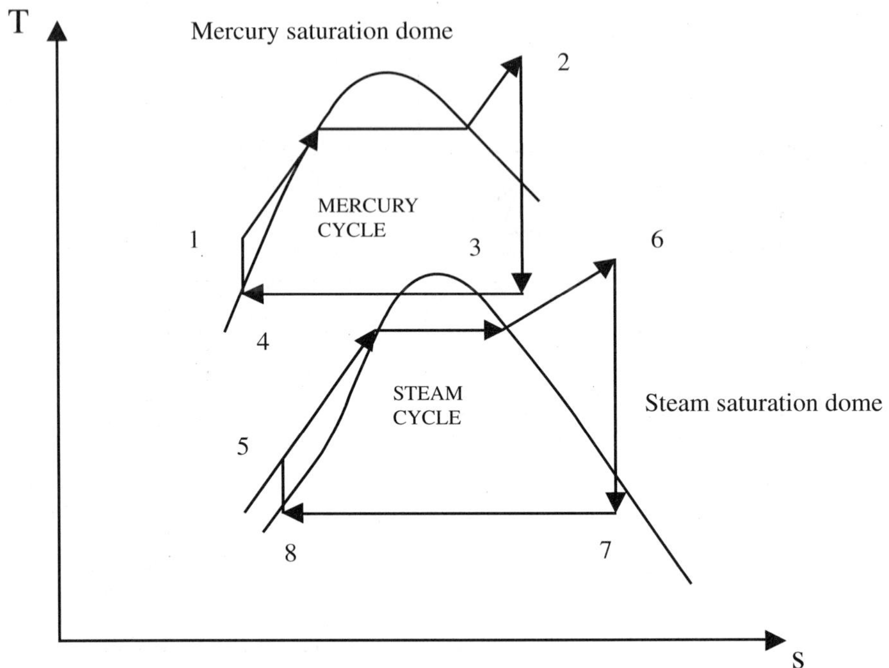

FIGURE 6.10(b)
Mercury-water vapor binary cycle.

than the increase of specific enthalpy of the water in the heat exchanger, the mass flow rate of the mercury has to be greater than that of the water.

The mercury-water binary cycle is more efficient than the water vapor cycle alone because the first one more closely approximates the Carnot cycle. However, they are not attractive for power generation on land because of their high capital cost and the competition from combined gas-steam power plants.

Example 6.9

Problem: Consider a mercury-and-steam binary cycle comprising two ideal Rankine cycles. In the mercury cycle, saturated mercury vapor leaves the boiler at 850 K, and saturated liquid leaves the condenser at 600 K. Relevant property data for saturated mercury are given in the table below.

The heat rejected from the mercury cycle is transferred to the water cycle, which produces saturated water vapor at 500 K. The steam is then superheated

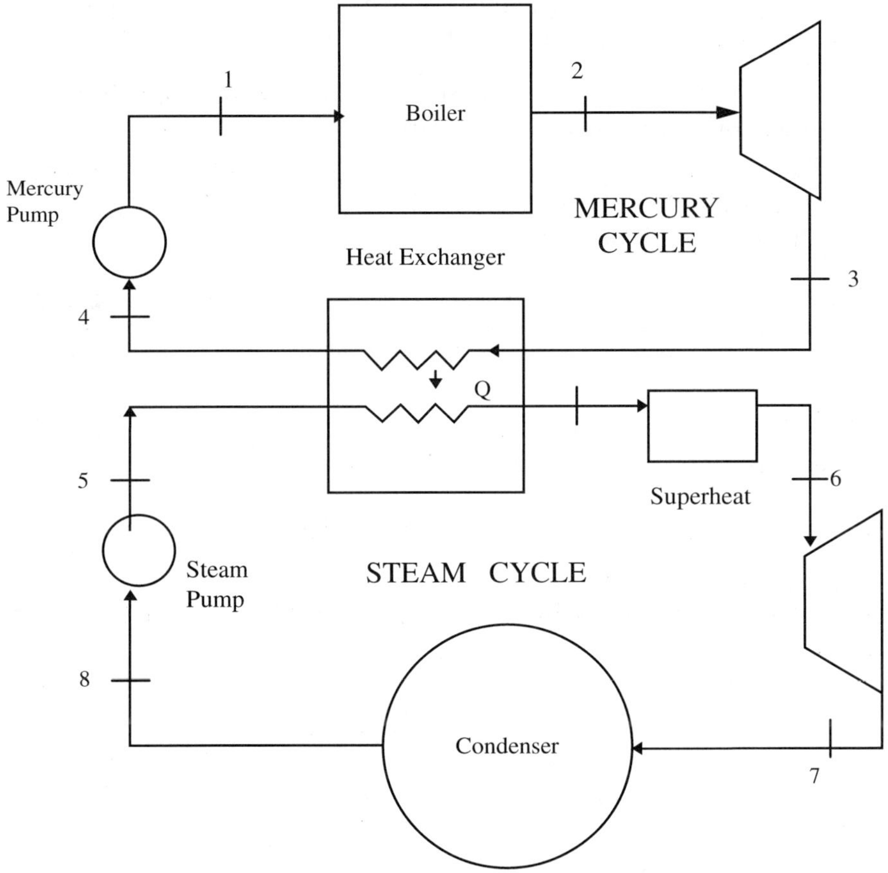

1
Boiler
2
Mercury
Pump
MERCURY
CYCLE
Heat Exchanger
3
4
Q
5
6
Superheat
Steam
Pump
STEAM CYCLE
8
Condenser
7

Schematic for Example 6.9.

T
2
1
3
6
4
5'
5
8
7
s

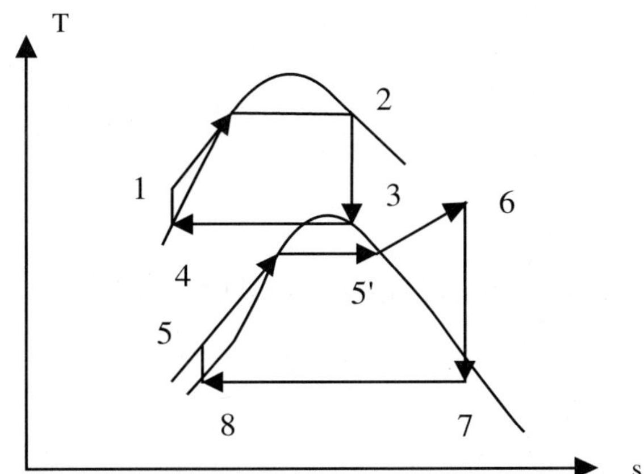

T-s Diagram for Example 6.9.

at that pressure to 750 K using another source of heat. Saturated liquid water leaves the condenser at 8 kPa. Determine:

(a) The mass in kilograms of mercury condensed per mass in kilograms of water converted into vapor.
(b) The rate of heat addition to the overall binary cycle, in kilojoules per kilogram of water.
(c) The thermal efficiency.

T(K)	P(MPa)	v_f(m³/kg)	h_f(kJ/kg)	h_g(kJ/kg)	s_f(kJ/kg.K)	s_g(kJ/kg.K)
600	0.05757	0.0000780	26.19	322.53	0.0532	0.5471
850	1.890	0.0000821	61.06	346.36	0.1014	0.4370

Solution

The entropy at state 3 is the same as that at 2:

$$s_3 = s_2 = 0.4370 \text{ kJ/(kg.K)}.$$

Mercury cycle

Hence, state 3 is in the two-phase region, and the dryness fraction may be evaluated as

$$x_3 = \frac{s_3 - s_f}{s_g - s_f} = \frac{0.4370 - 0.0532}{0.5471 - 0.0532} = 0.7771.$$

Hence,

$$h_3 = x_3 h_g + (1 - x_3) h_f = 256.5 \text{ kJ/kg of mercury.}$$

Work done by the turbine is

$$w_T = h_2 - h_3 = 89.86 \text{ kJ/kg}.$$

Work done on the pump is

$$w_P \approx v(P_1 - P_4)$$

$$w_p \approx 0.00078 \frac{m^3}{kg}(1.890 - 0.5757)1000 \text{ kPa} \frac{1000 \text{ N.m}^{-2}}{1 \text{ kPa}} \cdot \frac{1 \text{ kJ}}{1000 \text{ J}}$$

$$= 0.143 \frac{kJ}{kg}.$$

From the first law, it may be shown that the enthalpy at state 1 is equal to that at state 4 plus the pump work. Hence,

$$h_1 = h_4 + w_P = 26.37 \text{ kJ/kg of mercury.}$$

Heat added to the mercury boiler is

$$q_b = h_2 - h_1 = 346.36 - 26.37 = 320 \text{ kJ/kg of mercury.}$$

Steam cycle

From the thermodynamic tables,

$$\begin{aligned}
&h_{5'} = 2804 \text{ kJ/kg}, && && P_{5'} = 2.644 \text{ MPa}\\
&h_6 = 3409 \text{ kJ/kg}, && && s_6 = 7.228 \text{ kJ/(kg.K)}\\
&h_7 = 2262 \text{ kJ/kg}, && && s_7 = s_6 = 7.228 \text{ kJ/(kg.K)}\\
&h_8 = 173.86 \text{ kJ/kg}, && s_8 = 0.5925 \text{ kJ/(kg.K)}, && v_8 = 0.0010084 \text{ m}^3\text{/kg.}
\end{aligned}$$

Since

$$s_5 = s_8 = 0.5925 \text{ kJ/(kg.K)}, \; h_5 \approx h_8 + (P_5 - P_8)\, v_8 = 176.5 \text{ kJ/kg.}$$

Work done by the steam turbine is

$$w_T = h_6 - h_7 = 1147 \text{ kJ/kg of water.}$$

Work done on the pump is

$$w_P = h_5 - h_8 = 2.64 \text{ kJ/kg of water.}$$

Net work done by the steam cycle = 1144.4 kJ/kg of water.

Select as control volume the mercury condenser-cum-boiler for the water. The first law for the control volume gives

$$m_{Hg}\left(h_3 - h_4\right) = m_{water}\left(h_{5'} - h_5\right).$$

The mass of mercury condensed per mass of water converted into vapor is

$$\frac{m_{Hg}}{m_{water}} = \frac{\left(h_{5'} - h_5\right)}{\left(h_3 - h_4\right)} = 11.4.$$

For 1 kg of water, the rate of heat addition to the overall binary cycle is

$$q_{overall} = \frac{m_{Hg}}{m_{water}} = \left(h_2 - h_1\right) + 1.\left(h_6 - h_{5'}\right)$$

$$q_{overall} = 4252.9 \text{ kJ/kg of water.}$$

All heat lost is from the water condenser,

$$q_{loss} = h_7 - h_8 = 2088.1 \text{ kJ/kg of water.}$$

The thermal efficiency is

$$\eta_{th} = \frac{w_{net}}{q_{overall}} = \frac{q_{overall} - q_{loss}}{q_{overall}} = 50.9\%.$$

6.10 Combined Gas-Vapor Power Cycles

A popular binary cycle involves a gas power topping cycle with a vapor bottoming cycle, called the combined gas-vapor cycle or the combined cycle. Of the various possible combinations, the gas-turbine (Brayton) cycle topping a steam cycle (Rankine cycle) is most popular. Recent advancements in gas turbines have made this combined gas-steam cycle economically attractive. The combined cycle increases the efficiency without much increase in the initial cost. As a result, many new power plants operate on combined cycles, and many existing steam power plants are being converted to combined-cycle plants. Thermal efficiencies of over 40% are attainable from the combined cycle.

The combined cycle makes use of the desirable characteristics of the gas-turbine cycle at high temperatures and uses the high-temperature exhaust gases as the energy input for the bottoming steam cycle. The schematic and the T-s diagram of the combined cycle is shown in Figure 6.11, Parts (a) and (b). In general, more than one gas turbine is needed to supply enough heat for the steam in the bottoming cycle. In addition, the steam cycle may include regeneration or reheating. Energy for the reheating may come from burning additional fuel.

Example 6.10

Problem: A combined gas turbine-vapor cycle power plant produces 37 MW net power. The exhaust gases from the gas turbine leaves the heat exchanger at 149.85°C (423 K). Steam leaves the heat exchanger at 5000 kPa and 399.85°C (673 K). Air enters the compressor of the gas turbine system at 21.85°C (295 K) and 100 kPa and exits the compressor at 1000 kPa. After isobaric heat addition, air leaves the combustor at 1126.85°C (1400 K). Calculate the flow rates in each cycle, the overall thermal efficiency, and the availability utilization of the gas turbine exhaust. The surrounding temperature is 21.85°C (295 K). The condenser pressure of the Rankine cycle is 10 kPa. Assume that air is the working fluid in the gas turbine cycle, so that the mass of the fuel may be neglected. Both the Rankine cycle and the gas turbine gas cycle may be assumed to be ideal.

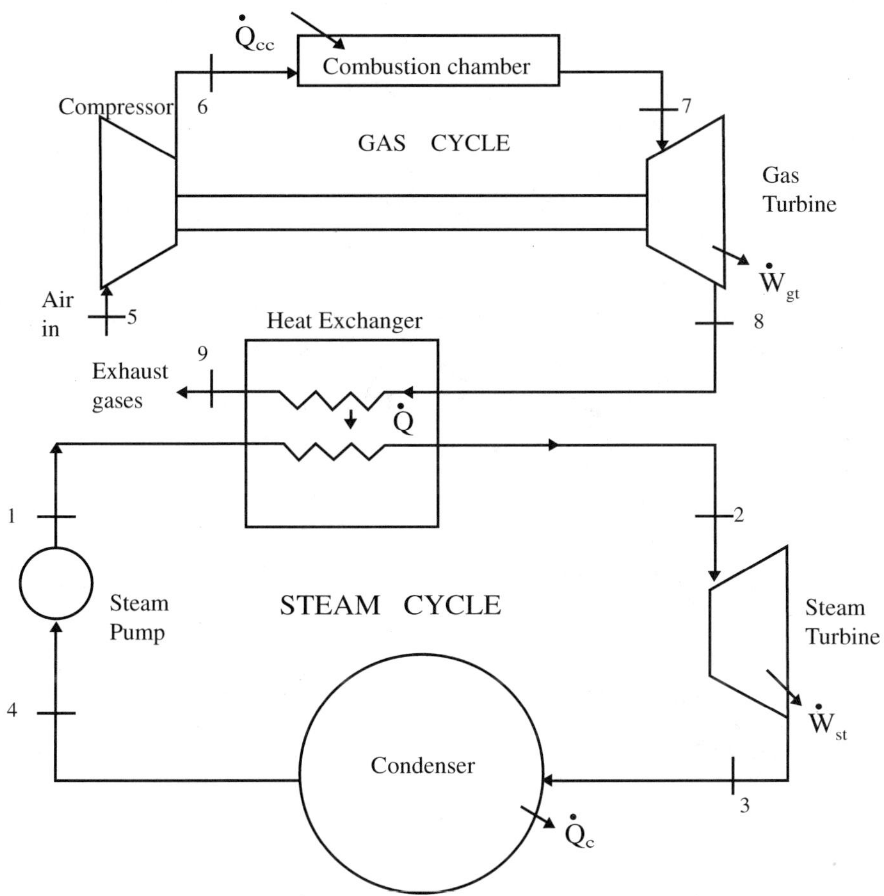

FIGURE 6.11(a)
Combined gas-steam power cycle.

Solution

Assumptions: (1) The water leaves the condenser as a saturated liquid, and (2) changes in kinetic and potential energy can be neglected.

Analysis: First, identify the state points on the temperature-entropy T-s diagram. Find the properties for air at each state, starting at state 5 since both T_5 and P_5 are known.

$$h_5 = 295.45 \text{ kJ/kg}, \ s_5 = 6.852 \text{ kJ/(kg.K)}, \ p_{r5} = 1.0509.$$

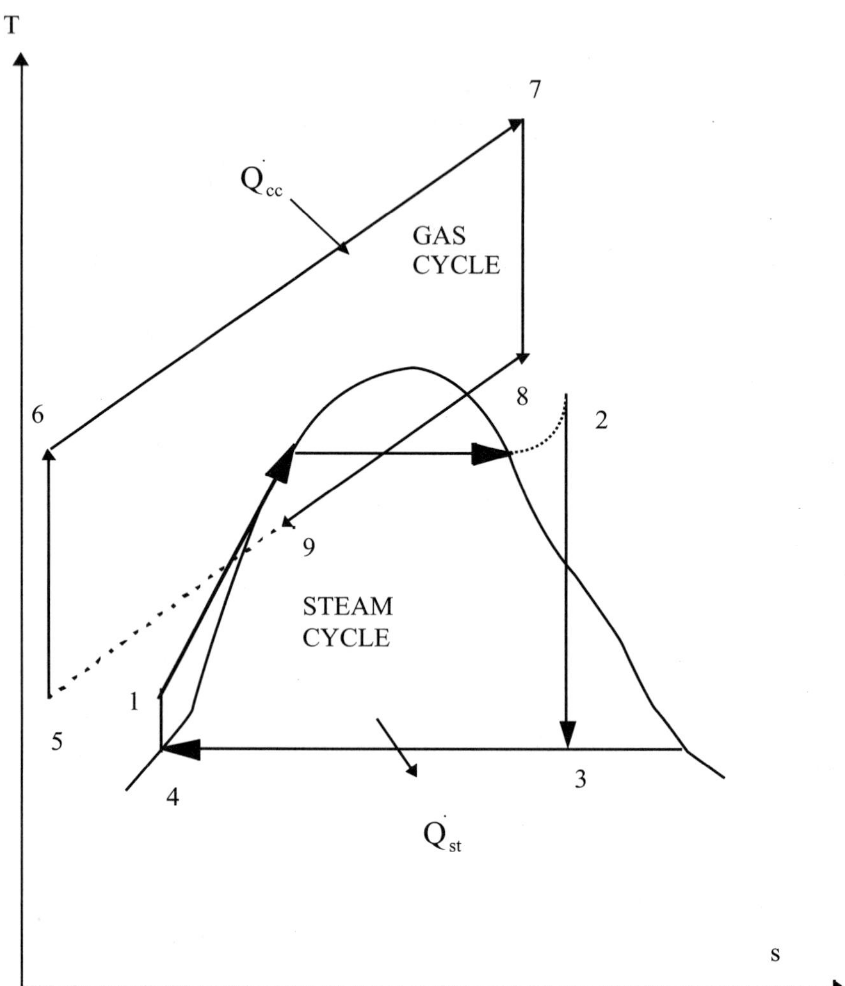

FIGURE 6.11(b)
Combined gas-steam power cycle.

Since

$$p_{r6} = 10.509 \text{ and } s_6 = s_5 ,$$
$$h_6 = 570.58 \text{ kJ/kg}, \; s_6 = 6.852 \text{ kJ/(kg.K)}, \; T_6 = 564.9 \text{ K}.$$

Since

$$p_7 = 1000 \text{ kPa and } T_7 = 1126.85 \text{ °C} = 1400 \text{ K},$$
$$h_7 = 1515.27 \text{ kJ/kg}, \; s_7 = 7.868 \text{ kJ/(kg.K)}, \; p_{r7} = 361.62.$$

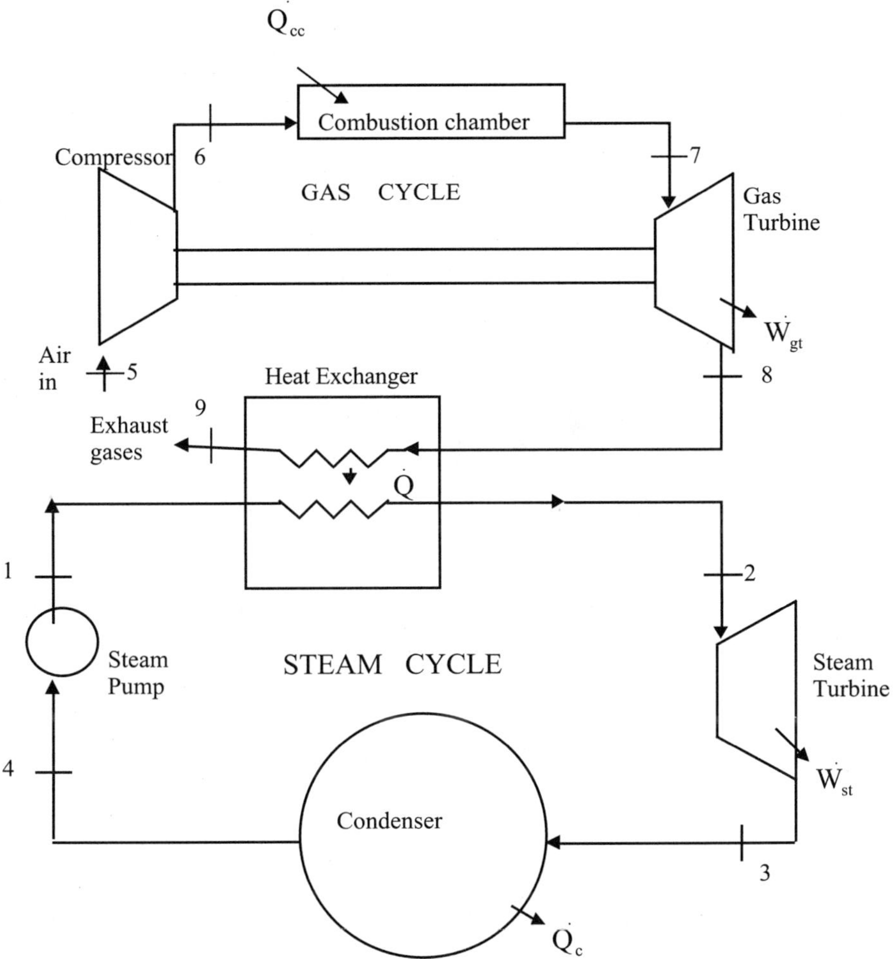

Schematic for Example 6.10.

The gas is expanded in the turbine to a pressure

$$p_8 = 100 \text{ kPa, so } p_{r8} = 36.162$$
$$h_8 = 808.6 \text{ kJ/kg, } s_8 = 7.868 \text{ kJ/(kg.K), } T_8 = 787.6 \text{ K.}$$

Since

$$p_9 = 100 \text{ kPa and } T_9 = 149.85°\text{C} = 423 \text{ K,}$$
$$h_9 = 424.64 \text{ kJ/kg, } s_9 = 7.216 \text{ kJ/(kg.K).}$$

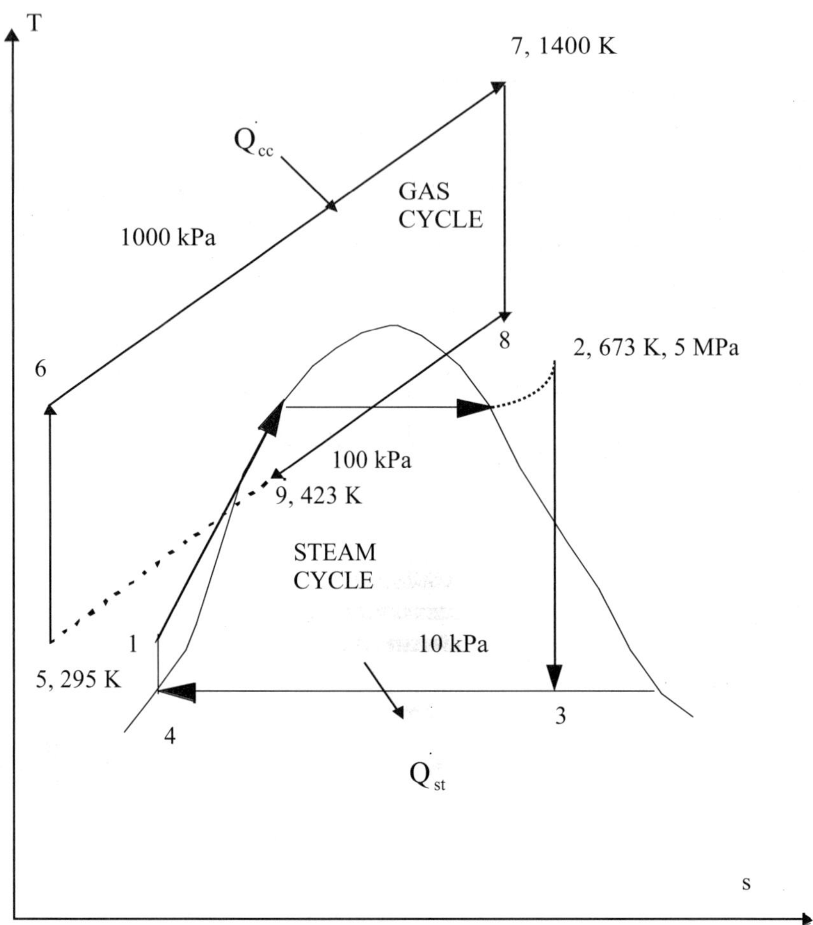

T-s Diagram for Example 6.10.

Now, find the state points for the steam cycle, starting with state 2 since we know T_2 and P_2.

$$h_2 = 3196 \text{ kJ/kg}, \; s_2 = 6.646 \text{ kJ/(kg.K)}.$$

Since

$$s_3 = s_2 \text{ and } p_3 = 10 \text{ kPa},$$
$$h_3 = 2105 \text{ kJ/kg}, \; s_3 = 6.646 \text{ kJ/(kg.K)}.$$

Since

$$p_4 = 10 \text{ kPa and state 4 is assumed to be saturated liquid,}$$
$$h_4 = 191.81 \text{ kJ/kg, } s_4 = 0.6492 \text{ kJ/(kg.K).}$$

Since

$$s_1 = s_4 \text{ and } p_1 = 5000 \text{ kPa,}$$
$$h_1 = 196.84 \text{ kJ/kg, } s_1 = 0.6492 \text{ kJ/(kg.K).}$$

The change in enthalpy and in entropy of the air across the heat exchanger are

$$h_8 - h_9 = 383.96 \text{ kJ/kg air}$$
$$s_8 - s_9 = 0.652 \text{ kJ/(kg.K).}$$

The change of steam enthalpy across the heat exchanger is

$$h_2 - h_1 = 2999.16 \text{ kJ/kg air.}$$

The first law applied to the heat exchanger gives us

$$\dot{m}_{air}(h_8 - h_9) = \dot{m}_s(h_2 - h_1)$$
$$\dot{m}_s / \dot{m}_{air} = 383.96 / 2999.16 = 0.128 \text{ kg steam/kg air.}$$

The net total power produced is 37 MW. This net total power is the sum of the net power from the gas cycle and the net power from the steam cycle. The net power from the gas cycle is the power produced by the gas turbine minus the power required to drive the compressor. Thus, the net power from the gas cycle is

$$\dot{W}_{net\ air} = \dot{m}_{air}\left[(h_7 - h_8) - (h_6 - h_5)\right]$$
$$= \dot{m}_{air}(431.54).$$

The net power produced by the steam cycle is the power produced by the steam turbine minus the power used to drive the pump. Thus, the net power from the steam cycle is

$$\dot{W}_{net\ s} = \dot{m}_s\left[(h_2 - h_3) - (h_1 - h_4)\right]$$
$$= \dot{m}_s(1085.97)$$
$$= \dot{m}_{air}(139).$$

The total net power is thus

$$\begin{aligned}\dot{W}_{net} &= \dot{W}_{net\ air} + \dot{W}_{net\ s} \\ &= \dot{m}_{air}(570.54) = 37\ \text{MW} \\ \dot{m}_{air} &= 37{,}000 / 570.54 = 64.85\ \text{kg/s} = 3891\ \text{kg/min}.\end{aligned}$$

The steam flow rate is

$$\dot{m}_s = (0.128)\ 3891\ \text{kg/min} = 498\ \text{kg/min}.$$

The overall thermal efficiency is

$$\begin{aligned}\eta_{th} &= \dot{W}_{net} / \dot{Q}_{in} = \dot{W}_{net} / \dot{Q}_{cc} \\ &= \dot{W}_{net} / \dot{m}_{air}(h_7 - h_6) \\ &= 570.54 / 944.69 \\ &= 0.604 = 60.4\%.\end{aligned}$$

The availability of the gas turbine exhaust gas is

$$\begin{aligned}a_{f8\ air} - a_{f9\ air} &= (h_8 - h_9) - T_o(s_8 - s_9) \\ &= 383.96\ \text{kJ/kg} - 295(0.652)\ \text{kJ/kg} \\ &= 191.6\ \text{kJ/kg}.\end{aligned}$$

The work produced by the steam cycle per kilogram mass of air is

$$\begin{aligned}&(0.128)(1085.97)\ \text{kJ/kg air} \\ &= 139.0\ \text{kJ/kg air}.\end{aligned}$$

Therefore, the percentage of availability converted to work by the Rankine cycle is

$$\begin{aligned}(\Delta a_f)_{used} &= 139.0 / 191.6 \\ &= 0.725 = 72.5\%.\end{aligned}$$

Problems

Rankine Cycle

6.1 The low pressure in an ideal Rankine cycle is 0.007 MPa. Steam enters the turbine in a superheated condition at 700°C (973.15 K) and 10 MPa, at a rate of 73.56 kg/s. Determine:

(a) The net power output of the cycle, in kilowatts,

(b) The rate of heat transfer in the condenser, $\dot{Q}_c$, in megawatts,

(c) The temperature of the cooling water leaving the condenser if it enters at 15°C (288.15 K), at a rate of 2134 kg/s,

(d) The thermal efficiency of the cycle.

6.2 The condenser pressure in an ideal Rankine cycle is 0.7 $lb_f/in.^2$ Steam leaves the boiler in a superheated condition at 1200°F (1659.67°R) and 1300 $lb_f/in.^2$ The net power output of the cycle is 100×10^3 Btu/s. Determine:

(a) Mass flow rate of the steam, in pounds per second,

(b) The rate of heat transfer in the condenser, $\dot{Q}_c$, in British thermal units per second,

(c) The mass flow rate of the condenser cooling water in kilograms per second, if cooling water enters the condenser at 60°F (519.67°R) and leaves at 86°F (545.67°R),

(d) The thermal efficiency of the cycle.

6.3 An ideal Rankine cycle has a fixed boiler exit state. The condenser pressure is decreased. State whether each of the following quantities increase, remain the same, or decrease: turbine work output, pump work input, heat supplied, heat rejected, thermal efficiency, and moisture content at turbine outlet.

6.4 An ideal Rankine cycle has fixed boiler and condenser pressures. The steam is superheated to a higher temperature. State whether each of the following quantities increase, remain the same, or decrease: turbine work output, pump work input, heat supplied, heat rejected, thermal efficiency, and moisture content at turbine outlet.

6.5 An ideal Rankine cycle has fixed boiler outlet temperature and condenser pressure. The boiler pressure is increased. State whether each of the following quantities increase, remain the same, or decrease: turbine work output, pump work input, heat supplied, heat rejected, thermal efficiency, and moisture content at turbine outlet.

6.6 The inlet state of steam to an isentropic turbine is 2 MPa and 400°C (673.15 K), and the outlet pressure is 5 kPa. The power output of the turbine is 1 MW. Determine the mass flow rate of steam through the turbine. Neglect changes in potential energy and kinetic energy.

6.7 A Rankine power cycle employing steam is being operated between 2 MPa and 0.01 MPa. The highest temperature in the boiler is 505°C (778.15 K). The isentropic efficiency of the turbine is 85%, and that of the pump is 95%. Calculate the thermal efficiency of the cycle.

6.8 In the Rankine cycle using steam, the state points indicated in the figure are as follows:

State 2 is at 25 MPa, 500°C (773.15 K)
State 1 is at 25 MPa, $s_1 = 0.4764$ kJ/(kg.K)
State 4 is at 0.005 MPa, $s_4 = 0.4764$ kJ/(kg.K)

Calculate the thermal efficiency of the cycle.

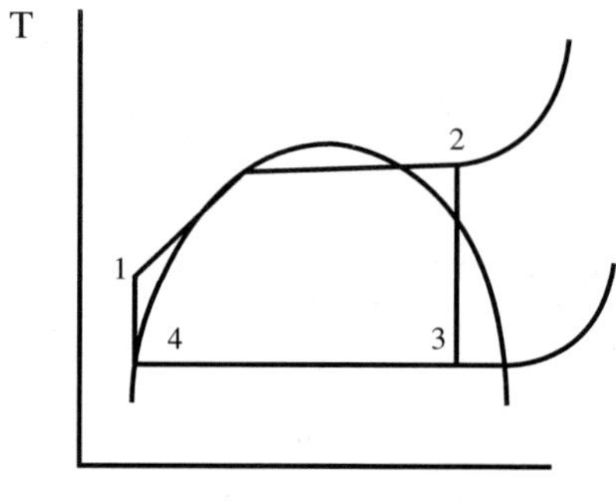

Schematic for Problem 6.8.

The Reheat Rankine Cycle

6.9 A steam power plant operates on an ideal Rankine cycle with reheat. The steam leaves the boiler at 18 MPa and 600°C (873.15 K) and the lowest pressure in the cycle is 0.006 MPa. If the moisture content of the steam at the outlet of the low-pressure (LP) turbine is not to exceed 20%, find (a) the pressure at which the steam has to be reheated, and (b) the thermal efficiency of the cycle. The steam is reheated to 600°C (873.15 K).

6.10 Consider a steam power plant operating on an ideal Rankine cycle with reheat. The steam leaves the steam generator at 2700 $lb_f/in.^2$ and 1150°F (1609.67°R) and the condenser pressure is 1 $lb_f/in.^2$ If the dryness fraction of the steam at the outlet of the low-pressure (LP) turbine should not drop below 0.8, determine (a) the pressure at which the steam has to be reheated, and (b) the cycle thermal efficiency. The steam is reheated to 1150°F (1609.67°R).

6.11 In Problem 6.9, compute the thermal efficiency of the corresponding ideal Rankine cycle without reheat. Determine (a) the moisture content at the turbine outlet, and (b) the thermal efficiency of the cycle. Comment on the effects of reheating.

6.12 In Problem 6.10, compute the thermal efficiency of the corresponding ideal Rankine cycle without reheat. Calculate (a) the quality of the steam at the turbine outlet, and (b) the thermal efficiency of the cycle. Comment on the effects of reheating.

6.13 Sketch an ideal Rankine cycle with four stages of reheat on a T-s diagram. The turbine inlet temperature is the same after each reheat stage. Discuss what happens to the average temperature of heat addition to the working fluid, as the number of reheat stages increase. Deduce the effect on the thermal efficiency of the cycle.

The Regenerative Rankine Cycle

6.14 Consider an ideal cycle where steam enters the turbine at 16 MPa, 610°C (883.15 K) and the lowest pressure in the cycle is 80 kPa. Calculate the thermal efficiency for the cases described below.

(a) No feedwater heater.

(b) One feedwater heater, working at 1 MPa.

(c) Two feedwater heaters, one at 5 MPa and the other at 500 kPa.

6.15 An ideal cycle is such that steam enters the turbine at 2700 $lb_f/in.^2$, 1200°F (1659.67°R) and the pressure in the condenser is 1 $lb_f/in.^2$ Find the thermal efficiency for the cases described below.

(a) No feedwater heater.

(b) One feedwater heater, working at 110 $lb_f/in.^2$

(c) Two feedwater heaters, one at 600 $lb_f/in.^2$ and the other at 60 $lb_f/in.^2$

6.16 An ideal regenerative steam power cycle operates between the pressure limits of 10 MPa and 0.12 MPa. The steam leaves the boiler at 525°C (798.15 K). An open feedwater heater that operates at 5 MPa is used. Calculate the cycle thermal efficiency.

6.17 An ideal Rankine cycle has a fixed mass flow rate of working fluid through the boiler; the cycle is modified with regeneration. State whether each of the following quantities increase, remain the same, or decrease: turbine work output, heat supplied, heat rejected, and moisture content at turbine outlet.

6.18 Design an ideal Rankine cycle with regeneration that approaches the thermal efficiency of the corresponding Carnot cycle. Sketch such a cycle on a T-s diagram.

Air Preheater and Economizer

6.19 The exhaust gases in a boiler before venting are at 900°C (1173.15 K) and at a pressure of about 1 atm. Compute the flow availability of the gases in kilojoules per kilogram. Assume the gases to have the properties of air. The environment is at 25°C (298.15 K) and 1 bar. If the gases are used in an air preheater, they are vented at 450°C (723.15 K) instead of 900°C (1173.15 K). Compute the percentage of availability used in this case.

In Chapter 5, Section 5.5.1, a second law ratio to measure thermal environmental impact was introduced, where

$$r_{II} = \frac{\left(1 - \frac{T_o}{T_j}\right)\dot{Q}_1}{\left(1 - \frac{T_o}{T_s}\right)\dot{Q}_s}.$$

Determine the percentage change of this ratio owing to the use of the preheater.

6.20 In Problem 6.19, the exhaust gases are used in an economizer instead of the air preheater. The water entering the economizer is at 10 MPa and 39.29°C (312.44 K). It leaves at 10 MPa and 45°C (318.15 K). Calculate the ratio $\dot{m}_{water} / \dot{m}_{gases}$ if the gases are vented at 450°C (723.15 K), after the economizer.

Availability Analysis of Vapor Power Cycles

6.21 Compute the irreversibility of each of the four major components of the Rankine cycle described in Problem 6.1. Heat is transferred to the working fluid in the furnace of the boiler at 1300°C (1573.15 K) and heat is transferred to the cooling water in the condenser at 15°C (288.15 K) and 100 kPa. Calculate the availability of the steam before and after the turbine.

6.22 Calculate the irreversibility of each of the four major components of the Rankine cycle described in Problem 6.2. Heat transfer to the working fluid in the boiler furnace takes place at 2300°F (2759.67°R) and heat transfer to the cooling water in the condenser occurs at 60°F (519.67°R) and 1 atm. Determine the availability of the steam entering and leaving the turbine.

6.23 A turbine expands steam from 600°C (873.15 K) and 13 MPa to 10 kPa, at an isentropic efficiency of 85%. Calculate the entropy production and the irreversibility. The environmental conditions are at 25°C (298.15 K) and 1 atm.

6.24 Steam expands in a turbine from 1110°F (1569.67°R) and 1900 $lb_f/in.^2$ to 1.5 $lb_f/in.^2$, with an isentropic efficiency of 85%. Compute the entropy production and the irreversibility. The environment is at 77°F (536.67°R) and 1 atm.

6.25 Steam converts from 500°C (773.15 K) and 10 MPa to 5 kPa in a turbine, and the isentropic efficiency is 75%. Compute the entropy production and the irreversibility. The environmental conditions are at 25°C (298.15 K) and 1 atm.

6.26 A turbine expands steam from 930°F (1389.67°R) and 1470 $lb_f/in.^2$ to 0.735 $lb_f/in.^2$, at an isentropic efficiency of 75%. Calculate the entropy production and the irreversibility. The environment is at 77°F (536.67°R) and 1 atm.

6.27 Steam at 500°C (773.15 K), 4.5 MPa is expanded in a turbine to the saturated steam state at 0.01 MPa. Calculate the work delivered by the turbine, and the availability change within the turbine.

Cogeneration

6.28 Explain the difference between cogeneration and regeneration.

6.29 The conditions of steam entering a turbine are 2 MPa and 400°C (673.15 K). The steam leaves the turbine to a process steam line at 0.15 MPa. The turbine produces 850 kW at an isentropic efficiency of 68%. Determine the mass flow rate of the steam.

6.30 The highest and lowest pressures in a cogeneration plant, shown in the schematic below, are 9 MPa and 6 kPa, respectively. The superheated steam leaves the boiler at 575°C (848.15 K). The steam is bled from the turbine at 0.8 MPa for process heating. The water enters the boiler at a rate of 12 kg/s. In periods of a large demand for process heat, some steam is expanded through the pressure-reducing valve (PRV) to 0.8 MPa and directed to the process heater. The bled fractions are manipulated so that the working fluid leaves the process heater as a saturated liquid at 0.8 MPa. Find:

(a) The maximum power produced

(b) The maximum rate at which process heat is supplied

(c) The rate of process heat supply when 15% of the steam is directed through the PRV, and 60% of the steam is bled from the turbine at 0.8 MPa for process heating

6.31 The condenser pressure of a cogeneration plant, as shown in the schematic below, is 0.8 $lb_f/in.^2$ The superheated steam leaves the boiler at

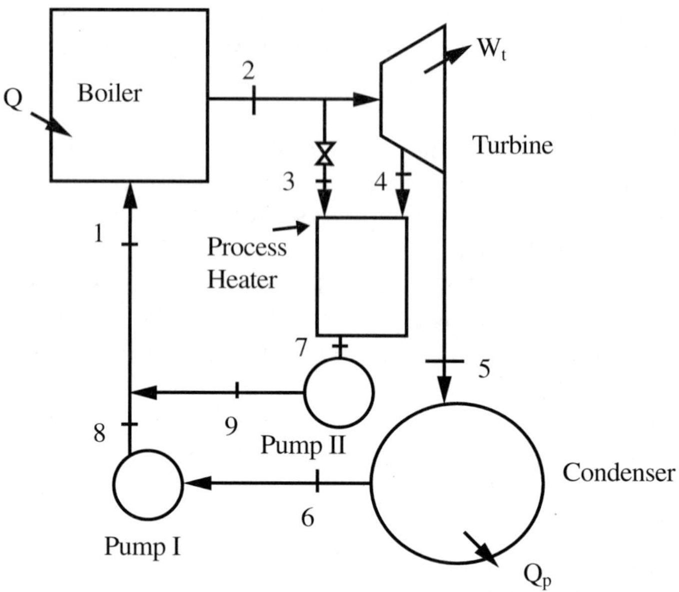

Schematic for Problem 6.30 and Problem 6.31.

1200 $lb_f/in.^2$ and 1100°F (1559.67°R). The steam is bled from the turbine at 120 $lb_f/in.^2$ for process heating. When there is a large need for process heat, some steam is expanded through the pressure-reducing valve (PRV) to 120 $lb_f/in.^2$ and directed to the process heater. The bleeding is adjusted so that the working fluid leaves the process heater as a saturated liquid at 120 $lb_f/in.^2$ The mass flow rate of the working fluid through the boiler is 25 lb_m/s. Find:

(a) The maximum power produced

(b) The maximum rate at which process heat is supplied

(c) The rate of process heat supply when 15% of the steam is directed through the PRV, and 60% of the steam is bled from the turbine at 120 $lb_f/in.^2$ for process heating

6.32 A steam turbine, rated at 800 kW power output, is to operate from a geothermal supply of hot water. The geothermal hot water is at 5 MPa, 200°C (473.15 K); it is throttled into an adiabatic flash chamber, forming liquid and vapor at a pressure of 900 kPa. The liquid is used for process heating, and the vapor is used for the steam turbine. If the exit pressure of the turbine is 40 kPa, and the isentropic efficiency is 90%, calculate the hot water requirement from the geothermal source. Calculate the corresponding rate of process heat supplied, if the water leaves the process heater as saturated liquid at 100 kPa.

Binary Vapor and Combined Gas-Vapor Power Cycles

6.33 A binary cycle uses cesium for the topping cycle and water for the other. What is the heat source for the water?

6.34 Mercury and steam are the two fluids used in a binary cycle comprising two ideal Rankine cycles. In the mercury cycle, saturated mercury vapor leaves the boiler at 900 K, and saturated liquid leaves the condenser at 575 K. Relevant property data for saturated mercury are given in the table below.

T(K)	P(MPa)	v_f(m^3/kg)	h_f(kJ/kg)	h_g(kJ/kg)	s_f(kJ/(kg.K)	s_g(kJ/(kg.K)
575	0.03424	0.0000777	22.97	319.98	0.0478	0.5643
900	2.999	0.0000830	68.92	350.53	0.1102	0.4231

The heat rejected from the mercury cycle is transferred to the water cycle, which produces saturated water vapor at 559 K. The steam is then superheated at that pressure to 773 K using another source of heat. Saturated liquid water leaves the condenser at 5 kPa. Calculate:

(a) The mass in kilograms of mercury condensed per mass in kilograms of water converted into vapor.

(b) The rate of heat addition to the overall binary cycle, in kilojoules of water.

(c) The thermal efficiency.

6.35 The two fluids used in a binary cycle consisting of two ideal Rankine cycles, are mercury and water. Saturated mercury vapor leaves the boiler at 2080°R, and saturated liquid leaves the condenser at 1495°R. The property data for saturated mercury are listed in the table below.

T(°R)	P(psia)	v_f(ft^3/lb$_m$)	h_f(Btu/lb)	h_g(Btu/lb)	s_f(Btu/lb.°R)	s_g(Btu/lb.°R)
1495	4.9834	0.001245	9.875	137.57	0.0114	0.1348
2080	436.49	0.001330	29.63	150.70	0.0263	0.1011

The heat rejected from the mercury cycle is transferred to the water cycle, which produces saturated water vapor at 1006.87°R. The steam is then superheated at that pressure to 1392°R using another source of heat. Saturated liquid water leaves the condenser at 0.7277 $lb_f/in.^2$ Find:

(a) The mass in pounds of mercury condensed per mass in pounds of water converted into vapor.

(b) The rate of heat addition to the overall binary cycle, in British thermal units per pound of water.

(c) The thermal efficiency.

6.36 Give the reasons why mercury is a suitable working fluid for the topping cycle in a binary vapor cycle, and not for the bottoming cycle.

6.37 The net power produced by a combined gas turbine-vapor cycle power plant is 40 MW. Steam leaves the heat exchanger at 6000 kPa and 427°C (700.15 K). Incoming air into the compressor of the gas turbine system is at 25°C (298.15 K) and 100 kPa and it leaves the compressor at 1100 kPa. The gas leaves the cycle at 160°C (433.15 K). Heat is added at constant pressure, after which the air leaves the combustor at 1177°C (1450.15 K). Both the gas turbine cycle and the Rankine cycle are ideal. Compute the flow rates in each cycle, the overall thermal efficiency, and the availability utilization of the gas turbine exhaust. The condenser pressure of the Rankine cycle is 7 kPa. The environment is at 25°C (298.15 K). Assume that air is the working fluid in the gas turbine cycle, thus neglecting the mass of the fuel.

6.38 A combined gas turbine-vapor cycle power plant produces 42×10^3 Btu/s net power. The exhaust gases from the gas turbine leaves the heat exchanger at 300°F (759.67°R). Steam leaves the heat exchanger at 900 $lb_f/in.^2$ and 800°F (1259.67°R). Air enters the compressor of the gas turbine system at 77°F (536.67°R) and 1 atm and exits the compressor at 147 $lb_f/in.^2$ Air leaves the combustor at 2150°F (2609.67°R), after heat is added at constant pressure. Calculate the flow rates in each cycle, the overall thermal efficiency, and the availability utilization of the gas turbine exhaust. The surrounding temperature is 77°F (536.67°R). The condenser pressure of the Rankine cycle is 1 $lb_f/in.^2$ Assume that air is the working fluid in the gas turbine cycle, so that the mass of the fuel may be neglected. In addition, assume that both the Rankine cycle and the gas turbine cycle are ideal.

6.39 Explain the difference between the binary vapor power cycle and the combined gas-vapor power cycle.

6.40 Explain the reasons for the increased efficiency of the combined gas-vapor cycle, as compared to that of either cycle operated separately.

6.41 A steam power plant operates in a Rankine cycle between 4 MPa and 12 kPa, with the highest temperature in the boiler being at 723.15 K. The high temperature source is 200 kg/s of air at 903.15 K from the exhaust of a gas turbine. The boiler is a parallel flow heat exchanger where the maximum flow rate of water is 10 kg/s. Determine the smallest temperature difference in the boiler between the water and the air.

Computer, Design, and General Problems

6.42 Fuel cells employing solar energy are being used to provide power. Research the subject, and write an essay on their potential for more widespread use.

6.43 Research literature on fuel cells used to replace power supplied by batteries (e.g., lead-acid cells).

6.44 Large quantities of cooling water leave the condensers of huge power plants (especially nuclear power plants) at over 10°C above the environmental temperature. If they are discharged directly into the lake, river, or ocean, there is concern about thermal pollution. Discuss ways and means to utilize this wasted energy.

6.45 Research articles involving conflict or strife brought about by thermal pollution caused by power plants in the past.

6.46 The incineration of solid waste is an energy source for steam generation. Investigate their use and their potential throughout the world. Discuss the additional considerations, if any, that are required for such steam generators as compared to those using fossil fuels.

6.47 The incineration of medical wastes provides the heat energy source for cogeneration plants that provide district heating / cooling, and generate electricity. This incineration may be carried out at each hospital, or all the medical wastes from many hospitals may be trucked to a single commercial facility to be incinerated at one place. Taking capital costs, trucking costs, distribution costs, environmental concerns, and other factors into consideration, compare the pros and cons of each hospital having its own incinerator to one centralized commercial incinerator used by many hospitals.

6.48 Incineration of solid waste as an energy source may produce large quantities of harmful products, depending on the composition of the waste. For instance, the incineration of polyvinyl chloride (PVC) plastic produces hydrochloric acid vapor which is very harmful. Design an equipment to separate out substances like PVC, heavy metals, etc., before the waste is incinerated.

6.49 A fabric filter is recommended for use to control emissions when solid waste is incinerated. Design such a fabric filter, based on the volume of exhaust gases handled, the temperature and pressure of these gases, and the amount and nature of particulate matter expected.

6.50 Design a power plant to be used in a space station. Explain the reasons for the system design. In addition, elaborate on the suitability of the working fluid chosen.

7

Principles of Energy (Heat) Transfer

CONTENTS

Transfer of energy by heat takes place only when there is a temperature difference. The temperature acts as a measure of potential for heat energy to move from a higher potential to a lower potential.

Experimental methods are used to measure energy transfer by heat. Two basic transfer mechanisms are commonly recognized: conduction and radiation. A third classification, convection, is often used.

7.1 Conduction

Conduction is the heat transfer mode that takes place in stationary media. It can occur in solids, gases, and liquids. It may be perceived as the transfer of energy from the more energetic particles of a substance to adjacent particles that are less energetic owing to particle interactions. The conduction heat transfer phenomenon is described macroscopically by **Fourier's law**, which is

$$Q = -kA\nabla T \tag{7.1}$$

where k is a property of the medium (substance) called the **thermal conductivity**, and A is the area through which the heat is flowing. Fourier's law states that the driving influence for conduction is the temperature gradient. The negative sign indicates that the energy flow is in the direction of decreasing temperature. Substances with large values of thermal conductivity like gold and copper are good thermal conductors, and those with small conductivities like wood and cement are good thermal insulators.

In one dimension, Fourier's law reduces to

$$Q|_x = -kA\frac{dT}{dx}\bigg|_x. \tag{7.2}$$

The application of this law to a plane wall is shown in Figure 7.1. The heat flow $Q|_x$ is the heat energy transfer in the x-direction. The area is the cross-sectional area of the control volume normal to the heat flow, in this case, the area normal to the x-direction. In the special case when the temperature gradient is constant, the equation may be written as

$$Q|_x = -kA\frac{\Delta T}{\Delta x}\bigg|_x. \tag{7.3}$$

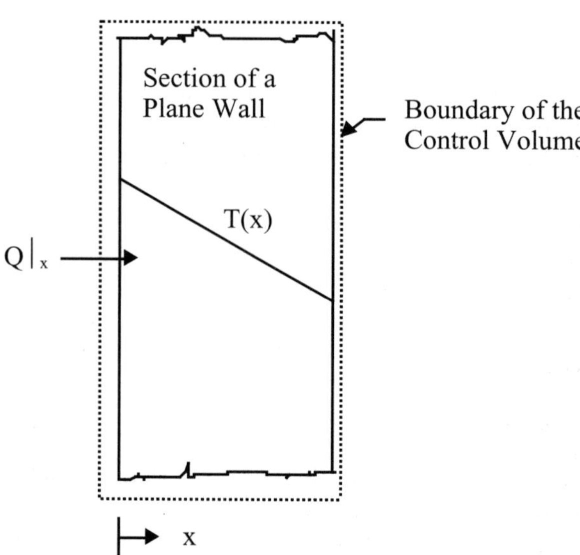

FIGURE 7.1
Fourier's law applied to a plane wall.

In Figure 7.2 a control volume drawn around a plane wall with three layers is shown. The three layers are made of three different materials, A, B, and C, and are of different thicknesses, Δx_A, Δx_B, and Δx_C. The thermal conductivities of the three substances are k_A, k_B, and k_C, respectively. The heat conducted through each of the three layers has to be equal, by the conservation of energy. Hence, application of Fourier's law yields

$$Q|_x = -k_A A\frac{\Delta T_A}{\Delta x_A} = -k_B A\frac{\Delta T_B}{\Delta x_B} = -k_C A\frac{\Delta T_C}{\Delta x_C}. \tag{7.4}$$

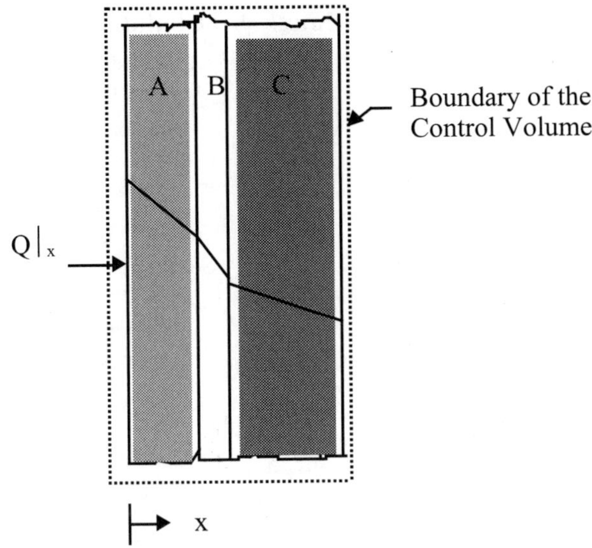

FIGURE 7.2
Fourier's law applied to a plane wall with three layers.

For the particular illustration given in Figure 7.2, the temperature gradient in B is observed to be larger than that in A, which is in turn larger than that in C. It can then be deduced from Equation (7.4) that $k_B < k_A < k_C$. In other words, material C is the best conductor and material B is the worst conductor of the three.

Example 7.1

Problem: A silver plate is 5 cm thick. One surface is at 500°C and the other is at 0°C. The thermal conductivity for silver is 369 W/(m.°C) at 150°C. Determine the heat conducted through the plate.

Solution

From Fourier's law,

$$q|_x = \frac{Q|_x}{A} = -k\frac{\Delta T}{\Delta x}$$
$$= \frac{-(369)(0-500)}{5\times 10^{-2}} \frac{W}{m.°C}.°C.\frac{1}{m}$$
$$= 3.69\ MW/m^2.$$

Example 7.2

Problem: The wall has the same thickness as the building insulation. The thermal conductivities of the wall and the building insulation are in the ratio 3:1. If the temperature drop across the wall is 6°C, compute the temperature drop across the insulation.

Solution

Fourier's law yields

$$Q|_x = -k_A A \frac{\Delta T_A}{\Delta x_A} = -k_B A \frac{\Delta T_B}{\Delta x_B}.$$

Hence,

$$-k_{wall} A \frac{\Delta T_{wall}}{\Delta x_{wall}} = -k_{insulation} A \frac{\Delta T_{insulation}}{\Delta x_{insulation}}$$

$$\Delta T_{insulation} = \frac{k_{wall}}{k_{insulation}} \Delta T_{wall} = \frac{3}{1}(6) = 18°C.$$

7.2 Radiation

Thermal radiation takes place without the need of a medium. Thermal radiation may be thought of as being emitted by matter resulting from changes in the electronic configurations of its atoms or molecules. Solid surfaces, gases, and liquids all emit, absorb, and transmit thermal radiation to different degrees. The radiation heat transfer phenomenon is described macroscopically by a modified form of the **Stefan-Boltzmann law**, which is

$$Q = \varepsilon \sigma A T_b^4 \tag{7.5}$$

where σ is the Stefan-Boltzmann constant and ε is a property of the surface that describes how effectively the surface radiates ($0 \le \varepsilon \le 1$). This property is called the **emissivity** of the surface. The Stefan-Boltzmann constant σ is 5.669 $\times 10^{-8}$ W/(m^2.K^4) or 0.1714 $\times 10^{-8}$ Btu/(ft^2.°R^4.h). It can be seen that thermal radiation takes place according to the fourth power of the absolute temperature of the surface, T_b. In general, however, the net thermal radiation heat transfer between two surfaces involves complex relationships among the properties of the surfaces, their orientations with respect to each other, and the extent to which the medium in between scatters, emits, and absorbs thermal radiation, and other factors.

For a simple two-body radiation problem with a nonparticipating intervening medium, the radiation equation gives

$$Q = \varepsilon_1 \sigma A_1 \left(T_1^4 - T_2^4\right) \tag{7.6}$$

where Q is the net radiation heat transfer from body 1 with the higher temperature T_1 to body 2 of the lower temperature T_2. The parameters ε_1 and A_1 are the emissivity of body 1 and the effective area of radiation for body 1. Similarly, the net radiation can be expressed as

$$Q = \varepsilon_2 \sigma A_2 \left(T_2^4 - T_1^4\right). \tag{7.7}$$

Here, ε_2 and A_2 are the emissivity of body 2 and the effective area of radiation for body 2. From the conservation of energy principle, the magnitude of the net heat radiated in Equation (7.6) has to equal the magnitude of the heat radiated in Equation (7.7).

When there are more than two bodies, a further refinement to the radiation equation may be made according to this equation:

$$Q_{1-2} = \varepsilon_1 \sigma F_{1-2} A_1 \left(T_1^4 - T_2^4\right) \tag{7.8}$$

where F_{1-2} is called the view-factor of body 2 from body 1. This **view-factor** F_{1-2} is the fraction of the thermal radiation from body 1 that falls on body 2. If body 1 also sees body 3, then F_{1-3} is the fraction of the thermal radiation from body 1 that falls on body 3. Then the net radiation from body 1 to body 3 will be given by

$$Q_{1-3} = \varepsilon_1 \sigma F_{1-3} A_1 \left(T_1^4 - T_3^4\right). \tag{7.9}$$

It follows that the sum of the view-factors from body 1 is equal to unity, i.e., $F_{1-2} + F_{1-3} = 1$.

Writing the net radiation from body 2 to the other bodies, we get

$$Q_{2-1} = \varepsilon_2 \sigma F_{2-1} A_2 \left(T_2^4 - T_1^4\right) \tag{7.10}$$

and

$$Q_{2-3} = \varepsilon_2 \sigma F_{2-3} A_2 \left(T_2^4 - T_3^4\right). \tag{7.11}$$

Doing the same for body 3, we get

$$Q_{3-1} = \varepsilon_3 \sigma F_{3-1} A_3 \left(T_3^4 - T_1^4\right) \tag{7.12}$$

and

$$Q_{3-2} = \varepsilon_3 \sigma F_{3-2} A_3 \left(T_3^4 - T_2^4\right). \tag{7.13}$$

It follows that the sum of the view-factors from body 2 is equal to unity, i.e., $F_{2-1} + F_{2-3} = 1$, and also that the sum of the view-factors from body 3 is equal to unity, i.e., $F_{3-1} + F_{3-2} = 1$. In the simple three-body radiation problem described, it has been assumed that none of the bodies can radiate to itself. In other words, the view factor of body 1 to itself is zero, and this is the case for bodies 2 and 3 also, that is, $F_{1-1} = F_{2-2} = F_{3-3} = 0$.

Example 7.3

Problem: Two very large parallel plates at 700°C and 500°C exchange heat via radiation. Determine the heat transfer per unit area. Assume that $\varepsilon = 1$ for the plates.

Solution

Assumptions: (1) The medium in between does not participate in the heat transfer.

Analysis: From the Stefan-Boltzmann law,

$$\begin{aligned}
\frac{q}{A} &= \sigma\varepsilon\left(T_1^4 - T_2^4\right) \\
&= \left(5.669 \times 10^{-8}\right)\left(973.15^4 - 773.15^4\right)\frac{W}{m^2K^4}.K^4 \\
&= 30.59\,\frac{kW}{m^2}.
\end{aligned}$$

Example 7.4

Problem: The view-factor of body 1 to 2 is 0.5 and that from body 1 to itself is 0.2. If the temperatures of bodies 1, 2, and 3 are 400, 300, and 200°C, respectively, calculate the heat radiated from body 1 to body 3. Assume all the emissivities are 0.8. The surface area of body 1 is 1 m^2.

Solution

Assumptions: (1) The medium inbetween does not participate in the heat transfer.

Analysis: From the Stefan-Boltzmann law,

$$\begin{aligned}
&Q_{1-3} = \varepsilon_1 \sigma F_{1-3} A_1 \left(T_1^4 - T_3^4\right) \\
&F_{1-1} + F_{1-2} + F_{1-3} = 1.
\end{aligned}$$

Thus,

$$F_{1\text{-}3} = 1 - 0.2 - 0.5 = 0.3, \text{ and}$$

$$Q_{1\text{-}3} = 0.8\left(5.669 \times 10^{-8}\right)(0.3)(1.0)\left(673.15^4 - 473.15^4\right)\frac{W}{m^2K^4}.m^2.K^4 = 2112 \text{ W}.$$

7.3 Convection

When heat is transferred by a moving medium, it is commonly referred to as convection. For instance, heat energy can be transferred from a solid plane surface at one temperature to an adjacent moving gas or liquid at another temperature. Consider the situation shown in Figure 7.3. Heat energy is conducted from the plane to the adjacent moving fluid, where the energy is carried away by the combined effects of conduction within the fluid and the bulk motion of the fluid. The heat transfer from the solid system to the fluid can be described by the empirical equation:

$$Q = h_{conv} A\left(T_b - T_f\right) \tag{7.14}$$

known as **Newton's law of cooling.** In this equation, A is the surface area, T_b is the temperature of the surface, and T_f is the fluid temperature away from the surface (bulk or mean temperature of the fluid). For $T_b > T_f$, heat energy flows from the solid to the fluid. The proportionality factor h_{conv} is called the **heat transfer coefficient.** This coefficient is not a thermodynamic property. It is a parameter that may be found experimentally, and incorporates into the heat transfer relationship the nature of the fluid flow pattern near the surface, the geometry of the system, and the fluid properties. When pumps or fans make the fluid flow, the value of the heat transfer coefficient is normally greater than when relatively slow buoyancy-driven motion takes place. These two categories are called **forced** and **free** (or **natural**) **convection**, respectively. The further study of convection requires an understanding of fluid mechanics.

In Figure (7.3), it can be seen that the heat transfer between the solid plane and the fluid may be given by Fourier's conduction law as

$$Q\Big|_y = -k_f A \frac{dT}{dy}\bigg|_{wall} \cong -k_f A \frac{\Delta T}{\Delta y}\bigg|_{wall} = \frac{k_f}{\Delta y} A\left(T_b - T_f\right). \tag{7.15}$$

Comparing Equations (7.14) and (7.15), one can see that the heat transfer coefficient h_{conv} is an approximation of the quantity $k_f/\Delta y$. The thermal conductivity of the fluid k_f is a thermodynamic property, but Δy is a function of the fluid flow pattern near the surface, the fluid properties, and the geometry

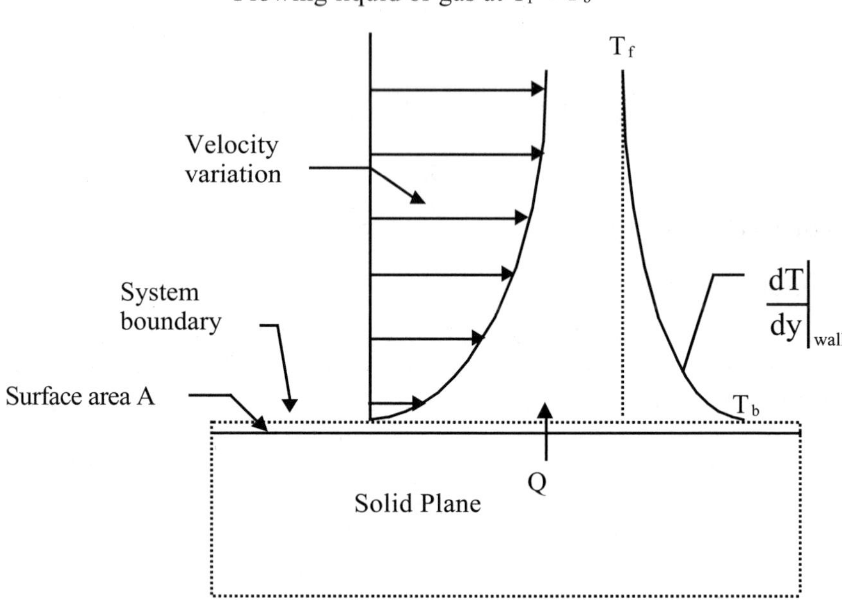

FIGURE 7.3
Newton's law of cooling.

of the system. Hence, as stated before, the heat transfer coefficient is not a thermodynamic property, but an empirical parameter. It can also be seen that Newton's law of cooling is a special case of the conduction law. This explains why only two basic heat transfer mechanisms are recognized generally. However, convection is complicated because of the complexity of fluid motion, and is often treated separately.

Example 7.5

Problem: Air at 25°C blows over a hot plate at 200°C. The convection heat transfer coefficient is 20 W/(m^2.°C). The dimensions of the plate is 20 by 30 cm. Determine the heat transfer.

Solution

From Newton's law of cooling,

$$\begin{aligned} Q &= h_{conv} A(T_b - T_f) \\ &= 20(0.06)(200-20)\frac{W}{m^2.°C}.m^2.°C \\ &= 216\ W. \end{aligned}$$

7.4 Combined Convection and Radiation

The heat transferred via convection may be added to the heat transferred via radiation. So the total heat transferred from a body 1 to the surrounding fluid f is the sum of the convective heat transferred and the radiative heat transferred between say body 1 and body 2. The convective heat transferred from body 1 to the fluid f is

$$Q_{conv} = h_{conv} A_{conv} (T_1 - T_f). \tag{7.16}$$

The radiative heat transferred from body 1 to body 2 is

$$Q_{rad} = \varepsilon\sigma F_{1-2} A_{rad} (T_1^4 - T_2^4). \tag{7.17}$$

The total heat transferred from body 1 from convection and radiation is thus

$$Q_{total} = h_{conv} A_{conv} (T_1 - T_f) + \varepsilon\sigma F_{1-2} A_{rad} (T_1^4 - T_2^4). \tag{7.18}$$

Note that the area available for convective heat transfer is not necessarily the same as that available for radiative heat transfer between bodies 1 and 2. In addition, the temperature of the surrounding fluid T_f is in general not the same as the fluid of the body 2, T_2.

Under the special condition that T_1 is close to T_2, Equation (7.16) may be simplified as follows:

$$\begin{aligned} Q_{rad} &= \varepsilon\sigma F_{1-2} A_{rad} (T_1^2 + T_2^2)(T_1^2 - T_2^2) \\ &= \varepsilon\sigma F_{1-2} A_{rad} (T_1^2 + T_2^2)(T_1 - T_2)(T_1 + T_2) \\ &\approx \varepsilon\sigma F_{1-2} A_{rad} T_1^3 (T_1 - T_2) \\ &= h_{rad} A_{rad} (T_1 - T_2). \end{aligned} \tag{7.19}$$

In other words, the radiative heat exchange has been approximated and written to look like Newton's law of cooling, with a heat transfer coefficient due to radiation. Like the convection heat transfer coefficient, the radiative heat transfer coefficient is not a property of either of the bodies. The total heat transferred from body 1 may then be written as

$$Q_{total} = h_{conv} A_{conv} (T_1 - T_f) + h_{rad} A_{rad} (T_1 - T_2). \tag{7.20}$$

Example 7.6

Problem: For a certain body, the convection heat transfer coefficient to the surrounding air is 26 W/(m^2.°C), and the radiative heat transfer coefficient from this body to another body is approximately 28 W/(m².°C). If the temperature of body under study is 200°C and that of the surrounding air is 30°C, determine the temperature of the second body so that the heat transferred by convection is equal in magnitude to the heat transferred via radiation. Assume that the area ratio A_{conv}:A_{rad} is 1:1.5.

Solution

Let T_2 be the temperature of the second body.
The requirement is for

$$Q_{conv} = Q_{rad}$$

$$h_{conv} A_{conv} (T_1 - T_f) = h_{rad} A_{rad} (T_1 - T_2)$$

$$\frac{(T_1 - T_2)}{(T_1 - T_f)} = \frac{h_{conv} A_{conv}}{h_{rad} A_{rad}}$$

$$T_2 = T_1 - (T_1 - T_f)\frac{h_{conv} A_{conv}}{h_{rad} A_{rad}}$$

$$T_2 = 200°C - 170°C\frac{26(1)}{28(15)} = 94.8°C.$$

The temperature of the second body is 94.8°C.

Problems

Conduction

7.1 A copper plate is 5 cm thick. One surface is at 400°C (673.15 K) and the other is at 200°C (473.15 K). The thermal conductivity for copper is 369 W/(m.°C) at 300°C (573.15 K). Determine the heat conducted through the plate in megawatts per square meter.

7.2 A silver plate is 10 cm thick. The higher temperature surface is at 400°C (673.15 K). The thermal conductivity for silver can be taken to be 362 W/(m.°C). The heat conducted through the plate is 1.35 MW/m². Calculate the temperature of the other surface.

7.3 A copper plate is 2 in. thick. The cooler surface is at 392°F (851.67°R). The thermal conductivity for copper can be taken to be 213.2 Btu/(ft.°F.h). The heat conducted through the plate is 460×10^3 Btu/(ft².h). Calculate the hotter surface temperature.

7.4 A silver plate is 4 in. thick. One surface is at 752°F (1211.67°R) and the other is at 32°F (491.67°R). The heat conducted through the plate is

452×10^3 Btu/(ft^2.h). Compute the thermal conductivity of silver for this range of temperature, if it is assumed to have a constant value.

7.5 The wall is half as thick as the building insulation. The thermal conductivities of the wall and the building insulation are in the ratio 4:1. If the temperature drop across the wall is 2°C, compute the temperature drop across the insulation.

7.6 The wall has the same thickness as the building insulation. The thermal conductivities of the wall and the building insulation are in the ratio 3.5:1. If the temperature drop across the insulation is 28°F, compute the temperature drop across the wall.

Radiation

7.7 Two very large parallel plates at 600°C (873.15 K) and 450°C (723.15 K) exchange heat via radiation. Determine the heat transfer per unit area. Assume that $\varepsilon = 1$ for the plates.

7.8 Two very large parallel plates exchange heat via radiation at the rate of 103 kW/m^2. The hotter plate is at 900°C (1173.15 K). Calculate the temperature of the cooler plate. Assume that $\varepsilon = 1$ for the plates.

7.9 Two very large parallel plates at 1650°F (2109.67°R) and 480°F (939.67°R) exchange heat via radiation. The cooler plate is at 850°F (1309.67°R). Assume that $\varepsilon = 1$ for the plates. Determine the area of each plate if the total heat transfer is to be 1×10^8 Btu.

7.10 Two very large parallel plates at 1650°F (2109.67°R) and 480°F (939.67°R) exchange heat via radiation. Assume that $\varepsilon = 1$ for the plates. Determine the area of each plate if the total heat transfer is to be 1×10^8 Btu.

7.11 Body 1 views bodies 2 and 3, besides itself. The view-factor of body 1 to 3 is 0.3. The temperature of body 1 is 450°C (723.15 K), and the heat radiated from body 1 to body 3 is 2110 W. Calculate the temperature of body 3. All the emissivities are 0.78. The surface area of body 1 is 0.8 m^2.

7.12 The view-factor of body 1 to 2 is 0.35. When the temperatures of bodies 1, 2, and 3 are 600, 300, and 150°C (873.15, 573.15, and 423.15 K), respectively, the heat radiated from body 1 to body 3 is 4667 W. It is given that all the emissivities are 0.6. The surface area of body 1 is 0.5 m^2. Determine the view-factor of body 1 to itself.

7.13 The view-factor of body 1 to 2 is 0.45 and that from body 1 to itself is 0.25. The temperatures of bodies 1, 2, and 3 are 840, 660, and 480°F (1299.67, 1119.67, and 939.67°R), respectively; the heat radiated from body 1 to body 3 is 582.7 Btu/h. The surface area of body 1 is 0.7 ft^2. If all the emissivities are the same, calculate this emissivity.

7.14 The view-factor of body 1 to 2 is 0.35. If the temperatures of bodies 1, 2, and 3 are 1110, 570, and 300°F (1569.67, 1029.67, and 759.67°R), respectively, calculate the heat radiated from body 1 to body 2. Assume all the emissivities are 0.65. The surface area of body 1 is 4.5 ft^2.

Convection

7.15 Air at 22°C (295.15 K) blows over a hot plate at 300°C (573.15 K). The convection heat transfer coefficient is 28 W/(m^2.°C). The dimensions of the plate are 40 by 80 cm. Compute the heat transfer.

7.16 Air at 30°C (303.15 K) moves over a plate at 150°C (423.15 K). The plate is 10 cm by 100 cm. If the heat transfer is 216 W, calculate the convection heat transfer coefficient between the air and the plate.

7.17 Carbon dioxide at 18°C (291.15 K) flows over a hot plate at 320°C (593.15 K). The velocity of the gas is such that the convection heat transfer coefficient is 25 W/(m^2.°C). If the heat transfer is 2548 W, determine the area of the plate.

7.18 A gas (carbon dioxide) at 35°C (308.15 K) flows over a hot plate at 150°C (423.15 K) such that the convection heat transfer coefficient is 19 W/(m^2.°C). The plate is 100 by 100 cm. If the heat gained by the gas is 2185 W, calculate the temperature of the plate.

7.19 Air moves over a hot plate at 572°F (1031.67°R) such that the convection heat transfer coefficient is 4.93 Btu/(ft^2.°F.h). The dimensions of the plate are 16 by 32 in. If the heat transfer is 8766 Btu/h, determine the temperature of the air.

7.20 Air at 86°F (545.67°R) blows over a plate at 302°F (761.67°R). The dimensions of the plate are 4 by 40 in. If the heat transfer is 761 Btu/h, calculate the convection heat transfer coefficient.

7.21 Carbon dioxide at 64°F (523.67°R) blows over a hot plate at 608°F (1067.67°R). The velocity of the gas is such that the convection heat transfer coefficient is 4.40 Btu/(ft^2.°F.h). The dimensions of the plate are 18 by 30 in. Compute the heat transfer in British thermal units per hour.

7.22 Carbon dioxide at 95°F (554.67°R) blows over a plate at 302°F (761.67°R). The flow of the carbon dioxide is such that the convection heat transfer coefficient is 3.346 Btu/(ft^2.°F.h). If the heat gained by the carbon dioxide is 7696 Btu/h, compute the length of one side of the square plate.

Combined Convection and Radiation

7.23 For an oven, the convection heat transfer coefficient to the surrounding air is 29 W/(m^2.°C), and the radiative heat transfer coefficient from this oven to another oven is 33 W/(m^2.°C). If the temperature of the oven under consideration is 180°C (453.15 K), and that of the second oven is 62.8°C (335.95 K) determine the surrounding air temperature when the heat transferred by convection is equal in magnitude to the heat transferred via radiation. Assume that the area ratio A_{conv}:A_{rad} is 1:1.2.

7.24 For a certain body, the convection heat transfer coefficient to the surrounding air is 20 W/(m^2.°C), and the radiative heat transfer coefficient from this body to another body is approximately 20 W/(m^2.°C). The temperature of the body under study is 180°C (453.15 K) and that of the surrounding air is 20°C (293.15 K). Determine the area ratio A_{conv}:A_{rad} such that the magnitude of the heat transferred by convection is 110% of that transferred via radiation.

7.25 The temperature of an oven under study is 360°F (819.67°R), that of the surrounding environment is 68°F (527.67°R), and that of a second oven is 170°F (629.67°R). If the magnitude of the heat transferred by convection is 90% of that transferred via radiation, calculate the ratio of the convection heat transfer coefficient to the radiative heat transfer coefficient. Assume that the area ratio A_{conv}:A_{rad} is 1:1.2.

7.26 The convection heat transfer coefficient for body A to the surrounding air is 3.0 Btu/(ft^2.°F.h), and the radiative heat transfer coefficient from body A to body B is 3.2 Btu/(ft^2.°F.h). If the temperature of body A is 360°F (819.67°R), that of the surrounding air is 68°F (527.67°R), determine the temperature of the body B so that the magnitude of the heat transfer via convection is equal to 95% of that via radiation for body A. The area ratio A_{conv}:A_{rad} is 1:1.05.

Computer, Design, and General Problems

7.27 Discuss the reasons for metals to be better heat conductors than nonmetals.

7.28 An uneducated person stated that heat cannot be transferred in a vacuum. Support or defend this statement.

7.29 Only heat radiation with a nonparticipating medium in between has been discussed. Discuss the phenomena when the medium in between participates in the heat transfer.

7.30 Find some references that provide the variation of thermal conductivity of some metals with temperature. Write a computer program that provides the thermal conductivity of some selected metals, when the temperature is specified.

Appendices

Table A-1 SI Critical Constants
Table A-2 SI Properties of Selected Ideal Gases at 25°C, 1 MPa
Table A-3 SI Ideal-Gas Properties of Air, Standard Entropy at 0.1 MPa
Table A-4 SI Ideal-Gas Properties of Nitrogen, Entropies at 0.1 MPa
Table A-5 SI Ideal-Gas Properties of Oxygen, Entropies at 0.1 MPa
Table A-6 SI Beattie-Bridgeman Constants
Figure A-1 Nelson-Obert Generalized Compressibility Chart - low pressures
Figure A-2 Nelson-Obert Generalized Compressibility Chart - intermediate pressures
Figure A-3 Nelson-Obert Generalized Compressibility Chart - high pressures
B-Series Tables SI Thermodynamic Properties of Water
- Table B-1 SI Saturated Water
- Table B-2 SI Saturated Water Pressure First Independent Variable
- Table B-3 SI Superheated Water Vapor
- Table B-4 SI Compressed Liquid Water

C-Series Tables SI Thermodynamic Properties of Ammonia
- Table C-1 SI Saturated Ammonia
- Table C-2 SI Superheated Ammonia
- Table C-3 SI Compressed Liquid Ammonia

D-Series Tables SI Thermodynamic Properties of Carbon Dioxide
- Table D-1 SI Saturated Carbon Dioxide
- Table D-2 SI Superheated Carbon Dioxide
- Table D-3 SI Compressed Liquid Carbon Dioxide

E-Series Tables SI Thermodynamic Properties of R-12
- Table E-1 SI Saturated R-12
- Table E-2 SI Superheated R-12
- Table E-3 SI Compressed Liquid R-12

F-Series Tables SI Thermodynamic Properties of R-134a
- Table F-1 SI Saturated R-134a
- Table F-2 SI Superheated R-134a
- Table F-3 SI Compressed Liquid R-134a

Table AA-1 U.S. Critical Constants
Table AA-2 U.S. Properties of Selected Ideal Gases at 77°F, 14.504 psia
Table AA-3 U.S. Ideal-Gas Properties of Air, Standard Entropy at 14.504 psia
Table AA-4 U.S. Ideal-Gas Properties of Nitrogen, Entropies at 14.504 psia
Table AA-5 U.S. Ideal-Gas Properties of Oxygen, Entropies at 14.504 psia
Table AA-6 U.S. Beattie-Bridgeman Constants

A-Series Tables

TABLE A-1 SI

Critical Constants

Substance	Formula	Molec. Mass	Vol. $m^3/kmol$	Temp. K	Pressure MPa
Ammonia	NH_3	17.031	0.0725	405.5	11.35
Argon	Ar	39.948	0.0749	150.8	4.87
Bromine	Br_2	159.808	0.1272	588	10.3
Carbon dioxide	CO_2	44.01	0.0939	304.1	7.38
Carbon monoxide	CO	28.01	0.0932	132.9	3.5
Chlorine	Cl_2	70.906	0.1238	416.9	7.98
Fluorine	F_2	37.997	0.0663	144.3	5.22
Helium	He	4.003	0.0574	5.19	0.227
Hydrogen	H_2	2.016	0.0651	33.2	1.3
Krypton	Kr	83.8	0.0912	209.4	5.5
Neon	Ne	20.183	0.0416	44.4	2.76
Nitric oxide	NO	30.006	0.0577	180	6.48
Nitrogen	N_2	28.013	0.0898	126.2	3.39
Nitrogen dioxide	NO_2	46.006	0.1678	431	10.1
Nitrous oxide	N_2O	44.013	0.0974	309.6	7.24
Oxygen	O_2	31.999	0.0734	154.6	5.04
Sulfur dioxide	SO_2	64.063	0.1222	430.8	7.88
Water	H_2O	18.015	0.0571	647.3	22.12
Xenon	Xe	131.3	0.1184	289.7	5.84
Benzene	C_6H_6	78.114	0.259	562.2	4.89
n-Butane	C_4H_{10}	58.124	0.255	425.2	3.8
Chlorodifluoromethane (r-22)	$CHClF_2$	86.469	0.1656	369.3	4.97
Dichlorodifluoromethane (r-12)	CCl_2F_2	120.914	0.2167	385	4.14
Ethane	C_2H_6	30.07	0.1483	305.4	4.88
Ethyl alcohol	C_2H_5OH	46.069	0.1671	513.9	6.14
Ethylene	C_2H_4	28.054	0.1304	282.4	5.04
n-Heptane	C_7H_{16}	100.205	0.432	540.3	2.74
n-Hexane	C_6H_{14}	86.178	0.37	507.5	3.01
Methane	CH_4	16.043	0.0992	190.4	4.6
Methyl alcohol	CH_3OH	32.042	0.118	512.6	8.09
n-Octane	C_8H_{18}	114.232	0.492	568.8	2.49
n-Pentane	C_5H_{12}	72.151	0.304	469.7	3.37
Propane	C_3H_8	44.094	0.203	369.8	4.25
Tetrafluoroethane (r-134a)	CF_3CH_2F	102.03	0.2008	374.2	4.06

Adapted from Sonntag, R. E., Borgnakke, C., and Van Wylen, G. J., *Fundamentals of Thermodynamics*, John Wiley & Sons, New York, 1998. With permission.

TABLE A-2 SI

Properties of Selected Ideal Gases at 25°C, 0.1 MPa (or Saturation Pressure if It Is < 0.1 MPa)

Gas	Formula	Molecular Mass	R kJ/(kg.K)	v m^3/kg	C_{vo} kJ/(kg.K)	C_{po} kJ/(kg.K)	k
Air	—	28.97	0.287	0.855432	0.717	1.004	1.400
Ammonia	NH_3	17.031	0.4882	1.440922	1.642	2.130	1.297
Argon	Ar	39.948	0.2081	0.619963	0.312	0.520	1.667
Butane	C_4H_{10}	58.124	0.143	0.415455	1.573	1.716	1.091
Carbon monoxide	CO	28.01	0.2968	0.884956	0.744	1.041	1.399
Carbon dioxide	CO_2	44.01	0.1889	0.563380	0.653	0.842	1.289
Ethane	C_2H_6	30.07	0.2765	0.818331	1.490	1.766	1.186
Ethanol	C_2H_5OH	46.069	0.1805	0.531067	1.246	1.427	1.145
Ethylene	C_2H_4	28.054	0.2964	0.878735	1.252	1.548	1.237
Helium	He	4.003	2.0071	6.191950	3.116	5.193	1.667
Hydrogen	H_2	2.016	4.1243	12.300123	1.008	14.209	1.409
Methane	CH_4	16.043	0.5183	1.543210	1.736	2.254	1.299
Methanol	CH_3OH	32.042	0.2595	0.763359	1.146	1.405	1.227
Neon	Ne	20.183	0.412	1.228501	0.618	1.030	1.667
Nitric oxide	NO	30.006	0.2771	0.826446	0.716	0.993	1.387
Nitrogen	N_2	28.013	0.2968	0.884956	0.745	1.042	1.400
Nitrous oxide	N_2O	44.013	0.1889	0.563380	0.690	0.879	1.274
n-Octane	C_8H_{18}	114.23	0.07279	10.869565	1.638	1.711	1.044
Oxygen	O_2	31.999	0.2598	0.773994	0.662	0.922	1.393
Propane	C_3H_8	44.094	0.1886	0.553097	1.490	1.679	1.126
R-12	CCl_2F_2	120.914	0.06876	0.200803	0.547	0.616	1.126
R-134a	CF_3CH_2F	102.03	0.08149	0.238095	0.771	0.852	1.106
Steam	H_2O	18.015	0.4615	43.290043	1.410	1.872	1.327

Adapted from Sonntag, R. E., Borgnakke, C., and Van Wylen, G. J., *Fundamentals of Thermodynamics,* John Wiley & Sons, New York, 1998. With permission.

TABLE A-3 SI

Ideal-Gas Properties of Air, Standard Entropy at 0.1 MPa

T K	v m^3/kg	h kJ/kg	s^o kJ/(kg.K)	P_r	v_r
200	0.573973	200.174	6.46260	0.27027	493.466
220	0.631370	220.218	6.55812	0.37700	389.150
240	0.688768	240.267	6.64535	0.51088	313.274
260	0.746165	260.323	6.72562	0.67573	256.584
280	0.803562	280.390	6.79998	0.87556	213.257
290	0.832261	290.430	6.83521	0.98990	195.361
298.15	0.855650	298.615	6.86305	1.09071	182.288
300	0.860960	300.473	6.86926	1.11458	179.491
320	0.918357	320.576	6.93413	1.39722	152.728
340	0.975754	340.704	6.99515	1.72814	131.200
360	1.033151	360.863	7.05276	2.11226	113.654
380	1.090549	381.060	7.10735	2.55479	99.1882
400	1.147946	401.299	7.15926	3.06119	87.1367
420	1.205343	421.589	7.20875	3.63727	77.0025
440	1.262741	441.934	7.25607	4.28916	68.4088
460	1.320138	462.340	7.30142	5.02333	61.0658
480	1.377535	482.814	7.34499	5.84663	54.7479
500	1.434933	503.360	7.38692	6.76629	49.2777
520	1.492330	523.982	7.42736	7.78997	44.5143
540	1.549727	544.686	7.46642	8.92569	40.3444
560	1.607124	565.474	7.50422	10.18197	36.6765
580	1.664522	585.350	7.54084	11.56771	33.4358
600	1.721919	607.316	7.57638	13.09232	30.5609
620	1.779316	628.375	7.61090	14.76564	28.0008
640	1.836714	649.528	7.64448	16.59801	25.7132
660	1.894111	670.776	7.67717	18.60025	23.6623
680	1.951508	692.120	7.70903	20.78367	21.8182
700	2.008906	713.561	7.74010	23.16010	20.1553
720	2.066303	735.098	7.77044	25.74188	18.6519
740	2.123700	756.731	7.80008	28.54188	17.2894
760	2.181097	778.460	7.82905	31.57347	16.0518
780	2.238495	800.284	7.85740	34.85061	14.9250
800	2.295892	822.202	7.88514	38.38777	13.8972
850	2.439385	877.397	7.95207	48.46828	11.6948
900	2.582879	933.152	8.01581	60.51977	9.91692
950	2.726372	989.436	8.07667	74.81519	8.46770
1000	2.869865	1046.221	8.13493	91.65077	7.27604
1050	3.013358	1103.478	8.19081	111.3467	6.28845
1100	3.156852	1161.18	8.24449	134.2478	5.46408
1150	3.300345	1219.298	8.29616	160.7245	4.77141
1200	3.443838	1277.805	8.34596	191.1736	4.18586
1250	3.587331	1336.677	8.39402	226.0192	3.68804
1300	3.730825	1395.892	8.44046	265.7145	3.26257
1350	3.874318	1455.429	8.48539	310.7426	2.89711
1400	4.017811	1515.270	8.52891	361.6192	2.58171
1450	4.161304	1575.398	8.57111	418.8942	2.30831
1500	4.304798	1635.800	8.61208	483.1554	2.07031
1550	4.448291	1696.446	8.65185	554.9577	1.86253
1600	4.591784	1757.329	8.69051	634.9670	1.68035

TABLE A-3 SI (CONTINUED)

Ideal-Gas Properties of Air, Standard Entropy at 0.1 MPa

T K	v m^3/kg	h kJ/kg	s^o kJ/(kg.K)	P_r	v_r
1650	4.735277	1818.436	8.72811	723.8560	1.52007
1700	4.878771	1879.755	8.76472	822.3320	1.37858
1750	5.022264	1941.275	8.80039	931.1376	1.25330
1800	5.165757	2002.987	8.83516	1051.051	1.14204
1850	5.309250	2064.882	8.86908	1182.888	1.04294
1900	5.452744	2126.951	8.90219	1327.498	0.95445
1950	5.596237	2189.186	8.93452	1485.772	0.87521
2000	5.739730	2251.581	8.96611	1658.635	0.80410
2050	5.883223	2314.128	8.99699	1847.077	0.74012
2100	6.026717	2376.823	9.02721	2052.109	0.68242
2150	6.170210	2439.659	9.05678	2274.789	0.63027
2200	6.313703	2502.630	9.08573	2516.217	0.58305
2250	6.457196	2565.733	9.11409	2777.537	0.54020
2300	6.600690	2628.962	9.14189	3059.939	0.50124
2350	6.744183	2692.313	9.16913	3364.658	0.46576
2400	6.887676	2755.782	9.19586	3692.974	0.43338
2450	7.031169	2819.366	9.22208	4046.215	0.40378
2500	7.174663	2883.059	9.24781	4425.759	0.37669
2550	7.318156	2946.859	9.27308	4833.031	0.35185
2600	7.461649	3010.763	9.29790	5269.505	0.32903
2650	7.605142	3074.767	9.32228	5736.707	0.30805
2700	7.748636	3138.868	9.34625	6236.215	0.28872
2750	7.892129	3203.064	9.36980	6769.657	0.27089
2800	8.035622	3267.351	9.39297	7338.715	0.25443
2850	8.179115	3331.726	9.41576	7945.124	0.23921
2900	8.322609	3396.188	9.43818	8590.676	0.22511
2950	8.466102	3460.733	9.46025	9277.216	0.21205
3000	8.609595	3525.359	9.48198	10006.645	0.19992

Adapted from Sonntag, R. E., Borgnakke, C., and Van Wylen, G. J., *Fundamentals of Thermodynamics*, John Wiley & Sons, New York, 1998. With permission.

TABLE A-4 SI

Ideal-Gas Properties of Nitrogen, Entropies at 0.1 MPa

		Diatomic Nitrogen (N_2) M = 28.013		Monatomic Nitrogen (N) M = 14.007	
T K	v m³/kmol	$(\bar{h} \pm \bar{h}^o_{298})$ kJ/kmol	$\bar{s}^o$ kJ/(kmol.K)	$(\bar{h} \pm \bar{h}^o_{298})$ kJ/kmol	$\bar{s}^o$ kJ/(kmol.K)
0	0.0000	–8670.0	0.000	–6197.0	0.000
100	8.3140	–5768.0	159.812	–4119.0	130.593
200	16.6280	–2857.0	179.985	–2040.0	145.001
298	24.7757	0.0	191.609	0.0	153.300
300	24.9420	54.0	191.789	38.0	153.429
400	33.2560	2971.0	200.181	2117.0	159.409
500	41.5700	5911.0	206.740	4196.0	164.047
600	49.8840	8894.0	212.177	6274.0	167.837
700	58.1980	11937.0	216.856	8353.0	171.041
800	66.5120	15046.0	221.016	10431.0	173.816
900	74.8260	18223.0	224.757	12510.0	176.265
1000	83.1400	21463.0	228.171	14589.0	178.455
1100	91.4540	24760.0	231.314	16667.0	180.436
1200	99.7680	28109.0	234.227	18746.0	182.244
1300	108.0820	31503.0	236.943	20825.0	183.908
1400	116.3960	34936.0	239.487	22903.0	185.448
1500	124.7100	38405.0	241.881	24982.0	186.883
1600	133.0240	41904.0	244.139	27060.0	188.224
1700	141.3380	45430.0	246.276	29139.0	189.484
1800	149.6520	48979.0	248.304	31218.0	190.672
1900	157.9660	52549.0	250.234	33296.0	191.796
2000	166.2800	56137.0	252.075	35375.0	192.863
2200	182.9080	63362.0	255.518	39534.0	194.845
2400	199.5360	70640.0	258.684	43695.0	196.655
2600	216.1640	77963.0	261.615	47860.0	198.322
2800	232.7920	85323.0	264.342	52033.0	199.868
3000	249.4200	92715.0	266.892	56218.0	201.311
3200	266.0480	100134.0	269.286	60420.0	202.667
3400	282.6760	107577.0	271.542	64646.0	203.948
3600	299.3040	115042.0	273.675	68902.0	205.164
3800	315.9320	122526.0	275.698	73194.0	206.325
4000	332.5600	130027.0	277.622	77532.0	207.437
4400	365.8160	145078.0	281.209	86367.0	209.542
4800	399.0720	160188.0	284.495	95457.0	211.519
5200	432.3280	175352.0	287.530	104813.0	213.397
5600	465.5840	190572.0	290.349	114550.0	215.195
6000	498.8400	205848.0	292.984	124590.0	216.926

Adapted from Sonntag, R. E., Borgnakke, C., and Van Wylen, G. J., *Fundamentals of Thermodynamics*, John Wiley & Sons, New York, 1998. With permission.

TABLE A–5 SI

Ideal-Gas Properties of Oxygen, Entropies at 0.1 MPa

		Diatomic Oxygen (O_2) M = 31.999		Monatomic Oxygen (O) M = 16.00	
T K	v m³/kmol	$(\bar{h} \pm \overline{h^o_{298}})$ kJ/kmol	$\bar{s}^o$ kJ/(kmol.K)	$(\bar{h} \pm \overline{h^o_{298}})$ kJ/kmol	$\bar{s}^o$ kJ/(kmol.K)
0	0	–8683	0	–6725	0
100	8.314	–5777	173.308	–4518	135.947
200	16.628	–2868	193.483	–2186	152.153
298	24.775	0	205.148	0	161.059
300	24.942	54	205.329	41	161.194
400	33.256	3027	213.873	2207	167.431
500	41.570	6086	220.693	4343	172.198
600	49.884	9245	226.450	6462	176.060
700	58.198	12499	231.465	8570	179.310
800	66.512	15833	235.920	10671	182.116
900	74.826	19241	239.931	12767	184.585
1000	83.140	22703	243.579	14860	186.790
1100	91.454	26212	246.923	16950	188.783
1200	99.768	29761	250.011	19039	190.600
1300	108.082	33345	252.878	21126	192.270
1400	116.396	36958	255.556	23212	193.816
1500	124.710	40600	258.068	25296	195.254
1600	133.024	44267	260.434	27381	196.599
1700	141.338	47959	262.673	29464	197.862
1800	149.652	51674	264.797	31547	199.053
1900	157.966	55414	266.819	33630	200.179
2000	166.280	59716	268.748	35713	201.247
2200	182.908	66770	272.366	39878	203.232
2400	199.536	74453	275.708	44054	205.045
2600	216.164	82225	278.818	48216	206.714
2800	232.792	90080	281.729	52391	208.262
3000	249.420	98013	284.466	56574	209.705
3200	266.048	106022	287.050	60767	211.058
3400	282.676	114101	289.499	64971	212.332
3600	299.304	122245	291.826	69190	213.538
3800	315.932	130447	294.043	73424	214.682
4000	332.560	138705	296.161	77675	215.773
4400	365.816	155374	300.133	86234	217.812
4800	399.072	172240	303.801	94873	219.691
5200	432.328	189312	307.217	103592	221.435
5600	465.584	206618	310.423	112391	223.066
6000	498.840	224210	313.457	121264	224.597

Adapted from Sonntag, R. E., Borgnakke, C., and Van Wylen, G. J., *Fundamentals of Thermodynamics*, John Wiley & Sons, New York, 1998. With permission.

TABLE A-6 SI

Beattie-Bridgeman Constants

Beattie-Bridgeman Equation of State: $$\mathbf{P} = \frac{\mathbf{R}_u T}{\bar{v}^2}\left(1 - \frac{c}{\bar{v}T^3}\right)(\bar{v} + B) - \frac{A}{\bar{v}^2}$$

where $$A = A_0\left(1 - \frac{a}{\bar{v}}\right) \text{ and } B = B_0\left(1 - \frac{b}{\bar{v}}\right)$$

The constants in the Beattie–Bridgeman equation are as listed below for various gases, when T is in K, P is in kPa, $\bar{v}$ is in $m^3/kmol$, and $R_u = 8.314$ $kPa.m^3/(kmol.K)$.

Gas	A_0	a	B_0	b	c
Air	131.8441	0.01931	0.04611	−0.001101	43,400
Carbon dioxide, CO_2	507.2836	0.07132	0.10476	0.07235	660,000
Hydrogen, H_2	20.0117	−0.00506	0.02096	−0.04359	504
Nitrogen, N_2	136.2315	0.02617	0.05046	−0.00691	42,000
Oxygen, O_2	151.0857	0.02562	0.04624	0.004208	48,000

Adapted from Van Wylen, G. J. and Sonntag, R. E., *Fundamentals of Classical Thermodynamics*, E/SI Vr, 3E, John Wiley & Sons, New York, 1986, p. 46. With permission.

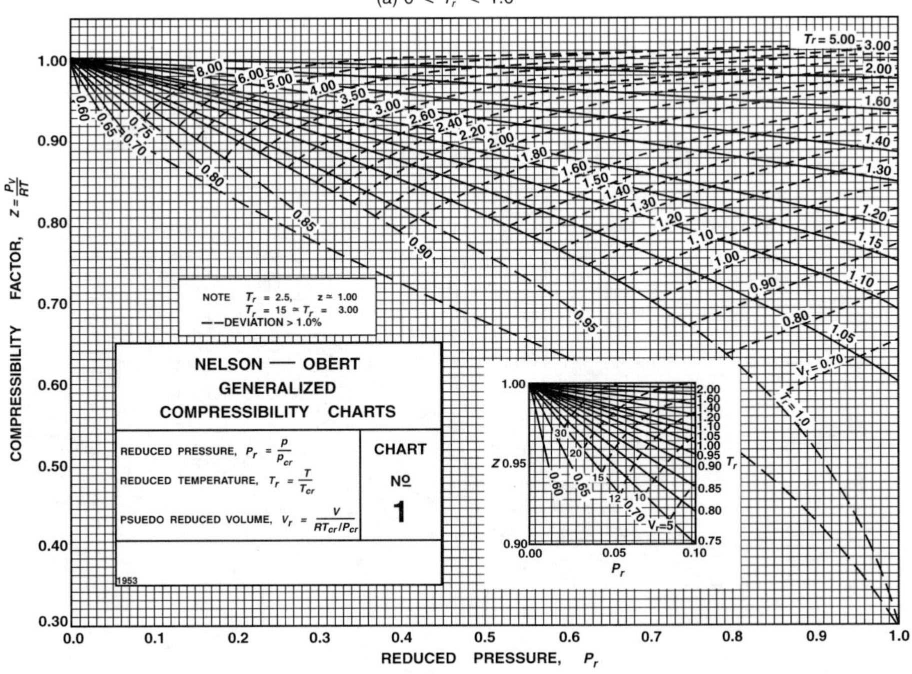

FIGURE A-1

Nelson-Obert generalized compressibility chart — low pressures. (From Cengel, Y. A. and Boles, M., *Thermodynamics*, McGraw-Hill, New York, 1994. With permission.)

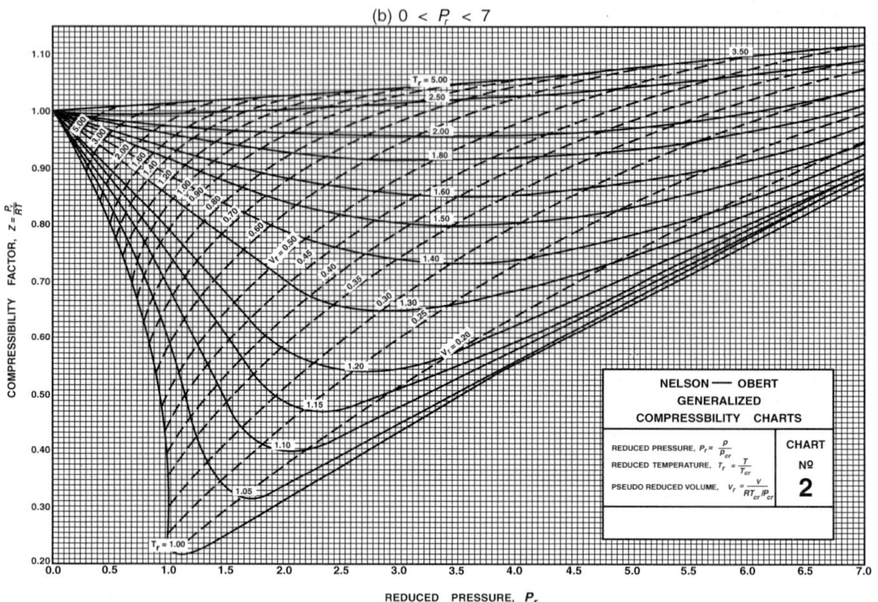

FIGURE A-2
Nelson-Obert generalized compressibility chart — intermediate pressures. (From Cengel, Y. A. and Boles, M., *Thermodynamics*, McGraw-Hill, New York, 1994. With permission.)

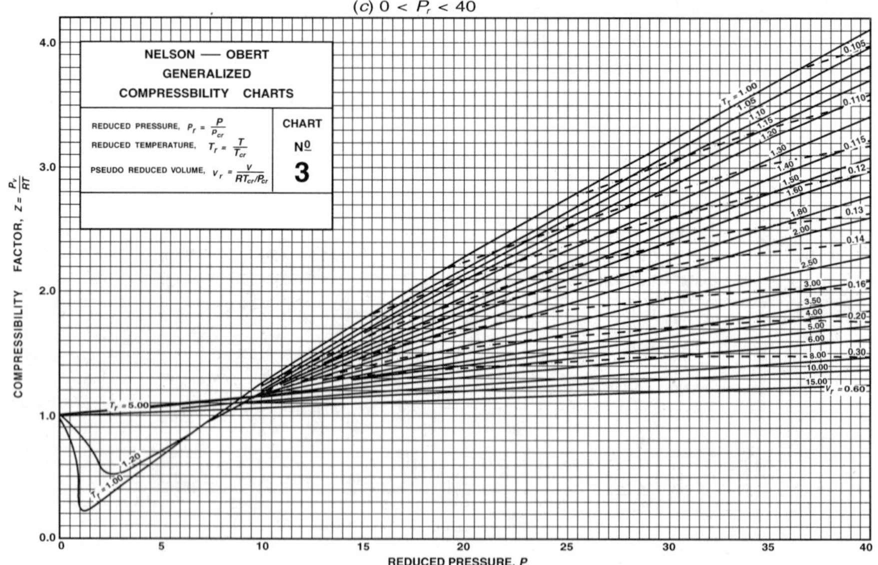

FIGURE A-3
Nelson-Obert generalized compressibility chart — high pressures. (From Cengel, Y. A. and Boles, M., *Thermodynamics*, McGraw-Hill, New York, 1994. With permission.)

B-Series Tables

Thermodynamic Properties of Water

TABLE B-1 SI

Saturated Water

		Specific Volume m^3/kg		Enthalpy kJ/kg		Entropy kJ/(kg.K)	
Temp. K T	Press. MPa P	Sat. Liquid v_f	Sat. Vapor v_g	Sat. Liquid h_f	Sat. Vapor h_g	Sat. Liquid s_f	Sat. Vapor s_g
273.16	0.0006113	0.00100	206.132	0	2501.35	0	9.1562
278.15	0.0008721	0.00100	147.118	20.98	2510.54	0.0761	9.0257
283.15	0.0012276	0.00100	106.376	41.99	2519.74	0.1510	8.9004
288.15	0.001705	0.001001	77.9250	62.98	2528.91	0.2245	8.7813
293.15	0.002339	0.001002	57.7897	83.94	2538.06	0.2966	8.6671
298.15	0.003169	0.001003	43.3593	104.87	2547.17	0.3673	8.5579
303.15	0.004246	0.001004	32.8932	125.77	2556.25	0.4369	8.4533
308.15	0.005628	0.001006	25.2158	146.66	2565.28	0.5052	8.3530
313.15	0.007384	0.001008	19.5229	167.54	2574.26	0.5724	8.2569
318.15	0.009593	0.001010	15.2581	188.42	2583.19	0.6386	8.1647
323.15	0.012350	0.001012	12.0318	209.31	2592.06	0.7037	8.0762
328.15	0.015758	0.001015	9.56835	230.20	2600.86	0.7679	7.9912
333.15	0.019941	0.001017	7.67071	251.11	2609.59	0.8311	7.9095
338.15	0.02503	0.001020	6.19656	272.03	2618.24	0.8934	7.8309
343.15	0.03119	0.001023	5.04217	292.96	2626.8	0.9548	7.7552
348.15	0.03858	0.001026	4.13123	313.91	2635.28	1.0154	7.6824
353.15	0.04739	0.001029	3.40715	334.88	2643.66	1.0752	7.6121
358.15	0.05783	0.001032	2.82757	355.88	2651.93	1.1342	7.5444
363.15	0.07014	0.001036	2.36056	376.90	2660.09	1.1924	7.4790
368.15	0.08455	0.001040	1.98186	397.94	2668.13	1.2500	7.4158
373.15	0.1013	0.001044	1.67290	419.02	2676.05	1.3068	7.3548
378.15	0.1208	0.001047	1.41936	440.13	2683.83	1.3629	7.2958
383.15	0.1433	0.001052	1.21014	461.27	2691.47	1.4184	7.2386
388.15	0.1691	0.001056	1.03658	482.46	2698.96	1.4733	7.1832
393.15	0.1985	0.001060	0.89186	503.69	2706.30	1.5275	7.1295
398.15	0.2321	0.001065	0.77059	524.96	2713.46	1.5812	7.0774
403.15	0.2701	0.001070	0.66850	546.29	2720.46	1.6343	7.0269
408.15	0.3130	0.001075	0.58217	567.67	2727.26	1.6869	6.9777
413.15	0.3613	0.001080	0.50885	589.11	2733.87	1.7390	6.9298
418.15	0.4154	0.001085	0.44632	610.61	2740.26	1.7906	6.8832
423.15	0.4759	0.001090	0.39278	632.18	2746.44	1.8417	6.8378
428.15	0.5431	0.001096	0.34676	653.82	2752.39	1.8924	6.7934
433.15	0.6178	0.001102	0.30706	675.53	2758.09	1.9426	6.7501
438.15	0.7005	0.001108	0.27269	697.32	2763.53	1.9924	6.7075
443.15	0.7917	0.001114	0.24283	719.20	2768.70	2.0418	6.6663
448.15	0.8920	0.001121	0.21680	741 16	2773.58	2.0909	6.6256
453.15	1.0022	0.001127	0.19405	763.21	2778.16	2.1395	6.5857
458.15	1.1227	0.001134	0.17409	785.36	2782.43	2.1878	6.5464
463.15	1.2544	0.001141	0.15654	807.61	2786.37	2.2358	6.5078

TABLE B-1 SI (CONTINUED)

Saturated Water

		Specific Volume m^3/kg		Enthalpy kJ/kg		Entropy kJ/(kg.K)	
Temp. K T	Press. MPa P	Sat. Liquid v_f	Sat. Vapor v_g	Sat. Liquid h_f	Sat. Vapor h_g	Sat. Liquid s_f	Sat. Vapor s_g
468.15	1.3978	0.001149	0.14105	829.96	2789.96	2.2835	6.4697
473.15	1.5538	0.001156	0.12620	852.43	2793.18	2.3308	6.4322
478.15	1.7230	0.001164	0.11521	875.03	2796.03	2.3779	6.3951
483.15	1.9063	0.001173	0.10444	897.75	2798.48	2.4247	6.3584
488.15	2.1042	0.001181	0.09479	920.61	2800.51	2.4713	6.3221
493.15	2.3178	0.001190	0.08619	943.61	2802.12	2.5177	6.2860
498.15	2.5477	0.001199	0.07849	966.77	2803.27	2.5639	6.2502
503.15	2.7949	0.001209	0.07158	990.10	2803.95	2.6099	6.2146
508.15	3.0601	0.001219	0.06536	1013.61	2804.13	2.6557	6.1791
513.15	3.3442	0.001229	0.05976	1037.31	2803.81	2.7015	6.1436
518.15	3.6482	0.00124	0.05470	1061.21	2802.95	2.7471	6.1083
523.15	3.9730	0.001251	0.05013	1085.34	2801.52	2.7927	6.0729
528.15	4.3195	0.001263	0.04598	1109.72	2799.51	2.8382	6.0374
533.15	4.6886	0.001276	0.04220	1134.35	2796.89	2.8837	6.0018
538.15	5.0813	0.001289	0.03877	1159.27	2793.61	2.9293	5.9661
543.15	5.4987	0.001302	0.03564	1184.49	2789.65	2.9750	5.9301
548.15	5.9418	0.001317	0.03279	1210.05	2784.97	3.0208	5.8937
553.15	6.4117	0.001332	0.03017	1235.97	2779.53	3.0667	5.8570
558.15	6.9094	0.001348	0.02777	1262.29	2773.27	3.1129	5.8198
563.15	7.4340	0.001366	0.02557	1289.04	2766.13	3.1593	5.7821
568.15	7.9928	0.001384	0.02354	1316.27	2758.05	3.2061	5.7436
573.15	8.5810	0.001404	0.02167	1344.01	2748.94	3.2533	5.7044
578.15	9.2018	0.001425	0.01995	1372.33	2738.72	3.3009	5.6642
583.15	9.8566	0.001447	0.01835	1401.29	2727.27	3.3492	5.6229
588.15	10.547	0.001472	0.01687	1430.97	2714.44	3.3981	5.5803
593.15	11.274	0.001499	0.01549	1461.45	2700.08	3.4479	5.5361
598.15	12.040	0.001528	0.01420	1492.84	2683.97	3.4987	5.4900
603.15	12.845	0.001561	0.01300	1525.29	2665.85	3.5506	5.4416
608.15	13.694	0.001597	0.01186	1558.98	2645.35	3.6040	5.3903
613.15	14.586	0.001638	0.01080	1594.15	2622.01	3.6593	5.3356
618.15	15.525	0.001685	0.00978	1631.17	2595.19	3.7169	5.2763
623.15	16.514	0.00174	0.00881	1670.54	2563.92	3.7776	5.2111
628.15	17.554	0.001807	0.00787	1713.13	2526.72	3.8427	5.1378
633.15	18.651	0.001892	0.00694	1760.48	2481.00	3.9146	5.0525
638.15	19.807	0.002011	0.00599	1815.96	2421.40	3.9983	4.9470
643.15	21.028	0.002213	0.00493	1890.37	2332.12	4.1104	4.7972
647.25	22.089	0.003155	0.00315	2099.26	2099.26	4.4297	4.4297

TABLE B-2 SI

Saturated Water Pressure First Independent Variable

Press. MPa P	Temp. K T	Specific Volume m³/kg Sat. Liquid v_f	Sat. Vapor v_g	Enthalpy kJ/kg Sat. Liquid h_f	Sat. Vapor h_g	Entropy kJ/(kg.K) Sat. Liquid s_f	Sat. Vapor s_g
0.000611	273.16	0.001000	206.1320	0	2501.30	0	9.1562
0.000100	280.13	0.001000	129.2080	29.290	2514.18	0.1059	8.9756
0.001500	286.18	0.001001	87.98013	54.700	2525.30	0.1956	8.8278
0.000200	290.65	0.001001	67.00385	73.470	2533.49	0.2607	8.7236
0.002500	294.23	0.001002	54.25385	88.470	2540.03	0.3120	8.6431
0.003000	297.23	0.001003	45.66502	101.03	2545.50	0.3545	8.5775
0.004000	302.11	0.001004	34.80015	121.44	2554.37	0.4226	8.4746
0.005000	306.03	0.001005	28.19251	137.79	2561.45	0.4763	8.3950
0.007500	313.44	0.001008	19.23775	168.77	2574.79	0.5763	8.2514
0.010000	318.96	0.001010	14.67355	191.81	2584.63	0.6492	8.1501
0.015000	327.12	0.001014	10.02218	225.91	2599.06	0.7548	8.0084
0.020000	333.21	0.001017	7.64937	251.38	2609.70	0.8319	7.9085
0.025000	338.12	0.001020	6.20424	271.90	2618.19	0.8930	7.8313
0.030000	342.25	0.001022	5.22918	289.21	2625.28	0.9439	7.7686
0.040000	349.02	0.001026	3.99345	317.55	2636.74	1.0258	7.6700
0.050000	354.48	0.001030	3.24034	340.47	2645.87	1.0910	7.5939
0.075000	364.92	0.001037	2.21711	384.36	2662.96	1.2129	7.4563
0.1000	372.77	0.001043	1.69400	417.44	2675.46	1.3025	7.3593
0.1250	379.14	0.001048	1.37490	444.30	2685.35	1.3739	7.2843
0.1500	384.52	0.001053	1.15933	467.08	2693.54	1.4335	7.2232
0.1750	389.21	0.001057	1.00363	486.97	2700.53	1.4848	7.1717
0.2000	393.38	0.001061	0.88573	504.68	2706.63	1.5300	7.1271
0.2250	397.15	0.001064	0.79325	520.69	2712.04	1.5705	7.0878
0.2500	400.58	0.001067	0.71871	535.34	2716.89	1.6072	7.0526
0.2750	403.75	0.001070	0.65731	548.87	2721.29	1.6407	7.0208
0.3000	406.70	0.001073	0.60582	561.45	2725.30	1.6717	6.9918
0.3250	409.45	0.001076	0.56201	573.23	2728.99	1.7005	6.9651
0.3500	412.03	0.001079	0.52425	584.31	2732.40	1.7274	6.9404
0.3750	414.47	0.001081	0.79137	594.79	2735.58	1.7527	6.9174
0.4000	416.78	0.001084	0.46246	604.73	2738.53	1.7766	6.8958
0.4500	421.08	0.001088	0.41398	623.24	2743.91	1.8206	6.8565
0.5000	425.01	0.001093	0.37489	640.21	2748.67	1.8606	6.8212
0.5500	428.63	0.001097	0.34268	655.91	2752.94	1.8972	6.7892
0.6000	432.00	0.001101	0.31567	670.54	2756.80	1.9311	6.7600
0.6500	435.16	0.001104	0.29268	684.26	2760.30	1.9627	6.7330
0.7000	438.12	0.001108	0.27286	697.20	2763.50	1.9922	6.7080
0.7500	440.92	0.001111	0.25560	709.45	2766.43	2.0199	6.6846
0.8000	443.58	0.001115	0.24043	721.10	2769.13	2.0461	6.6627
0.8500	446.11	0.001118	0.22698	732.20	2771.63	2.0709	6.6421
0.9000	448.53	0.001121	0.21497	742.82	2773.94	2.0946	6.6225
0.9500	450.84	0.001124	0.20419	753.00	2776.08	2.1171	6.6040
1.0000	453.06	0.001127	0.19444	762.79	2778.08	2.1386	6.5864
1.1000	457.24	0.001133	0.17753	781.32	2781.68	2.1791	6.5535
1.2000	461.14	0.001139	0.16333	798.64	2784.82	2.2165	6.5233
1.3000	464.79	0.001144	0.15125	814.91	2787.58	2.2514	6.4953
1.4000	468.22	0.001149	0.14084	830.290	2790.00	2.2842	6.4692

TABLE B-2 SI (CONTINUED)

Saturated Water Pressure First Independent Variable

		Specific Volume m^3/kg		Enthalpy kJ/kg		Entropy kJ/(kg.K)	
Press. MPa P	Temp. K T	Sat. Liquid v_f	Sat. Vapor v_g	Sat. Liquid h_f	Sat. Vapor h_g	Sat. Liquid s_f	Sat. Vapor s_g
1.5000	471.47	0.001154	0.13177	844.870	2792.15	2.3150	6.4448
1.7500	478.91	0.001166	0.11349	878.480	2796.43	2.3851	6.3895
2.0000	485.57	0.001177	0.09963	908.770	2799.51	2.4473	6.3408
2.2500	491.60	0.001187	0.08875	936.480	2801.67	2.5034	6.2971
2.5000	497.14	0.001197	0.07998	962.090	2803.07	2.5546	6.2574
2.7500	502.27	0.001207	0.07275	985.970	2803.86	2.6018	6.2208
3.0000	507.05	0.001216	0.06668	1008.41	2804.14	2.6456	6.1869
3.2500	511.53	0.001226	0.06152	1029.60	2803.97	2.6886	6.1551
3.5000	515.75	0.001235	0.05707	1049.73	2803.43	2.7252	6.1252
4.0000	523.55	0.001252	0.04978	1087.29	2801.38	2.7963	6.0700
5.0000	537.14	0.001286	0.03944	1154.21	2794.33	2.9201	5.9733
6.0000	548.79	0.001319	0.03244	1213.32	2784.33	3.0266	5.8891
7.0000	559.03	0.001351	0.02737	1266.97	2772.07	3.1210	5.8132
8.0000	568.21	0.001384	0.02352	1316.61	2757.94	3.2067	5.7431
9.0000	576.55	0.001418	0.02048	1363.23	2742.11	3.2857	5.6771
10.000	584.21	0.001452	0.01803	1407.53	2724.67	3.3595	5.6140
11.000	591.30	0.001489	0.01599	1450.05	2705.60	3.4294	5.5527
12.000	597.90	0.001527	0.01426	1491.24	2684.83	3.4961	5.4923
13.000	604.08	0.001567	0.01278	1531.46	2662.22	3.5604	5.4323
14.000	609.90	0.001611	0.01149	1571.08	2637,55	3.6231	5.3716
15.000	615.39	0.001658	0.01034	1610.45	2610.49	3.9847	5.3097
16.000	620.58	0.001711	0.00931	1650.00	2580.59	3.7460	5.2454
17.000	625.52	0.001770	0.00836	1690.25	2547.15	3.8078	5.1776
18.000	630.21	0.001840	0.00749	1731.97	2509.09	3.8713	5.1044
19.000	634.69	0.001924	0.00666	1776.43	2464.54	3.9387	5.0227
20.000	638.96	0.002035	0.00583	1826.18	2409.74	4.0137	4.9269
21.000	643.04	0.002206	0.00495	1888.30	2334.72	4.1073	4.8015
22.000	646.95	0.002808	0.00353	2034.92	2158.97	4.3307	4.5224
22.089	647.29	0.003155	0.00315	2099.26	2099.26	4.4297	4.4297

TABLE B-3 SI

Superheated Water Vapor

Temp. K	Specific Volume m³/kg	Enthalpy kJ/kg	Entropy kJ/(kg.K)	Temp. K	Specific Volume m³/kg	Enthalpy kJ/kg	Entropy kJ/(kg.K)
	P = 0.01 MPa (318.96)				*P = 0.05 MPa (354.48)*		
318.96	14.67355	2584.63	8.1501	354.48	3.24034	2645.87	7.5939
323.15	14.86920	2592.56	8.1749	373.15	3.41833	2682.52	7.6947
373.15	17.19561	2687.46	8.4479	423.15	3.88937	2780.08	7.9400
473.15	21.82507	2879.52	8.9037	473.15	4.35595	2877.64	8.1579
573.15	26.44508	3076.51	9.2812	573.15	5.28391	3075.52	8.5372
673.15	31.06252	3279.51	9.6076	673.15	6.20929	3278.89	8.8641
773.15	35.67896	3489.05	9.8977	773.15	7.13364	3488.62	9.1545
873.15	40.29488	3705.40	10.1608	873.15	8.05748	3705.10	9.4178
973.15	44.91052	3928.73	10.4028	973.15	8.98104	3928.51	9.6599
1073.15	49 52599	4159.10	10.6281	1073.15	9.90444	4158.92	9.8852
1173.15	54.14137	4396.44	10.8395	1173.15	10.82773	4396.30	10.0967
1273.15	58.75669	4640.58	11.0392	1273.15	11.75097	4640.46	10.2964
1373.15	63.37198	4891.19	11.2287	1373.15	12.64742	4891.08	10.4858
1473.15	67.98724	5147.78	11.4090	1473.15	13.59737	5147.69	10.6662
1573.15	72.60250	5409.70	11.5810	1573.15	14.52054	5409.61	10.8382
1673.15	77.22000	5676.00	11.7450	1573.15	14.52054	5409.61	10.8382
	P = 0.1 MPa (372.77)				*P = 0.2 MPa (393.38)*		
372.77	1.6940	2675.46	7.3593	393.38	0.88573	2706.63	7.1271
423.15	1.93636	2776.38	7.6133	423.15	0.95964	2768.80	7.2795
473.15	2.17226	2875.27	7.8342	473.15	1.08034	2870.46	7.5066
523.15	2.40604	2974.33	8.0332	523.15	1.19880	2970.98	7.7085
573.15	2.63876	3074.28	8.2157	573.15	1.31616	3071.79	7.8926
623.15	2.87100	3175.00	8.3850	623.15	1.43290	3174.00	8.0630
673.15	3.10263	3278.11	8.5434	673.15	1.54930	3276.55	8.2217
773.15	3.56547	3488.09	8.8341	773.15	1.78139	3487.03	8.5132
873.15	4.02781	3704.72	9.0975	873.15	2.01297	3703.96	8.7769
973.15	4.48986	3928.23	9.3398	973.15	2.24426	3927.66	9.0194
1073.15	4.95174	4158.71	9.5652	1073.15	2.47539	4158.27	9.2450
1173.15	5.41353	4396.12	9.7767	1173.15	2.70643	4395.77	9.4565
1273.15	5.87526	4640.31	9.9764	1273.15	2.93740	4640.01	9.6563
1373.15	6.33696	4890.95	10.1658	1373.15	3.16834	4890.68	9.8458
1473.15	6.78863	5147.56	10.3462	1473.15	3.39927	5147.32	10.0262
1573.15	7.26030	5409.49	10.5182	1573.15	3.63018	5409.26	10.1982
	P = 0.3 MPa (406.7)				*P = 0.4 MPa (416.78)*		
406.7	0.60582	2725.30	6.9918	416.78	0.46246	2738.53	6.8958
423.15	0.63388	2760.95	7.0778	423.15	0.47084	2752.82	6.9299
473.15	0.71629	2865.54	7.3115	473.15	0.53422	2860.51	7.1706
523.15	0.79636	2967.59	7.5165	523.15	0.59512	2964.16	7.3788
573.15	0.87529	3069.28	7.7022	573.15	0.65484	3066.75	7.5661
623.15	0.9536	3172.00	7.8730	623.15	0.71390	3170.00	7.7380
673.15	1.03151	3274.98	8.0329	673.15	0.77262	3273.41	7.8984
773.15	1.18669	3485.96	8.3250	773.15	0.88934	3484.89	8.1912

TABLE B-3 SI (CONTINUED)

Superheated Water Vapor

Temp. K	Specific Volume m^3/kg	Enthalpy kJ/kg	Entropy kJ/(kg.K)	Temp. K	Specific Volume m^3/kg	Enthalpy kJ/kg	Entropy kJ/(kg.K)
	P = 0.3 MPa (406.7)				*P = 0.4 MPa (416.78)*		
873.15	1.34136	3703.20	8.5892	873.15	1.00555	3702.44	8.4557
973.15	1.49573	3927.10	8.8319	973.15	1.12147	3926.53	8.6989
1073.15	1.64994	4157.83	9.0575	1073.15	1.23722	4157.40	8.9244
1173.15	1.80406	4395.42	9.2691	1173.15	1.35288	4395.06	9.1361
1273.15	1.95812	4639.71	9.4689	1273.15	1.46847	4639.41	9.3360
1373.15	2.11214	4890.41	9.6585	1373.15	1.58404	4890.15	9.5255
1473.15	2.26614	5147.09	9.8389	1473.15	1.69958	5146.83	9.7059
1573.15	2.42013	5409.03	10.0109	1573.15	1.81511	5408.80	9.8780
	P = 0.5 MPa (425.01)				*P = 0.6 MPa (432.00)*		
425.01	0.37489	2748.67	6.8212	432.00	0.31567	2756.80	6.7600
473.15	0.42492	2855.37	7.0592	473.15	0.35202	2850.12	6.9665
523.15	0.47436	2960.68	7.2703	523.15	0.39383	2957.16	7.1816
573.15	0.52256	3064.20	7.4598	573.15	0.43437	3061.63	7.3723
623.15	0.57012	3167.65	7.6326	623.15	0.47424	3165.66	7.5463
673.15	0.61728	3271.83	7.7937	673.15	0.51372	3270.25	7.7078
723.15	0.66420	3377.00	7.9450	723.15	0.55290	3376.00	7.8590
773.15	0.71093	3483.82	8.0872	773.15	0.59199	3482.75	8.0020
873.15	0.80406	3701.67	8.3521	873.15	0.66974	3700.91	8.2673
973.15	0.89691	3925.97	8.5952	973.15	0.74720	3925.41	8.5107
1073.15	0.98959	4156.96	8.8211	1073.15	0.82450	4156.52	8.7367
1173.15	1.08217	4394.71	9.0329	1173.15	0.90169	4394.36	8.9485
1273.15	1.17469	4639.11	9.2328	1273.15	0.97883	4638.81	9.1484
1373.15	1.26718	4889.88	9.4224	1373.15	1.05594	4889.61	9.3381
1473.15	1.35964	5146.58	9.6080	1473.15	1.13302	5146.34	9.5185
1573.15	1.45210	5408.57	9.7749	1573.15	1.21009	5408.34	9.6906
	P = 0.8 MPa (443.58)				*P = 1.0 MPa (453.06)*		
443.58	0.24043	2769.13	6.6627	453.06	0.19444	2778.08	6.5864
473.15	0.26080	2839.25	6.8158	473.15	0.20596	2827.86	6.6939
523.15	0.29314	2949.97	7.0384	523.15	0.23268	2942.59	6.9246
573.15	0.32411	3056.43	7.2327	573.15	0.25794	3051.15	7.1228
623.15	0.35439	3161.68	7.4088	623.15	0.28247	3157.65	7.3010
673.15	0.38426	3267.07	7.5715	673.15	0.30659	3263.88	7.4650
723.15	0.41390	3373.00	7.7240	723.15	0.33040	3371.00	7.6180
773.15	0.44331	3480.60	7.8672	773.15	0.35411	3478.44	7.7621
873.15	0.50184	3699.38	8.1332	873.15	0.40109	3697.85	8.0289
973.15	0.56007	3924.27	8.3770	973.15	0.44790	3923.14	8.2731
1073.15	0.61813	4155.65	8.6033	1073.15	0.49432	4154.78	8.4996
1173.15	0.67610	4393.65	8.8153	1173.15	0.54075	4392.94	8.7118
1273.15	0.73401	4638.20	9.0153	1273.15	0.58712	4637.60	8.9119
1373.15	0.79188	4889.08	9.2049	1373.15	0.63345	4888.55	9.1016
1473.15	0.84974	5145.85	9.3854	1473.15	0.37977	5145.36	9.2821
1573.15	0.90758	5407.89	9.5575	1573.15	0.72608	5407.41	9.4542

TABLE B-3 SI (CONTINUED)

Superheated Water Vapor

Temp. K	Specific Volume m³/kg	Enthalpy kJ/kg	Entropy kJ/(kg.K)	Temp. K	Specific Volume m³/kg	Enthalpy kJ/kg	Entropy kJ/(kg.K)
	P = 1.2 MPa (461.14)				*P = 1.4 MPa (468.22)*		
461.14	0.16333	2784.82	6.5233	468.22	0.14084	2790.00	6.4692
473.15	0.16930	2815.90	6.5898	473.15	0.14302	2803.32	6.4975
523.15	0.19235	2935.01	6.8293	523.15	0.16350	2927.22	6.7467
573.15	0.21382	3045.80	7.0316	573.15	0.18228	3040.35	6.9533
623.15	0.23452	3153.59	7.2120	623.15	0.20026	3149.49	7.1359
673.15	0.25480	3260.66	7.3773	673.15	0.21780	3257.42	7.3025
723.15	0.27480	3368.00	7.5310	723.15	0.23510	3365.00	7.4570
773.15	0.29463	3476.28	7.6758	773.15	0.25215	3474.11	7.6026
873.15	0.33393	3696.32	7.9434	873.15	0.28596	3694.78	7.8710
973.15	0.37294	3922.01	8.1881	973.15	0.31947	3920.87	8.1160
1073.15	0.41177	4153.90	8.4149	1073.15	0.35281	4153.03	8.3431
1173.15	0.45054	4392.23	8.6272	1173.15	0.38606	4391.53	8.5555
1273.15	0.48919	4637.00	8.8274	1273.15	0.41924	4636.41	8.7558
1373.15	0.52783	4888.02	9.0171	1373.15	0.45239	4887.49	8.9456
1473.15	0.56646	5144.87	9.1977	1473.15	0.48552	5144.38	9.1262
1573.15	0.60507	5406.95	9.3698	1573.15	0.51864	5406.49	9.2983
	P = 2.0 MPa (485.57)				*P = 2.5 MPa (497.14)*		
485.57	0.09963	2799.51	6.3408	497.14	0.07998	2803.07	6.2574
523.15	0.11144	2902.46	6.5452	523.15	0.08700	2880.06	6.4084
573.15	0.12547	3023.50	6.7663	573.15	0.09890	3008.81	6.6437
623.15	0.13857	3136.96	6.9562	623.15	0.10976	3126.24	6.8402
673.15	0.15120	3247.60	7.1270	673.15	0.12010	3239.28	7.0147
723.15	0.16353	3357.48	7.2844	723.15	0.13014	3350.77	7.1745
773.15	0.17568	3467.55	7.4316	773.15	0.13998	3462.04	7.3233
823.15	0.18770	3578.00	7.5700	823.15	0.14970	3574.00	7.4630
873.15	0.19960	3690.14	7.7023	873.15	0.15930	3686.25	7.5960
973.15	0.22323	3917.45	7.9487	973.15	0.17832	3914.59	7.8435
1073.15	0.24668	4150.40	8.1766	1073.15	0.19716	4148.20	8.0720
1173.15	0.27004	4389.40	8.3895	1173.15	0.21590	4387.64	8.2853
1273.15	0.29333	4634.61	8.5900	12i3.15	0.23458	4633.12	8.4860
1373.15	0.31659	4885.89	8.7800	1373.15	0.25322	4884.57	8.6761
1473.15	0.33984	5142.92	8.9606	1473.15	0.27185	5141.70	8.8569
1573.15	0.36306	5405.10	9.1328	1573.15	0.29046	5403.95	9.0291
	P = 3.0 MPa (507.05)				*P = 5.0 MPa (537.14)*		
507.05	0.06668	2804.14	6.1869	537.14	0.03944	2794.33	5.9733
523.15	0.07058	2855.75	6.2871	573.15	0.04532	2924.53	6.2083
573.15	0.08114	2993.48	6.5389	623.15	0.05194	3068.39	6.4492
623.15	0.09053	3115.25	6.7427	673.15	0.05781	3195.64	6.6458
673.15	0.09936	3230.82	6.9211	723.15	0.06330	3316.15	6.8185
723.15	0.10787	3344.00	7.0833	773.15	0.06857	3433.76	6.9758
773.15	0.11619	3456.48	7.2337	823.15	0.07368	3550.23	7.1217
823.15	0.12440	3569.00	7.3750	873.15	0.07869	3666.47	7.2588

TABLE B-3 SI (CONTINUED)

Superheated Water Vapor

Temp. K	Specific Volume m³/kg	Enthalpy kJ/kg	Entropy kJ/(kg.K)	Temp. K	Specific Volume m³/kg	Enthalpy kJ/kg	Entropy kJ/(kg.K)
	P = 3.0 MPa (507.05)				*P = 5.0 MPa (537.14)*		
873.15	0.13243	3682.34	7.5084	923.15	0.08362	3783.00	7.3890
973.15	0.14838	3911.72	7.7571	973.15	0.08849	3900.13	7.5122
1073.15	0.16414	4146.00	7.9862	1073.15	0.09811	4137.17	7 7440
1173.15	0.17980	4385.87	8.1999	1173.15	0.10762	4378.82	7.9593
1273.15	0.19541	4631.63	8.4009	1273.15	0.11707	4625.69	8.1612
1373.15	0.21098	4883.26	8.5911	1373.15	0.12648	4878.02	8.3519
1473.15	0.22652	5140.49	8.7719	1473.15	0.13587	5135.67	8.5330
1573.15	0.24206	5402.81	8.9442	1573.15	0.14526	5398.24	8.7055
	P = 7.0 MPa (559.03)				*P = 8.0 MPa (568.21)*		
559.03	0.02737	2772.07	5.8132	568.21	0.02352	2757.94	5.7431
573.15	0.02947	2838.40	5.9304	573.15	0.02426	2784.98	5.7905
623.15	0.03524	3016.02	6.2282	623.15	0.02995	2987.30	6.1300
673.15	0.03993	3158.07	6.4477	673.15	0.03432	3138.28	6.3633
723.15	0.04416	3287.04	6.6326	723.15	0.03817	3271.99	6.5550
773.15	0.04814	3410.29	6.7974	773.15	0.04175	3398.27	6.7239
823.15	0.05195	3530.87	6.9486	823.15	0.04516	3521.01	6.8778
873.15	0.05565	3650.26	7.0894	873.15	0.04845	3642.03	7.0205
923.15	0.05927	3769.00	7.2220	923.15	0.05166	3762.00	7.1540
973.15	0.06283	3888.39	7.3476	973.15	0.05481	3882.47	7.2512
1073.15	0.06981	4128.30	7.5822	1073.15	0.06097	4123.84	7.5173
1173.15	0.07669	4371.77	7.7991	1173.15	0.06702	4368.26	7.7350
1273.15	0.08350	4619.80	8.0020	1273.15	0.07301	4616.87	7.9384
1373.15	0.09027	4872.83	8.1933	1373.15	0.07896	4870.25	8.1299
1473.15	0.07030	5130.90	8.3747	1473.15	0.08490	5128.54	8.3115
1573.15	0.10377	5393.71	8.5472	1573.15	0.09080	5391.46	8.4842
	P = 9.0 MPa (576.55)				*P = 10.0 MPa (584.21)*		
576.55	0.02048	2742.11	5.6771	584.21	0.01803	2726.67	5.6140
623.15	0.02580	2956.55	6.0361	623.15	0.02242	2923.39	5.9442
673.15	0.02993	3117.76	6.2853	673.15	0.02641	3096.46	6.2119
723.15	0.03350	3256.59	6.4843	723.15	0.02975	3240.83	6.4189
773.15	0.03677	3386.05	6.6575	773.15	0.03279	3373.63	6.5965
823.15	0.03987	3511.02	6.8141	823.15	0.03564	3500.92	6.7561
873.15	0.04285	3633.73	6.9588	873.15	0.03837	3625.34	6.9028
923.15	0.04574	3755.32	7.0943	923.15	0.04101	3748.27	7.0397
973.15	0.04857	3876.51	7.2221	973.15	0.04358	3870.52	7.1687
1023.15	0.05135	3998.00	7.3440	1023.15	0.04611	3993.00	7.2910
1073.15	0.05409	4119.38	7.4597	1073.15	0.04859	4114.91	7.4077
1173.15	0.05950	4364.74	7.6782	1173.15	0.05349	4361.24	7.6272
1273.15	0.06485	4613.95	7.8821	1273.15	0.05832	4611.04	7.8315
1373.15	0.07016	4867.69	8.0739	1373.15	0.06312	4865.14	8.0236
1473.15	0.07544	5126.18	8.2556	1473.15	0.06789	5123.84	8.2054
1573.15	0.08072	5389.22	8.4283	1573.15	0.07265	5386.99	8.3783

TABLE B-3 SI (CONTINUED)

Superheated Water Vapor

Temp. K	Specific Volume m³/kg	Enthalpy kJ/kg	Entropy kJ/(kg.K)	Temp. K	Specific Volume m³/kg	Enthalpy kJ/kg	Entropy kJ/(kg.K)
	P = 12.5 *MPa (601.04)*				*P* = 15.0 *MPa (615.39)*		
601.04	0.01350	2673 77	5.4623	615.39	0.01034	2610.49	5.3097
623.15	0.01613	2826.15	5.7117	623.15	0.01147	2692.41	5.4420
673.15	0.02000	3039.30	6.0416	673.15	0.01565	2975.44	5.8810
723.15	0.02990	3199.78	6.2718	723.15	0.01845	3156.15	6.1403
773.15	0.02500	3341.72	6.4617	773.15	0.02080	3308.53	6.3442
823.15	0.02800	3475.13	6.6289	823.15	0.02293	3448.61	6.5198
873.15	0.03020	3604.05	6.7810	873.15	0.02491	3582.30	6.6775
923.15	0.03240	3730.44	6.9213	923.15	0.02680	3712.32	6.8223
973.15	0.03461	3855.41	7.0536	973.15	0.02861	3840.12	6.9572
1023.15	0.03667	3980.00	7.1780	1023.15	0.03037	3967.00	7.0840
1073.15	0.03861	4103.69	7.2965	1073.15	0.03210	4092.43	7.2040
1173.15	0.04200	4352.48	7.5181	1173.15	0.03546	4343.75	7.4279
1273.15	0.04600	4603.81	7.7237	1273.15	0.03875	4596.63	7.6347
1373.15	0.05000	4858.82	7.9165	1373.15	0.04200	4852.56	7.8282
1473.15	0.05430	5118.02	8.0987	1473.15	0.04523	5112.27	8.0108
1573.15	0.05800	5381.44	8.2717	1573.15	0.04845	5375.94	8.1839
	P = 17.5 *MPa (627.90)*				*P* = 20.0 *MPa (638.96)*		
627.90	0.0070	2528.79	5.1418	638.96	0.00583	2409.74	4.9269
673.15	0.0120	2902.82	5.7212	673.15	0.00994	2818.07	5.5539
723.15	0.0151	3109.69	6.0182	723.15	0.01270	3060.06	5.9016
773.15	0.0173	3274.02	6.2382	773.15	0.01477	3238.18	6.1400
823.15	0.0192	3421.37	6.4229	823.15	0.01656	3393.45	6.3347
873.15	0.0210	3560.13	6.5866	873.15	0.01818	3537.57	6.5048
923.15	0.0220	3693.94	6.7356	923.15	0.01969	3675.32	6.6582
973.15	0.0240	3824.67	6.8936	973.15	0.02113	3809.09	6.7993
1023.15	0.0250	3953.48	7.0026	1023.15	0.02251	3940.27	6.9308
1073.15	0.0270	4081.13	7.1245	1073.15	0.02385	4069.80	7.0544
1123.15	0.0289	4208.00	7.2400	1123.15	0.02516	4198.00	7.1710
1173.15	0.0300	4335.05	7.3507	1173.15	0.02645	4326.37	7.2830
1273.15	0.0330	4589.52	7.5588	1273.15	0.02897	4582.45	7.4925
1373.15	0.0350	4846.37	7.7530	1373.15	0.03145	4840.24	7.6874
1473.15	0.0380	5106.59	7.9359	1473.15	0.03391	5100.96	7.8706
1573.15	0.0415	5370.50	8.1093	1573.15	0.03636	5365.1	8.0441
	P = 25.0 *MPa*				*P* = 30.0 *MPa*		
648.15	0.00100	1847.93	4.0319	648.15	0.001789	1791.43	3.9303
673.15	0.00600	2580.16	5.1418	673.15	0.002790	2151.04	4.4728
698.15	0.00700	2806.25	5.4722	698.15	0.005304	2614.17	5.1503
723.15	0.00900	2949.70	5.6743	723.15	0.006735	2821.35	5.4423
773.15	0.01100	3162.39	5.9592	773.15	0.008679	3081.03	5.7904
823.15	0.01200	3335.62	6.1764	823.15	0.010168	3275.36	6.0342
873.15	0.01400	3491.36	6.3602	873.15	0.011446	3443.91	6.233
923.15	0.01500	3637.46	6.5229	923.15	0.012596	3598.93	6.4057

TABLE B-3 SI (CONTINUED)

Superheated Water Vapor

Temp. K	Specific Volume m³/kg	Enthalpy kJ/kg	Entropy kJ/(kg.K)	Temp. K	Specific Volume m³/kg	Enthalpy kJ/kg	Entropy kJ/(kg.K)
	P = 25.0 MPa				*P = 30.0 MPa*		
973.15	0.01600	3777.56	6.6707	973.15	0.013661	3745.67	6.5606
1073.15	0.01891	4047.08	6.9345	1073.15	0.015623	4024.31	6.8332
1123.15	0.01999	4179.00	7.0540	1123.15	0.016550	4159.00	6.9560
1173.15	0.021045	4309.09	7.1679	1173.15	0.017448	4291.93	7.0717
1273.15	0.023102	4568.47	7.3801	1273.15	0.019196	4554.68	7.2867
1373.15	0.025119	4828.15	7.5765	1373.15	0.020903	4816.28	7.4845
1473.15	0.027115	5089.86	7.7604	1473.15	0.022589	5078.97	7.6691
1573.15	0.029101	5354.44	7.9342	1573.15	0.024266	5343.95	7.8432
	P = 35.0 MPa				*P = 40.0 MPa*		
648.15	0.001700	1762.37	3.8721	648.15	0.001641	1742.71	3.8289
673.15	0.002100	1987.52	4.2124	673.15	0.001908	1930.83	4.1134
698.15	0.003428	2373.41	4.7747	698.15	0.002532	2198.11	4.5028
723.15	0.004962	2672.36	5.1962	723.15	0.003693	2512.79	4.9459
773.15	0.006927	2994.34	5.6281	773.15	0.005623	2903.26	5.4699
823.15	0.008345	3213.01	5.9025	823.15	0.006984	3149.05	5.7784
873.15	0.009527	3395.49	6.1178	873.15	0.008094	3346.38	6.0113
923.15	0.010575	3559.91	6.3010	923.15	0.009064	3520.58	6.2054
973.15	0.011533	3713.54	6.4631	973.15	0.009942	3681.29	6.3750
1073.15	0.013278	4001.54	6.7450	1073.15	0.011523	3978.80	6.6662
1123.15	0.014090	4139.00	6.8710	1123.15	0.012256	4120.00	6.7950
1173.15	0.014883	4274.87	6.9886	1173.15	0.012963	4257.93	6.9150
1273.15	0.016410	4541.05	7.2063	1273.15	0.014324	4527.59	7.1356
1373.15	0.017895	4804.59	7.4056	1373.15	0.015643	4793.09	7.3364
1473.15	0.01936	5068.26	7.5910	1473.15	0.016940	5057.72	7.5224
1573.15	0.020815	5333.62	7.7652	1573.15	0.018229	5323.45	7.6969
	P = 50.0 MPa				*P = 60.0 MPa*		
648.15	0.001559	1716.52	3.7638	648.15	0.001503	1699.51	3.7140
673.15	0.001731	1874.58	4.0030	673.15	0.001633	1843.35	3.9317
698.15	0.002007	2059.98	4.2733	698.15	0.001817	2001.65	4.1625
723.15	0.002486	2283.91	4.5883	723.15	0.002085	2178.96	4.4119
773.15	0.003892	2720.07	5.1725	773.15	0.002956	2567.88	4.9320
823.15	0.005118	3019.51	5.5485	823.15	0.003957	2896.16	5.3440
873.15	0.006112	3247.59	5.8177	873.15	0.004835	3151.21	5.6451
923.15	0.006966	3441.84	6.0342	923.15	0.005595	3364.55	5.8829
973.15	0.007727	3616.91	6.2189	973.15	0.006272	3553 56	6.0824
1073.15	0.009076	3933.62	6.5290	1073.15	0.007459	3889.12	6.4110
1123.15	0.009693	4081.00	6.6640	1123.15	0.007996	4043.00	6.5510
1173.15	0.010283	4224.41	6.7882	1173.15	0.008508	4191.47	6.6805
1273.15	0.011411	4501.09	7.0146	1273.15	0.009480	4475.16	6.9126
1373.15	0.012497	4770.55	7.2183	1373.15	0.010409	4748.61	7.1194
1473.15	0.013561	5037.15	7.4058	1473.15	0.011317	5017.19	7 3082
1573.15	0.014616	5303.56	7.5807	1573.15	0.012215	5284.28	7.4837

TABLE B-4 SI

Compressed Liquid Water

Temp. K	Specific Volume m³/kg	Enthalpy kJ/kg	Entropy kJ/(kg.K)	Temp. K	Specific Volume m³/kg	Enthalpy kJ/kg	Entropy kJ/(kg.K)
	P = 5 MPa (537.14)				*P = 10 MPa (584.21)*		
273.15	0.000998	5.02	0.0001	273.15	0.000995	10.05	0.0003
293.15	0.001000	88.64	0.2955	293.15	0.000997	93.32	0.2945
313.15	0.001006	171.95	0.5705	313.15	0.001003	176.36	0.5685
333.15	0.001015	255.28	0.8284	333.15	0.001013	259.47	0.8258
353.16	0.001027	338.83	1.0719	353.15	0.001025	342.81	1.0687
373.15	0.001041	422.71	1.3030	373.15	0.001039	426.48	1.2992
393.15	0.001058	507.07	1.5232	393.15	0.001055	510.61	1.5188
413.15	0.001077	592.13	1.7342	413.15	0.001074	595.40	1.7291
433.15	0.001099	678.10	1.9374	433.15	0.001195	681.07	1.9316
453.15	0.001124	765.24	2.1341	453.15	0.001120	767.83	2.1274
473.15	0.001153	853.85	2.3254	473.15	0.001148	855.97	2.3178
493.15	0.001187	944.36	2.5128	493.15	0.001181	945.88	2.5038
513.15	0.001226	1037.47	2.6978	513.15	0.001219	1038.13	2.6872
523.15	0.001249	1085.30	2.7900	533.15	0.001265	1133.68	2.8698
533.15	0.001275	1134.30	2.8829	553.15	0.001322	1234.11	3.0547
537.14	0.001286	1154.21	2.9201	584.21	0.001452	1407.53	3.3595
	P = 15 MPa (615.39)				*P = 20 MPa (638.96)*		
273.15	0.000993	15.04	0.0004	273.15	0.00099	20.00	0.0004
293.15	0.000995	97.97	0.2934	293.15	0.000993	102.61	0.2922
313.15	0.001001	180.75	0.5665	313.15	0.000999	185.14	0.5646
333.15	0.001011	263.65	0.8231	333.15	0.001008	267.82	0.8205
353.15	0.001022	346.79	1.0655	353.15	0.001020	350.78	1.0623
373.15	0.001036	430.26	1.2954	373.15	0.001034	434.04	1.2917
413.15	0.001071	598.70	1.7241	413.15	0.001068	602.03	1.7192
433.15	0.001092	684.07	1.9259	433.15	0.001089	687.11	1.9203
453.15	0.001116	770.48	2.1209	473.15	0.001139	860.47	2.3031
473.15	0.001143	858.18	2.3103	493.15	0.001169	949.27	2.4869
513.15	0.001211	1038.99	2.6770	513.15	0.001205	1040.04	2.6673
533.15	0.001255	1133.41	2.8575	553.15	0.001297	1230.62	3.0248
553.15	0.001308	1232.09	3.0392	573.15	0.001360	1333.29	3.2071
573.15	0.001377	1337.23	3.2259	613.15	0.001568	1571.01	3.6074
613.15	0.001631	1591.88	3.6545	633.15	0.001823	1739.23	3.8770
615.39	0.001658	1610.45	3.6847	638.96	0.002035	1826.18	4.0137
	P = 30 MPa				*P = 50 MPa*		
273.15	0.000986	29.82	0.0001	273.15	0.000977	49.03	–0.0014
293.15	0.000989	111.82	0.2898	293.15	0.000980	130.00	0.2847
313.15	0.000995	193.87	0.5606	313.15	0.000987	211.2	0.5526
333.15	0.001004	276.16	0.8153	333.15	0.000996	292.77	0.8051
353.15	0.001016	358.75	1.0561	353.15	0.001007	374.68	1.0439
373.15	0.001029	441.63	1.2844	373.15	0.001020	456.87	1.2703
393.15	0.001044	524.91	1.5017	413.15	0.001052	622.33	1.6915
433.15	0.001082	693.27	1.9095	433.15	0.001070	705.91	1.8890

TABLE B-4 SI (CONTINUED)

Compressed Liquid Water

Temp. K	Specific Volume m³/kg	Enthalpy kJ/kg	Entropy kJ/(kg.K)	Temp. K	Specific Volume m³/kg	Enthalpy kJ/kg	Entropy kJ/(kg.K)
	P = 30 *MPa*				*P* = 50 *MPa*		
453.15	0.001105	778.71	2.1024	453.15	0.001091	790.24	2.0793
473.15	0.001130	865.24	2.2892	473.15	0.001115	875.46	2.2634
493.15	0.001159	953.09	2.4710	513.15	0.001170	1049.2	2.6158
533.15	0.001230	1134.29	2.8242	533.15	0.001203	1138.23	2.7860
573.15	0.001330	1327.80	3.1740	573.15	0.001286	1322.95	3.1200
593.15	0.001400	1432.63	3.3538	593.15	0.001339	1420.17	3.2867
633.15	0.001627	1675.36	3.7492	633.15	0.001484	1630.16	3.6290
653.15	0.001869	1837.43	4.0010	653.15	0.001588	1746.54	3.8100

Adapted from Sonntag, R.E., Borgnakke, C., and Van Wylen, G. J., *Fundamentals of Thermodynamics,* John Wiley & Sons, New York, 1998. With permission.

C-Series Tables

Thermodynamic Properties of Ammonia

TABLE C-1 SI

Saturated Ammonia

Temp. K T	Press. MPa P	Specific Volume m^3/kg Sat. Liquid v_f	Sat. Vapor v_g	Enthalpy kJ/kg Sat. Liquid h_f	Sat. Vapor h_g	Entropy kJ/(kg.K) Sat. Liquid s_f	Sat. Vapor s_g
223.15	0.0409	0.001424	2.6270	–43.76	1372.6	–0.1916	6.1554
228.15	0.0545	0.001437	2.0063	–21.94	1380.8	–0.0950	6.0534
233.15	0.0710	0.001450	1.5526	0.00	1388.8	0.0000	5.9567
238.15	0.0932	0.001463	1.2161	22.06	1396.5	0.0935	5.8650
245.15	0.1195	0.001476	0.9634	44.26	1404.0	0.1856	5.7778
248.15	0.1516	0.001490	0.7712	66.58	1411.2	0.2763	5.6947
253.15	0.1902	0.001504	0.6233	89.05	1418.0	0.3657	5.6155
258.15	0.2363	0.001519	0.5084	111.66	1424.6	0.4538	5.5397
263.15	0.2909	0.001534	0.4181	134.41	1430.8	0.5408	5.4673
268.15	0.3549	0.001550	0.3465	157.31	1436.7	0.6266	5.3977
273.15	0.4296	0.001566	0.2892	180.36	1442.2	0.7114	5.3309
278.15	0.5159	0.001583	0.2430	203.58	1447.3	0.7951	5.2666
283.15	0.6152	0.001600	0.2054	226.97	1452.0	0.8779	5.2045
288.15	0.7286	0.001619	0.1746	250.54	1456.3	0.9598	5.1444
293.15	0.8575	0.001638	0.1492	274.30	1460.2	1.0408	5.0860
298.15	1.0032	0.001658	0.1281	298.25	1463.5	1.1210	5.0293
303.15	1.1670	0 001680	0.1105	322.42	1466.3	1.2005	4.9738
308.15	1.3504	0.001702	0.0957	346.80	1468.6	1.2792	4.9196
313.15	1.5549	0.001725	0.0831	371.43	1470.2	1.3574	4.8662
318.15	1.7820	0.001750	0.0725	396.31	1471.2	1.4350	4.8136
323.15	2.0331	0.001777	0.0634	421.48	1471.5	1.5121	4.7614
328.15	2.3101	0.001804	0.0556	446.96	1471.0	1.5888	4.7095
333.15	2.6144	0.001834	0.0488	472.79	1469.7	1.6652	4.6577
338.15	2.9478	0.001866	0.0430	499.01	1467.5	1.7415	4.6057
343.15	3.3120	0.001900	0.0379	525.69	1464.4	1.8178	4.5533
348.15	3.7090	0.001937	0.0334	552.88	1460.1	1.8943	4.5001
353.15	4.1405	0.001978	0.0295	580.69	1454.6	1.9712	4.4458
358.15	4.6086	0.002022	0.0261	609.21	1447.8	2.0488	4.3901
363.15	5.1153	0.002071	0.0230	638.59	1439.4	2.1273	4.3325
368.15	5.6629	0.002126	0.0203	668.99	1429.2	2.2073	4.2723
373.15	6.2537	0.002188	0.0178	700.64	1416.9	2.2893	4.2088
378.15	6.8904	0.002261	0.0156	733.87	1402.0	2.3740	4.1407
383.15	7.5757	0.002347	0.0136	769.15	1383.7	2.4625	4.0665
388.15	8.3133	0.002452	0.0118	807.21	1361.0	2.5566	3.9833
393.15	9.1072	0.002589	0.0100	849.36	1331.7	2.6593	3.8861
398.15	9.9635	0.002783	0.0083	898.42	1291.4	2.7775	3.7645
403.15	10.8916	0.003122	0.0065	963.29	1227.0	2.9326	3.5866
405.45	11.3332	0.004255	0.0043	1085.85	1085.9	3.2316	3.2316

TABLE C-2 SI

Superheated Ammonia Vapor

Temp. K	Specific Volume m³/kg	Enthalpy kJ/kg	Entropy kJ/(kg.K)	Temp. K	Specific Volume m³/kg	Enthalpy kJ/kg	Entropy kJ/(kg.K)
	P = 0.05 MPa				*P = 0.075 MPa*		
226.62	2.17521	1378.3	6.0839	233.99	1.48922	1390.1	5.9411
243.15	2.34484	1413.4	6.2333	243.15	1.55321	1410.1	6.0247
253.15	2.44631	1434.6	6.3187	253.15	1.62221	1431.7	6.1120
263.15	2.54711	1455.7	6.4006	263.15	1.69050	1453.3	6.1954
273.15	2.64736	1476.9	6.4795	273.15	1.75832	1474.8	6.2756
283.15	2.74716	1498.1	6.5556	283.15	1.82551	1496.2	6.3527
293.15	2.84661	1519.3	6.6293	293.15	1.89243	1517.7	6.4272
303.15	2.94578	1540.6	6.7008	303.15	1.95906–	1539.2	6.4993
313.15	3.04472	1562.0	6.7703	313.15	2.02547	1560.7	6.5693
323.15	3.14348	1583.5	6.8379	323.15	2.09168	1582.4	6.6373
333.15	3.24209	1605.1	6.9038	333.15	2.15775	1604.1	6.7036
343.15	3.34058	1626.9	6.9682	343.15	2.22369	1626.0	6.7683
353.15	3.43897	1648.8	7.0312	353.15	2.28954	1648.0	6.8315
373.15	3.63551	1693.2	7.1533	373.15	2.42099	1692.4	6.9539
393.15	3.83183	1738.2	7.2708	393.15	2.55221	1737.5	7.0716
413.15	4.02797	1783.9	7.3842	413.15	2.68326	1783 4	7.1853
433.15	4.22398	1830.4	7.4941	433.15	2.81418	1829.9	7.2953
453.15	4.41988	1877 7	7.6008	453.15	2.94499	1877.2	7.4021
473.15	4.61570	1925.7	7.7045	473.15	3.07571	1925.3	7.5059
	P = 0.1 MPa				*P = 0.125 MPa*		
239.55	1.13806	1398.7	5.8401	244.08	0.92365	1405.4	5.7620
243.15	1.15727	1406.7	5.8734	253.15	0.96271	1425.9	5.8446
253.15	1.21007	1428.8	5.9626	263.15	1.00506	1448.3	5.9314
263.15	1.26213	1450.8	6.0477	273.15	1.04628	1470.5	6.0141
273.15	1.31362	1472.6	6.1291	283.15	1.08811	1492.5	6.0933
283.15	1.36465	1494.4	6.2073	293.15	1.12903	1514.4	6.1694
293.15	1.41532	1516.1	6.2826	303.15	1.16964	1536.3	6.2428
303.15	1.46569	1537.7	6.3553	313.15	1.21003	1558.2	6.3138
313.15	1.51582	1559.5	6.4258	323.15	1.25002	1580.1	6.3827
323.15	1.56577	1581.2	6.4943	333.15	1.29026	1602.1	6.4496
333.15	1.61557	1603.1	6.5609	343.15	1.33017	1624.1	6.5149
343.15	1.66525	1625.1	6.6258	353.15	1.36998	1646.3	6.5785
353.15	1.71482	1647.1	6.6892	373.15	1.44937	1691.0	6.7017
373.15	1.81373	1691.7	6.8120	393.15	1.52852	1736.3	6.8199
393.15	1.91240	1736.9	6.9300	413.15	1.60749	1782.2	6.9339
413.15	2.01091	1782.8	7.0439	433.15	1.68633	1828.9	7.0443
433.15	2.10927	1829.4	7.1540	453.15	1.76507	1876.3	7.1513
453.15	2.20754	1876.8	7.2609	473.15	1.84371	1924.5	7.2553
473.15	2.30571	1924.9	7.3648	493.15	1.92229	1973.4	7.3566

TABLE C-2 SI (CONTINUED)

Superheated Ammonia Vapor

Temp. K	Specific Volume m^3/kg	Enthalpy kJ/kg	Entropy kJ/(kg.K)	Temp. K	Specific Volume m^3/kg	Enthalpy kJ/kg	Entropy kJ/(kg.K)
	$P = 0.15$ *MPa*				$P = 0.2$ *MPa*		
247.93	0.77870	1410.9	5.6983	254.29	0.59460	1419.6	5.5979
253.15	0.79774	1422.9	5.7465	263.15	0.61926	1440.6	5.6791
263.15	0.83364	1445.7	5.8349	263.15	0.61926	1440.6	5.6791
273.15	0.86892	1468.3	5.9189	273.15	0.64648	1463.8	5.7659
283.15	0.90373	1490.6	5.9992	283.15	0.67319	1486.8	5.8484
293.15	0.93815	1512.8	6.0716	293.15	0.69951	1506.4	5.9270
303.15	0.97227	1534.8	6.1502	303.15	0.72553	1531.9	6.0025
313.15	1.00615	1556.9	6.2217	313.15	0.75129	1554.3	6.0751
323.15	1.03984	1578.9	6.2910	323.15	0 77685	1576.6	6.1453
333.15	1.07338	1601.0	6.3583	333.15	0.80226	1598.9	6.2133
343.15	1.10678	1623.2	6.4238	343.15	0.82754	1621.3	6.2794
353.15	1.14009	1645.4	6.4877	353.15	0.85271	1643.7	6.3437
373.15	1.20646	1690.2	6.6112	373.15	0.90282	1688.8	6.4679
393.15	1.27259	1735.6	6.7297	393.15	0.95268	1734.4	6.5869
413.15	1.33855	1781.7	6.8439	413.15	1.00237	1780.6	6.7015
433.15	1.40437	1828.4	6.9544	433.15	1.05192	1827.4	6.8123
453.15	1.47009	1875.9	7.0615	453.15	1.10136	1875.0	6.9196
473.15	1.53572	1924.1	7.1656	473.15	1.15072	1923.3	7.0239
493.15	1.60127	1973.1	7.2670	493.15	1.20000	1972.4	7.1255
	$P = 0.25$ *MPa*				$P = 0.3$ *MPa*		
259.49	0.48213	1426.3	5.5201	263.91	0.40607	1431.7	5.4565
273.15	0.51293	1459.3	5.6441	273.15	0.42382	1454.7	5.5420
283.15	0.53481	1482.9	5.7288	283.15	0.44251	1478.9	5.6290
293.15	0.55629	1506.0	5.8093	293.15	0.46077	1502.6	5.7113
303.15	0.57745	1529.0	5.8861	303.15	0.47870	1525.9	5.7896
313.15	0.59835	1551.7	5.9599	313.15	0.49636	1549.0	5.8645
323.15	0.61904	1574.3	6.0309	323.15	0.51382	1571.9	5.9365
333.15	0.63958	1596.8	6.0997	333.15	0.53111	1594.7	6.0060
343.15	0.65998	1619.4	6.1663	343.15	0.54827	1617.5	6.0732
353.15	0.68028	1641.9	6.2312	353.15	0.56532	1640.2	6.1385
373.15	0.72063	1687.3	6.3561	373.15	0.59916	1685.8	6.2642
393.15	0.76073	1733.1	6.4756	39315	0.63276	1731.8	6.3842
413.15	0.80065	1779.4	6.5906	413.15	0.66618	1778.3	6.4996
433.15	0.84044	1826.4	6.7016	433.15	0.69946	1825.4	6.6109
453.15	0.88012	1874.1	6.8093	453.15	0.73263	1873.2	6.7188
473.15	0.91972	1922.5	6.9138	473.15	0.76572	1921.7	6.8235
493.15	0.95923	1971.6	7.0155	493.15	0.79872	1970.9	6.9254
513.15	0.99868	2021.5	7.1147	513.15	0.83167	2020.9	7.0247
533.15	1.03808	2072.2	7.2115	533.15	0.86455	2071.6	7.1217

TABLE C-2 SI (CONTINUED)

Superheated Ammonia Vapor

Temp. K	Specific Volume m³/kg	Enthalpy kJ/kg	Entropy kJ/(kg.K)	Temp. K	Specific Volume m³/kg	Enthalpy kJ/kg	Entropy kJ/(kg.K)
	P = 0.35 MPa				*P = 0.4 MPa*		
267.79	0.35108	1436.3	5.4026	271.26	0.30942	1440.2	5.3559
273.15	0.36011	1449 9	5.4532	283.15	0.32701	1470.7	5.4663
283.15	0.37654	1474.9	5.5427	293.15	0.34129	1495.6	5.5525
293.15	0.39251	1499.1	5.6270	303.15	0.35520	1519.8	5.6338
303.15	0.40814	1522.9	5.7068	313.15	0.36884	1543.6	5.7111
313.15	0.42360	1546.3	5.7828	323.15	0.38226	1567.1	5.7850
323.15	0.43865	1569.5	5.8557	333.15	0.39550	1590.4	5.8560
333.15	0.45362	1592.6	5.9259	343.15	0.40860	1613.6	5.9244
343.15	0.46846	1615.5	5.9938	353.15	0.42160	1636.7	5.9907
353.15	0.48319	1638.4	6.0596	373.15	0.44732	1682.8	6.1179
373.15	0.51240	1684.3	6.1860	393.15	0.47279	1729.2	6.2390
393.15	0.54135	1730.5	6.3066	413.15	0.49808	1776.0	6.3552
413.15	0.57012	1777.2	6.4223	433.15	0.52323	1823.4	6.4671
433.15	0.59876	1824.4	6.5340	453.15	0.54827	1871.4	6.5755
453.15	0.62728	1872.3	6.6421	473.15	0.57321	1920.1	6.6806
473.15	0.65571	1920.9	6.7470	493.15	0.59809	1969.5	6.7828
493.15	0.68407	1970.2	6.8491	513.15	0.62289	2019.6	6.8825
513.15	0.71237	2020.3	6.9486	533.15	0.64764	2070.5	6.9797
533.15	0.74060	2071.0	7.0456	553.15	0.67234	2122.1	7.0747
	P = 0.5 MPa				*P = 0.6 MPa*		
277.28	0.25035	1446.5	5.2776	282.43	0.21038	1451.4	5.2133
283.15	0.25757	1462.3	5.3340	283.15	0.21115	1453.4	5.2205
293.15	0.26949	1488.3	5.4244	293.15	0.22154	1480.8	5.3156
303.15	0.28103	1513.5	5.5090	303.15	0.23152	1507.1	5.4037
313.15	0.29227	1538.1	5.5889	313.15	0.24118	1532.5	5.4862
323.15	0.30328	1562.3	5.6647	323.15	0.25059	1557.3	5.5641
333.15	0.31410	1586.1	5.7373	333.15	0.25981	1581.6	5.6383
343.15	0.32478	1609.6	5.8070	343.15	0.26888	1605.7	5.7094
353.15	0.33530	1633.1	5.8744	353.15	0.27783	1629.5	5.7778
373.15	0.35621	1679.8	6.0310	373.15	0.29545	1676.8	5.9081
393.15	0.37681	1726.6	6.1253	393.15	0.31281	1724.0	6.0314
413.15	0.39722	1773.8	6.2422	413.15	0.32997	1771.5	6.1491
433.15	0.41748	1821.4	6.3548	433.15	0.34699	1819.4	6.2623
453.15	0.43764	1869.6	6.4636	453.15	0.36389	1867.8	6.3717
473.15	0.45771	1918.5	6.5691	473.15	0.38071	1916.0	6.4776
493.15	0.47770	1968.1	6.6717	493.15	0.39745	1966.6	6.5806
513.15	0.49763	2018.3	6.7717	513.15	0.41412	2017.1	6.6808
533.15	0.51749	2069.3	6.8692	533.15	0.43073	2068.2	6.7786
553.15	0.53731	2121.1	6.9644	553.15	0.44729	2120.1	6.8741

TABLE C-2 SI (CONTINUED)

Superheated Ammonia Vapor

Temp. K	Specific Volume m³/kg	Enthalpy kJ/kg	Entropy kJ/(kg.K)	Temp. K	Specific Volume m³/kg	Enthalpy kJ/kg	Entropy kJ/(kg.K)
	P = 0.8 MPa				*P = 1 MPa*		
291	0.15958	1458.6	5.1110	298.05	0.12852	1463.4	5.0304
293.15	0.16138	1464.9	5.1328	303.15	0.13206	1479.1	5.0826
303.15	0.16947	1493.5	5.2287	313.15	0.13868	1508.5	5.1778
313.15	0.17720	1520.8	5.3171	323.15	0.14499	1536.3	5.2654
323.15	0.18465	1547.0	5.3996	333.15	0.15106	1563.1	5.3471
333.15	0.19189	1572.5	5.4774	343.15	0.15659	1589.3	5.4240
343.15	0.19896	1597.5	5.5513	353.15	0.16270	1614.6	5.4971
353.15	0.20590	1622.1	5.6219	373.15	0.17389	1664.3	5.6342
373.15	0.21949	1670.6	5.7555	393.15	0.18477	1713.4	5.7622
393.15	0.23280	1718.7	5.8811	413.15	0.19545	1762.2	5.8834
413.15	0.24590	1766.9	6.0006	433.15	0.20597	1811.2	5.9992
433.15	0.25886	1815.3	6.1150	453.15	0.21638	1860.5	6.1105
453.15	0.27170	1864.2	6.2254	473.15	0.22669	1910.4	6.2182
473.15	0.28450	1913.6	6.3322	493.15	0.23693	1960.8	6.3226
493.15	0.29712	1963.7	6.4358	513.15	0.24710	2011.9	6.4241
513.15	0.30973	2014.5	6.5367	533.15	0.25720	2063.6	6.5229
533.15	0.32228	2065.9	6.6350	553.15	0.26726	2116.0	6.6194
553.15	0.33477	2118.0	6.7310	573.15	0.27726	2165.1	6.7137
573.15	0.34722	2170.9	6.8248	593.15	0.28723	2222.9	6.8059
	P = 1.2 MPa				*P = 1.6 MPa*		
304.09	0.10751	1466.8	4.9635	314.18	0.08079	1470.5	4.8553
308.15	0.10995	1480.0	5.0060	323.15	0.08506	1501.0	4.9510
313.15	0.11287	1495.4	5.0564	333.15	0.08951	1532.5	5.0472
323.15	0.11846	1525.1	5.1497	343.15	0.09372	1562.3	5.1351
333.15	0.12378	1553.3	5.2357	353.15	0.09774	1590.7	5.2167
343.15	0.12890	1580.5	5.3159	373.15	0.10539	1644.8	5.3659
353.15	0.13387	1606.8	5.3916	393.15	0.11268	1696.9	5.5018
373.15	0.14347	1658.0	5.5325	413.15	0.11974	1748.0	5.6286
393.15	0.15275	1708.0	5.6631	433.15	0.12662	1798.7	5.7485
413.15	0.16181	1757.5	5.7860	453.15	0.13339	1849.5	5.3631
433.15	0.17071	1807.1	5.9031	473.15	0.14005	1900.5	5.9734
453.15	0.17950	1856.9	6.0156	493.15	0.14663	1952.0	6.0800
473.15	0.18819	1907.1	6.1241	513.15	0.15314	2004.1	6.1834
493.15	0.19680	1957.9	6.2292	533.15	0.15959	2056.7	6.2839
513.15	0.20534	2009.3	6.3313	573.15	0.16599	2109.9	6.3819
533.15	0.21382	2061.3	6.4308	593.15	0.17234	2163.7	6.4775
553.15	0.22225	2114.0	6.5278	613.15	0.17865	2218.2	6.5710
573.15	0.23063	2167.3	6.6225	633.15	0.18492	2273.3	6.6624
593.15	0.23897	2221.3	6.7151	653.15	0.19115	2329.1	6.7519

TABLE C-2 SI (CONTINUED)

Superheated Ammonia Vapor

Temp. K	Specific Volume m³/kg	Enthalpy kJ/kg	Entropy kJ/(kg.K)	Temp. K	Specific Volume m³/kg	Enthalpy kJ/kg	Entropy kJ/(kg.K)
	P = 2 MPa				*P = 5 MPa*		
322.52	0.06444	1471.5	4.7680	362.05	0.02365	1441.4	4.3454
323.15	0.06471	1473.9	4.7754	373.15	0.02636	1501.5	4.5091
333.15	0.06875	1509.8	4.8848	393.15	0.03024	1586.3	4.7306
343.15	0.07246	1542.7	4.9821	413.15	0.03350	1657.3	4.9068
353.15	0.07595	1573.5	5.0707	433.15	0.03643	1721.7	5.0591
373.15	0.08248	1631.1	5.2294	453.15	0.03916	1782.7	5.1968
393.15	0.08861	1685.5	5.3714	473.15	0.04174	1841.8	5.3245
413.15	0.09447	1738.2	5.5022	493.15	0.04422	1900.0	5.4450
433.15	0.10016	1790.2	5.6251	513.15	0.04662	1957.9	5.5600
453.15	0.10571	1842.0	5.7420	533.15	0.04895	2015.6	5.6704
473.15	0.11116	1893.9	5.8540	553.15	0.05123	2073.6	5.7771
493.15	0.11652	1946.1	5.9621	573.15	0.05346	2131.8	5.8805
513.15	0.12182	1998.8	6.0668	593.15	0.05565	2190.3	5.9809
533.15	0.12705	2052.0	6.1685	613.15	0.05779	2249.2	6.0786
573.15	0.13224	2105.8	6.2675	633.15	0.05990	2308.6	6.1738
593.15	0.13737	2160.1	6.3641	653.15	0.06198	2368.4	6.2668
613.15	0.14246	2215.1	6.4583	673.15	0.06403	2428.6	6.3576
633.15	0.14751	2270.7	6.5505	693.15	0.06606	2489.3	6.4464
653.15	0.15253	2326.8	6.6406	713.15	0.06806	2550.4	6.5334
	P = 10 MPa				*P = 20 MPa*		
398.17	0.00826	1289.4	3.7587	413.15	0.00251	918.9	2.7630
413.15	0.01195	1461.3	4.1839	433.15	0.00323	1097.1	3.1838
433.15	0.01461	1578.3	4.4610	453.15	0.00490	1329.7	3.7087
453.15	0.01666	1667.2	4.6617	473.15	0.00653	1497.7	4.0721
473.15	0.01842	1744.5	4.8287	493.15	0.00782	1618.7	4.3228
493.15	0.02001	1816.0	4.9767	513.15	0.00891	1718.6	4.5214
513.15	0.02150	1884.2	5.1123	533.15	0.00988	1807.6	4.6916
533.15	0.02290	1950.6	5.2392	553.15	0.01077	1890.5	4.8442
553.15	0.02424	2015.9	5.3596	573.15	0.01159	1969.6	4.9847
573.15	0.02552	2080.7	5.4746	593.15	0.01237	2046.3	5.1164
593.15	0.02676	2145.2	5.5852	613.15	0.01312	2121.6	5.2412
613.15	0.02796	2209.6	5.6921	633.15	0.01382	2195.8	5.3603
633.15	0.02913	2274.1	5.7955	653.15	0.01450	2269.4	5.4748
653.15	0.03026	2338.7	5.8960	673.15	0.01516	2342.6	5.5851
673.15	0.03137	2403.5	5.9937	693.15	0.01579	2415.4	5.6917
693.15	0.03245	2468.5	6.0888	713.15	0.01641	2488.1	5.7950
713.15	0.03351	2533.7	6.1815	733.15	0.01700	2561.0	5.8950
733.15	0.03455	2599.0	6.2720	753.15	0.01758	2633.0	5.9930
753.15	0.03557	2665.0	6.3610	773.15	0.01814	2706.0	6.0880

TABLE C-3 SI

Compressed Liquid Ammonia

Temp. K	Specific Volume m³/kg	Enthalpy kJ/kg	Entropy kJ/(kg.K)	Temp. K	Specific Volume m³/kg	Enthalpy kJ/kg	Entropy kJ/(kg.K)
	P = 0.5 MPa				*P = 0.75 MPa*		
258.15	0.001518	111.96	0.4530	258.15	0.001518	112.140	0.4523
260.15	0.001524	121.02	0.4880	261.15	0.001527	125.750	0.5047
262.15	0.001530	130.11	0.5228	264.15	0.001536	139.400	0.5566
264.15	0.001537	139.22	0.5574	267.15	0.001543	148.530	0.5911
266.15	0.001543	148.35	0.5919	270.15	0.001555	166.860	0.6595
268.15	0.001549	157.51	0.6262	274.15	0.001569	185.290	0.7272
270.15	0.001556	166.69	0.6603	278.15	0.001582	203.800	0.7943
272.15	0.001562	175.90	0.6942	282.15	0.001600	222.500	0.8609
274.15	0.001569	185.13	0.7280	286.15	0.001611	241.200	0.9269
277.297	0.001580	199.71	0.7809	289.037	0.001622	254.800	0.9742
	P = 1 MPa				*P = 2 MPa*		
258.15	0.001518	112.330	0.4515	258.15	0.001516	113.080	0.4486
262.15	0.001530	130.470	0.5213	265.15	0.001537	144.850	0.5700
266.15	0.001542	148.700	0.5903	272.15	0.001560	176.890	0.6893
270.15	0.001555	167.030	0.6586	279.15	0.001583	209.200	0.8066
274.15	0.001568	185.450	0.7263	286.15	0.001609	241.900	0.9222
279.15	0.001585	208.600	0.8101	293.15	0.001635	274.900	1.0362
284.15	0.001603	232.000	0.8930	300.15	0.001664	308.300	1.1489
289.15	0.001622	255.500	0.9751	307.15	0.001695	342.200	1.2603
294.15	0.001640	279.200	1.0564	315.15	0.001734	381.500	1.3867
298.06	0.001658	297.900	1.1195	322.51	0.001773	418.400	1.5023
	P = 3 MPa				*P = 4 MPa*		
258.15	0.001515	113.830	0.4456	258.15	0.001513	114.590	0.4427
267.15	0.001542	154.670	0.6011	268.15	0.001543	159.940	0.6150
276.15	0.001571	195.980	0.7532	278.15	0.001576	205.800	0.7831
285.15	0.001603	237.800	0.9021	288.15	0.001612	252.400	0.9474
294.15	0.001637	280.100	1.0484	298.15	0.001651	299.600	1.1086
303.15	0.001674	323.100	1.1924	308.15	0.001694	347.600	1.2670
312.15	0.001715	366.900	1.3346	318.15	0.001742	396.600	1.4234
321.15	0.001761	411.500	1.4754	328.15	0.001796	446.800	1.5787
330.15	0.001813	457.200	1.6158	338.15	0.001859	498.600	1.7342
338.89	0.001871	503.000	1.7529	351.57	0.001965	572.000	1.9469
	P = 5 MPa				*P = 7.5 MPa*		
258.15	0.001512	115.340	0.4397	258.15	0.001508	117.240	0.4325
269.15	0.001545	165.200	0.6288	271.15	0.001547	176.050	0.6547
280.15	0.001581	215.700	0.8128	284.15	0.001590	235.800	0 8698
291.15	0.001621	267.000	0.9923	298.15	0.001642	301.200	1.0945
302.15	0.001664	319.100	1.1680	312.15	0.001701	368.000	1.3136

TABLE C-3 SI (CONTINUED)

Compressed Liquid Ammonia

Temp. K	Specific Volume m^3/kg	Enthalpy kJ/kg	Entropy kJ/(kg.K)	Temp. K	Specific Volume m^3/kg	Enthalpy kJ/kg	Entropy kJ/(kg.K)
	P = 5 *MPa*				*P* = 7.5 *MPa*		
314.15	0.001718	377.100	1.3562	326.15	0.001769	436.600	1.5285
326.15	0.001780	436.600	1.5421	340.15	0.001851	507.600	1.7417
338.15	0.001853	498.200	1.7274	354.15	0.001954	582.500	1.9574
350.15	0.001943	562.800	1.9153	368.15	0.002094	664.200	2.1840
362.06	0.002060	632.200	2.1100	382.64	0.002337	765.500	2.4530
	P = 10 *MPa*				*P* = 11 *MPa*		
258.15	0.001518	112.330	0.4515	258.15	0.001503	119.930	0.4225
273.15	0.001549	186.890	0.6803	274.15	0.001551	192.140	0.6938
288.15	0.001599	255.800	0.9258	290.15	0.001604	265.600	0.9543
303.15	0.001655	325.900	1.1632	306.15	0.001665	340.600	1.2057
318.15	0.001720	397.700	1.3942	322.15	0.001736	417.300	1.4501
334.15	0.001802	476.500	1.6359	338.15	0.001820	496.500	1.6900
350.15	0.001904	558.900	1.8766	354.15	0.001926	579.500	1.9297
366.15	0.002038	647.300	2.1230	370.15	0.002067	669.000	2.1770
382.15	0.002244	748.300	2.3930	387.15	0.002306	779.400	2.4680
398.37	0.002794	900.900	2.7830	403.73	0.003188	973.700	2.9570

Adapted from Sonntag, R. E., Borgnakke, C., and Van Wylen, G. J., *Fundamentals of Thermodynamics*, John Wiley & Sons, New York, 1998. With permission.

D-Series Tables

Thermodynamic Properties of Carbon Dioxide

TABLE D-1 SI

Saturated Carbon Dioxide

		Specific Volume m^3/kg		Enthalpy kJ/kg		Entropy kJ/(kg.K)	
Temp. K T	Press. MPa P	Sat. Liquid v_f	Sat. Vapor v_g	Sat. Liquid h_f	Sat. Vapor h_g	Sat. Liquid s_f	Sat. Vapor s_g
216.54	0.5173	0.000847	0.072780	0.00	351.87	0.0000	1.6250
220	0.6000	0.000857	0.063140	7.00	353.12	0.0318	1.6050
225	0.7366	0.000871	0.051810	17.52	354.73	0.0785	1.5772
230	0.8949	0.000885	0.042870	28.07	356.09	0.1243	1.5505
235	1.0770	0.000901	0.035730	38.52	357.19	0.1685	1.5246
240	1.2850	0.000918	0.029970	48.90	358.00	0.2114	1.4993
245	1.5210	0.000936	0.025270	59.32	358.48	0.2535	1.4745
250	1.7880	0.000956	0.021400	69.85	358.59	0.2950	1.4500
255	2.0870	0.000977	0.018200	80.57	358.29	0.3363	1.4254
260	2.4210	0.001001	0.015520	91.49	357.53	0.3774	1.4007
265	2.7920	0.001027	0.013250	102.66	356.23	0.4186	1.3754
270	3.2040	0.001056	0.011330	114.16	354.29	0.4500	1.3493
275	3.6590	0.001090	0.009675	126.12	351.57	0.5020	1.3219
280	4.1600	0.001129	0.008242	138.72	347.88	0.5455	1.2925
285	4.7110	0.001177	0.006982	152.21	342.90	0.5910	1.2601
290	5.3140	0.001239	0.005854	166.95	336.09	0.6397	1.2229
295	5.9770	0.001324	0.004808	183.67	326.28	0.6939	1.1774
300	6.7060	0.001471	0.003749	205.06	309.95	0.7624	1.1120
304.21	7.3830	0.002155	0.002155	257.31	257.31	0.9312	0.9312

TABLE D–2 SI

Superheated Carbon Dioxide Vapor

Temp. K	Specific Volume m^3/kg	Enthalpy kJ/kg	Entropy kJ/(kg.K)	Temp. K	Specific Volume m^3/kg	Enthalpy kJ/kg	Entropy kJ/(kg.K)
	P = 1 MPa				*P = 2 MPa*		
233	0.03844	356.78	1.5350	253.6	0.01903	358.42	1.4323
300	0.05379	419.95	1.7737	300	0.02535	409.41	1.6174
400	0.07418	513.65	2.0430	400	0.03640	508.45	1.9025
500	0.09376	613.22	2.2649	500	0.04654	610.01	2.1289
600	0.11300	718.90	2.4574	600	0.05638	716.73	2.3233
700	0.13220	829.90	2.6284	700	0.06608	828.37	2.4954
800	0.15130	945.34	2.7825	800	0.07571	944.25	2.6500
900	0.17030	1064.46	2.9228	900	0.08529	1063.68	2.7907
1000	0.18930	1186.61	3.0515	1000	0.09484	1186.07	2.9196
	P = 5 MPa				*P = 10 MPa*		
287.5	0.00641	339.84	1.2426	400	0.00620	463.19	1.5127
300	0.00779	366.98	1.3351	500	0.00885	584.73	1.7846
400	0.01374	492.26	1.6993	600	0.01112	700.24	1.9952
500	0.01824	600.43	1.9407	700	0.01325	816.97	2.1751
600	0.02241	710.37	2.1410	800	0.01530	936.20	2.3343
700	0.02643	823.93	2.3160	900	0.01731	1058.02	2.4777
800	0.03039	941.10	2.4724	1000	0.01929	1182.21	2.6086
900	0.03429	1061.45	2.6141				
1000	0.03817	1184.53	2.7438				
	P = 20 MPa						
400	0.00262	403.03	1.2634				
500	0.00426	555.30	1.6054				
600	0.00554	681.94	1.8366				
700	0.00670	804.67	2.0258				
800	0.00779	927.73	2.1901				
900	0.00885	1052.23	2.3367				
1000	0.00988	1178.42	2.4697				

TABLE D-3 SI

Compressed Liquid Carbon Dioxide

Temp. K	Specific Volume m³/kg	Enthalpy kJ/kg	Entropy kJ/(kg.K)
	P = 3.204 MPa		
220	0.0008523	7.677	0.0247
230	0.0008807	28.470	0.1720
240	0.0009129	49.030	0.2046
250	0.0009507	69.750	0.2892
260	0.0009672	78.950	0.3227
270	0.0010560	114.160	0.4600
	P = 5 MPa		
220	0.0008495	8.150	0.0199
230	0.0008771	28.810	0.1118
240	0.0009083	49.180	0.1984
250	0.0009442	69.640	0.2820
260	0.0009873	90.670	0.3644
270	0.0010415	112.850	0.4481
280	0.0011167	137.370	0.5373
	P = 10 MPa		
220	0.000842318	9.590	0.0073
230	0.000868056	29.940	0.0977
240	0.000896620	49.880	0.1826
250	0.000928764	69.750	0.2637
260	0.000965997	89.870	0.3426
270	0.001010132	110.590	0.4208
280	0.001064963	132.360	0.4999
290	0.001137708	155.970	0.5827
300	0.001246805	183.070	0.6746

Temp. K	Specific Volume m³/kg	Enthalpy kJ/kg	Entropy kJ/(kg.K)
	P = 4.16 MPa		
220	0.0008508	7.929	0.0221
230	0.0008788	28.650	0.1143
240	0.0009105	49.110	0.2013
250	0.0009472	69.690	0.2854
260	0.0009746	83.400	0.3384
270	0.0010120	98.990	0.3976
280	0.0011292	138.720	0.5455
	P = 6.706 MPa		
220	0.0008471	8.641	0.0156
230	0.0008740	29.200	0.1070
240	0.0009043	49.420	0.1930
250	0.0009389	69.680	0.2758
260	0.0009800	90.400	0.3570
270	0.0010310	112.100	0.4388
280	0.0010990	135.700	0.5245
290	0.0011150	144.100	0.5488
300	0.0014705	205.060	0.7624

Adapted from Reynolds, W. C., *Thermodynamic Properties in SI,* Stanford University Press, Stanford, CA. With permisison.

E-Series Tables

Thermodynamic Properties of R-12

TABLE E-1 SI

Saturated R-12

		Specific Volume m^3/kg		Enthalpy kJ/kg		Entropy kJ/(kg.K)	
Temp. K T	Press. MPa P	Sat. Liquid v_f	Sat. Vapor v_g	Sat. Liquid h_f	Sat. Vapor h_g	Sat. Liquid s_f	Sat. Vapor s_g
183.15	0.0028	0.000608	4.41494	–43.28	146.46	–0.2086	0.8273
193.15	0.0062	0.000617	2.13835	–34.72	151.02	–0.1631	0.7984
203.15	0.0123	0.000627	1.1273	–26.13	155.64	–0.1198	0.7749
213.15	0.0226	0.000637	0.63791	–17.49	160.29	–0.0783	0.7557
223.15	0.0391	0.000648	0.3831	–8.78	164.95	–0.0384	0.7401
228.15	0.0504	0.000654	0.30268	–4.40	167.28	–0.0190	0.7334
233.15	0.0642	0.000659	0.24191	0	169.59	0	0.7274
238.15	0.0807	0.000666	0.1954	4.42	171.9	0.0187	0.7219
243.15	0.1004	0.000672	0.15937	8.86	174.2	0.0371	0.7170
243.35	0.1013	0.000672	0.15803	9.05	174.29	0.0379	0.7168
248.15	0.1237	0.000679	0.13117	13.33	176.48	0.0552	0.7126
253.15	0.1509	0.000685	0.10885	17.85	178.74	0.0731	0.7087
258.15	0.1826	0.000693	0.09102	22.33	180.97	0.0906	0.7051
263.15	0.2191	0.000700	0.07665	26.78	183.19	0.1080	0.7019
268.15	0.2610	0.000708	0.06496	31.45	185.37	0.1251	0.6991
273.15	0.3086	0.000716	0.05539	36.05	187.53	0.1420	0.6965
278.15	0.3626	0.000724	0.04749	40.69	189.64	0.1587	0.6942
283.15	0.4233	0.000733	0.04091	45.37	191.74	0.1752	0.6921
288.15	0.4914	0.000743	0.03541	50.10	193.78	0.1915	0.6902
293.15	0.5673	0.000752	0.03078	54.87	195.78	0.2078	0.6884
298.15	0.6516	0.000763	0.02685	59.70	197.73	0.2239	0.6868
303.15	0.7449	0.000774	0.02351	64.59	199.62	0.2399	0.6853
308.15	0.8477	0.000786	0.02064	69.55	201.45	0.2559	0.6839
313.15	0.9670	0.000798	0.01817	74.59	203.20	0.2718	0.6825
318.15	1.0843	0.000811	0.01603	79.71	204.87	0.2877	0.6811
323.15	1.2193	0.000826	0.01417	84.94	206.45	0.3037	0.6797
328.15	1.3663	0.000841	0.01254	90.27	207.92	0.3197	0.6782
333.15	1.5259	0.000858	0.01111	95.74	209.26	0.3358	0.6765
338.15	1.6988	0.000877	0.00985	101.36	210.46	0.3552	0.6747
343.15	1.8858	0.000897	0.00873	107.15	211.48	0.3686	0.6726
348.15	2.0875	0.000920	0.00772	113.15	212.29	0.3854	0.6702
353.15	2.3046	0.000946	0.00682	119.39	212.83	0.4027	0.6672
358.15	2.5380	0.000976	0.00600	125.93	213.04	0.4204	0.6636
363.15	2.7885	0.001012	0.00526	132.84	212.80	0.4389	0.6590
368.15	3.0569	0.001056	0.00456	140.23	211.94	0.4583	0.6531
373.15	3.3441	0.001113	0.00391	148.61	210.12	0.4793	0.6449
378.15	3.6509	0.001197	0.00324	157.52	206.57	0.5028	0.6325
383.15	3.9785	0.001364	0.00246	169.55	197.99	0.5333	0.6076
385.15	4.1168	0.001792	0.00179	183.43	183.43	0.5689	0.5689

TABLE E-2 SI

Superheated R-12 Vapor

Temp. K T	Specific Volume m³/kg	Enthalpy kJ/kg	Entropy kJ/(kg.K)	Temp. K T	Specific Volume m³/kg	Enthalpy kJ/kg	Entropy kJ/(kg.K)
	P = 0.025 MPa				*P = 0.05 MPa*		
215.04	0.58130	161.10	0.7527	227.97	0.30515	167.19	0.7336
243.15	0.66179	176.19	0.8187	253.15	0.34186	181.17	0.7917
253.15	0.39001	181.74	0.8410	263.15	0.35623	186.89	0.8139
263.15	0.71811	187.40	0.8630	273.15	0.37051	192.70	0.8356
273.15	0.74613	193.17	0.8844	283.15	0.38472	198.61	0.8568
283.15	0.77409	199.03	0.9055	293.15	0.39886	204.62	0.8776
293.15	0.80198	204.99	0.9262	303.15	0.41296	210.71	0.8981
303.15	0.82982	211.05	0.9465	313.15	0.42701	216.89	0.9181
313.15	0.85762	217.20	0.9665	323.15	0.44103	223.16	0.9378
323.15	0.88539	223.45	0.9861	333.15	0.45502	229.51	0.9572
333.15	0.91312	229.77	1.0054	343.15	0.46898	235.95	0.9762
343.15	0.94083	236.19	1.0244	353.15	0.48292	242.46	0.9949
353.15	0.96852	242.68	1.0430	363.15	0.49684	249.05	1.0133
363.15	0.99618	249.26	1.0614	373.15	0.51074	255.71	1.0314
373.15	1.02384	255.91	1.0795	383.15	0.52463	262.45	1.0493
383.15	1.05148	262.63	1.0972	393.15	0.53851	269.26	1.0668
	P = 0.1 MPa				*P = 0.2 MPa*		
243.20	0.15999	174.15	0.7171	260.77	0.08354	182.07	0.7035
253.15	0.1677	179.99	0.7406	273.15	0.08861	189.80	0.7325
263.15	0.17522	185.84	0.7633	283.15	0.09255	196.02	0.7548
273.15	0.18265	191.77	0.7854	293.15	0.09642	202.28	0.7766
283.15	0.18999	197.77	0.8070	303.15	0.10023	208.60	0.7978
293.15	0.19728	203.85	0.8281	313.15	0.10399	214.97	0.8184
303.15	0.20451	210.02	0.8488	323.15	0.10771	221.41	0.8387
313.15	0.21169	216.26	0.8691	333.15	0.11140	227.90	0.8585
323.15	0.21884	222.58	0.8889	343.15	0.11506	234.46	0.8779
333.15	0.22596	228.98	0.9084	353.15	0.11869	241.09	0.8969
343.15	0.23305	235.46	0.9276	363.15	0.12230	247.77	0.9156
353.15	0.24011	242.01	0.9464	373.15	0.12590	254.53	0.9339
363.15	0.24716	248.63	0.9649	383.15	0.12948	261.34	0.9519
373.15	0.25419	255.32	0.9831	393.15	0.13305	268.21	0.8969
383.15	0.26121	262.08	1.0009	403.15	0.13661	275.15	0.9870
393.15	0.26821	268.91	1.0185	413.15	0.14016	282.14	1.0042
	P = 0.3 MPa				*P = 0.4 MPa*		
272.44	0.0569	187.16	0.6969	281.45	0.04321	190.97	0.6928
283.15	0.05998	194.17	0.7222	283.15	0.04363	192.21	0.6972
293.15	0.06273	200.64	0.7446	293.15	0.04584	198.91	0.7204
303.15	0.06542	207.12	0.7663	303.15	0.04797	205.58	0.7428
313.15	0.06805	213.64	0.7875	313.15	0.05005	212.25	0.7645
323.15	0.07064	220.19	0.8081	323.15	0.05207	218.94	0.7855

TABLE E-2 SI (CONTINUED)

Superheated R-12 Vapor

Temp. K T	Specific Volume m^3/kg	Enthalpy kJ/kg	Entropy kJ/(kg.K)	Temp. K T	Specific Volume m^3/kg	Enthalpy kJ/kg	Entropy kJ/(kg.K)
	P = 0.3 MPa				*P = 0.4 MPa*		
333.15	0.07319	226.79	0.8282	333.15	0.05406	225.65	0.8060
343.15	0.07571	233.44	0.8479	343.15	0.05601	232.40	0.8259
353.15	0.07820	240.15	0.8671	353.15	0.05794	239.19	0.8454
363.15	0.08067	246.90	0.8860	363.15	0.05985	246.02	0.8645
373.15	0.08313	253.72	0.9145	373.15	0.06173	252.79	0.8831
383.15	0.08557	260.58	0.9226	383.15	0.06360	259.81	0.9015
393.15	0.08799	267.5	0.9405	393.15	0.06546	266.79	0.9194
403.15	0.09041	274.48	0.9580	403.15	0.06730	273.81	0.9370
413.15	0.09281	281.51	0.9752	413.15	0.06913	280.88	0.9544
423.15	0.09520	288.59	0.9922	423.15	0.07095	287.99	0.9714
	P = 0.5 MPa				*P 0.75 MPa*		
288.75	0.03482	194.03	0.6899	303.56	0.02335	199.72	0.6852
303.15	0.03746	203.96	0.7235	313.15	0.02467	206.91	0.7086
323.15	0.04091	217.64	0.7672	323.15	0.02594	214.18	0.7314
333.15	0.04257	224.48	0.7881	333.15	0.02718	221.37	0.7533
343.15	0.04418	231.33	0.8083	343.15	0.02837	228.52	0.7745
353.15	0.04577	238.21	0.8281	353.15	0.02952	235.65	0.7949
363.15	0.04734	245.11	0.8473	363.15	0.03064	242.76	0.8148
373.15	0.04889	252.05	0.8662	373.15	0.03174	249.89	0.8342
383.15	0.05041	259.03	0.8847	383.15	0.03282	257.03	0.8530
393.15	0.05193	266.06	0.9028	393.15	0.03388	264.19	0.8715
403.15	0.05343	273.12	0.9205	403.15	0.03493	271.38	0.8895
413.15	0.05492	280.23	0.9379	413.15	0.03596	278.59	0.9072
423.15	0.05640	287.39	0.9550	423.15	0.03699	285.84	0.9246
433.15	0.05788	294.59	0.9718	433.15	0.03801	293.13	0.9416
443.15	0.05934	301.83	0.9884	443.15	0.03902	300.45	0.9583
453.15	0.06080	309.12	1.0046	453.15	0.04002	307.81	0.9747
	P = 1 MPa				*P = 1.5 MPa*		
314.94	0.01744	203.76	0.6820	332.37	0.01132	209.06	0.6768
323.15	0.01837	210.32	0.7026	353.15	0.01305	226.73	0.7284
333.15	0.01941	217.97	0.7259	363.15	0.01377	234.77	0.7508
343.15	0.0204	225.49	0.7481	373.15	0.01446	242.65	0.7722
353.15	0.02134	232.91	0.7695	383.15	0.01512	250.41	0.7928
363.15	0.02225	240.28	0.7900	393.15	0.01575	258.10	0.8126
373.15	0.02313	247.61	0.8100	403.15	0.01636	265.74	0.8318
383.15	0.02399	254.93	0.8293	413.15	0.01696	273.35	0.8504
393.15	0.02483	262.25	0.8482	423.15	0.01754	280.94	0.8686
403.15	0.02566	269.57	0.8665	433.15	0.01811	288.52	0.8863
413.15	0.02647	276.90	0.8845	443.15	0.01867	296.11	0.9036
423.15	0.02728	284.26	0.9021	453.15	0.01922	303.70	0.9205

TABLE E-2 SI (CONTINUED)

Superheated R-12 Vapor

Temp. K T	Specific Volume m^3/kg	Enthalpy kJ/kg	Entropy kJ/(kg.K)	Temp. K T	Specific Volume m^3/kg	Enthalpy kJ/kg	Entropy kJ/(kg.K)
	P = 1 MPa				*P = 1.5 MPa*		
433.15	0.02807	291.63	0.9193	463.15	0.01977	311.31	0.9371
443.15	0.02885	299.04	0.9362	473.15	0.02031	318.93	0.9534
453.15	0.02963	306.47	0.9528	483.15	0.02084	326.58	0.9694
	P = 2 MPa			493.15	0.02137	334.24	0.9851
346.03	0.00813	211.97	0.6713		*P = 4 MPa*		
353.15	0.00870	219.02	0.6914	383.62	0.00239	196.9	0.6046
363.15	0.00941	228.23	0.7171	393.15	0.00374	225.18	0.6777
373.15	0.01003	236.94	0.7408	403.15	0.00433	38.690	0.7116
383.15	0.01061	245.34	0.7630	413.15	0.00478	249.93	0.7392
393.15	0.01116	253.53	0.7841	423.15	0.00517	260.12	0.7636
403.15	0.01168	261.58	0.8043	433.15	0.00552	269.71	0.7860
413.15	0.01217	269.53	0.8238	443.15	0.00585	278.90	0.8069
423.15	0.01265	277.41	0.8426	453.15	0.00615	287.82	0.8269
433.15	0.01312	285.24	0.8609	463.15	0.00643	296.55	0.8459
443.15	0.01357	293.04	0.8787	473.15	0.00671	305.14	0.8642
453.15	0.01401	300.82	0.8961	483.15	0.00697	313.61	0.8820
463.15	0.01445	308.59	0.9131	493.15	0.00723	322.01	0.8992
473.15	0.01488	316.36	0.9297				
483.15	0.01530	324.14	0.9459				
493.15	0.01572	331.92	0.9619				

TABLE E-3 SI

Compressed Liquid R-12

Temp. K T	Press. MPa P	Specific Vol. m^3/kg	Enthalpy kJ/kg	Entropy kJ/(kg.K)
183.15	0.0028	0.000608	–43.28	–0.2086
213.15	0.0226	0.000637	–17.49	0.0783
243.35	0.1013	0.000672	9.05	0.0379
258.15	0.1826	0.000693	22.33	0.0906
273.15	0.308	0.000716	36.05	0.1420
288.15	0.491	0.000743	50.10	0.1915
303.15	0.744	0.000774	64.59	0.2399
318.15	1.084	0.000811	79.71	0.2877
333.15	1.525	0.000858	95.74	0.3358
348.15	2.087	0.000920	113.15	0.3854
363.15	2.788	0.001012	132.84	0.4389
373.15	3.344	0.001113	148.61	0.4793
385.15	4.116	0.001792	183.43	0.5689

Adapted from Sonntag, R. E., Borgnakke, C., and Van Wylen, G. J., *Fundamentals of Thermodynamics*, John Wiley & Sons, New York, 1998. With permission.

F-Series Tables

Thermodynamic Properties of R-134a

TABLE F-1 SI

Saturated R-134a

		Specific Volume m^3/kg		Enthalpy kJ/kg		Entropy kJ/(kg.K)	
Temp. K T	Press. MPa P	Sat. Liquid v_f	Sat. Vapor v_g	Sat. Liquid h_f	Sat. Vapor h_g	Sat. Liquid s_f	Sat. Vapor s_g
203.15	0.0083	0.000675	1.97274	119.47	354.62	0.6645	1.8220
208.15	0.0117	0.000679	1.42983	123.18	357.73	0.6825	1.8094
213.15	0.0163	0.000684	1.05268	127.53	360.86	0.7031	1.7978
218.15	0.0222	0.000689	0.78678	132.37	364.00	0.7256	1.7874
223.15	0.0299	0.000695	0.59657	137.62	367.16	0.7493	1.7780
228.15	0.0396	0.000701	0.45853	143.18	370.32	0.7740	1.7695
233.15	0.0518	0.000708	0.35696	148.98	373.48	0.7991	1.7820
238.15	0.0668	0.000715	0.28122	154.98	376.64	0.8245	1.7553
243.15	0.0851	0.000722	0.22402	161.12	379.80	0.8499	1.7493
246.85	0.1013	0.000728	0.19020	165.80	382.16	0.8690	1.7453
248.15	0.1072	0.000730	0.18030	167.38	382.95	0.8754	1.7441
253.15	0.1337	0.000738	0.14649	173.74	386.08	0.9007	1.7395
258.15	0.1650	0.000746	0.12007	180.19	389.20	0.9258	1.7354
263.15	0.2017	0.000755	0.09921	186.72	392.28	0.9507	1.7319
268.15	0.2445	0.000764	0.08257	193.32	395.34	0.9755	1.7288
273.15	0.2940	0.000773	0.06919	200.00	398.36	1.0000	1.7262
278.15	0.3509	0.000783	0.05833	206.75	401.32	1.0243	1.7239
283.15	0.4158	0.000794	0.04945	213.58	404.23	1.0485	1.7218
288.15	0.4895	0.000805	0.04213	220.49	407.07	1.0725	1.7200
293.15	0.5728	0.000817	0.03606	227.49	409.84	1.0963	1.7183
298.15	0.6663	0.000829	0.03098	234.59	412.51	1.1201	1.7168
303.15	0.7710	0.000843	0.02671	241.79	415.08	1.1437	1.7153
308.15	0.8876	0.000857	0.02310	249.10	417.52	1.1673	1.7139
313.15	1.0170	0.000873	0.02002	256.54	419.82	1.1909	1.7123
318.15	1.1602	0.000890	0.01739	264.11	421.96	1.2146	1.7106
323.15	1.3181	0.000908	0.01512	271.83	423.91	1.2391	1.7088
328.15	1.4916	0.000928	0.01316	279.72	425.65	1.2619	1.7066
333.15	1.6818	0.000951	0.01146	287.79	427.13	1.2857	1.7040
338.15	1.8899	0.000976	0.00997	296.09	428.30	1.3099	1.7008
343.15	2.1170	0.001005	0.00866	304.64	429.11	1.3343	1.6970
348.15	2.3644	0.001038	0.00749	313.51	429.45	1.3592	1.6923
353.15	2.6336	0.001078	0.00645	322.79	429.19	1.3849	1.6862
358.15	2.9262	0.001128	0.00550	332.65	428.10	1.4117	1.6782
363.15	3.2445	0.001195	0.00461	343.38	425.70	1.4404	1.6671
368.15	3.5915	0.001297	0.00373	355.83	420.81	1.4733	1.6498
373.15	3.9732	0.001557	0.00264	374.74	407.21	1.5228	1.6098
374.35	4.0640	0.001969	0.00197	390.98	390.98	1.5658	1.5658

TABLE F-2 SI

Superheated R-134a Vapor

Temp. K T	Specific Volume m^3/kg	Enthalpy kJ/kg	Entropy kJ/(kg.K)	Temp. K T	Specific Volume m^3/kg	Enthalpy kJ/kg	Entropy kJ/(kg.K)
	P = 0.05 MPa				*P = 0.1 MPa*		
232.63	0.36889	373.06	1.7629	246.61	0.19257	381.98	1.7456
253.15	0.40507	388.82	1.8279	253.15	0.19863	387.22	1.7665
263.15	0.42222	396.64	1.8582	263.15	0.20765	395.27	1.7978
273.15	0.43921	404.59	1.8878	273.15	0.21652	403.41	1.8281
283.15	0.45608	412.7	1.917	283.15	0.22527	411.67	1.8578
293.15	0.47287	420.96	1.9456	293.15	0.23392	420.05	1.8869
303.15	0.48958	429.38	1.9739	303.15	0.2425	428.56	1.9155
313.15	0.50623	437.96	2.0017	313.15	0.25101	437.22	1.9436
323.15	0.52284	446.7	2.0292	323.15	0.25948	446.03	1.9712
333.15	0.53941	455.6	2.0563	333.15	0.26791	454.99	1.9958
343.15	0.55595	464.66	2.0831	343.15	0.27631	464.1	2.0255
353.15	0.57247	473.88	2.1096	353.15	0.28468	473.36	2.0521
363.15	0.58896	483.26	2.1358	363.15	0.29302	482.78	2.0784
373.15	0.60544	492.81	2.1617	373.15	0.30135	492.35	2.1044
383.15	0.6219	502.5	2.1874	383.15	0.30967	502.07	2.1601
393.15	0.63835	512.36	2.2128	393.15	0.31797	511.95	2.1555
	P = 0.15 MPa				*P = 0.2 MPa*		
256.01	0.13139	387.77	1.7372	263.08	0.10002	392.15	1.7320
263.15	0.13602	393.84	1.7606	273.15	0.10501	400.91	1.7647
273.15	0.14222	402.19	1.7917	283.15	0.10974	409.50	1.7956
283.15	0.14828	410.60	1.8220	293.15	0.11436	418.15	1.8256
293.15	0.15424	419.11	1.8515	303.15	0.11889	426.87	1.8549
303.15	0.16011	427.73	1.8804	313.15	0.12335	435.71	1.8836
313.15	0.16592	436.47	1.9088	323.15	0.12776	444.66	1.9117
323.15	0.17168	445.35	1.9367	333.15	0.13213	453.74	1.9394
333.15	0.17740	454.37	1.9642	343.15	0.13646	462.95	1.9666
343.15	0.18308	463.53	1.9913	353.15	0.14076	472.30	1.9935
353.15	0.18874	472.83	2.018	363.15	0.14504	481.79	2.0200
363.15	0.19437	482.28	2.0444	373.15	0.14930	481.42	2.0461
373.15	0.19999	491.89	2.0705	383.15	0.15355	501.21	2.0720
383.15	0.20559	501.64	2.0963	393.15	0.15777	511.13	2.0976
393.15	0.21117	511.54	2.1218	403.15	0.16199	521.21	2.1229
403.15	0.21675	521.60	2.1470	413.15	0.16620	531.43	2.1479
	P = 0.4 MPa				*P = 0.5 MPa*		
282.14	0.05136	403.56	1.7223	288.96	0.04126	407.45	1.7198
283.15	0.05168	404.65	1.7261	303.15	0.04446	421.22	1.7663
293.15	0.05436	413.97	1.7584	313.15	0.04656	430.72	1.7971
303.15	0.05693	423.22	1.7895	323.15	0.04858	440.2	1.8270
313.15	0.05940	432.46	1.8195	333.15	0.05055	449.72	1.8560
323.15	0.05191	441.75	1.8487	343.15	0.05247	459.29	1.8843
333.15	0.06417	451.10	1.8772	353.15	0.05435	468.94	1.9120
343.15	0.06648	460.55	1.9051	363.15	0.05620	478.69	1.9392

TABLE F-2 SI (CONTINUED)

Superheated R-134a Vapor

Temp. K T	Specific Volume m³/kg	Enthalpy kJ/kg	Entropy kJ/(kg.K)	Temp. K T	Specific Volume m³/kg	Enthalpy kJ/kg	Entropy kJ/(kg.K)
	P = 0.4 MPa				*P = 0.5 MPa*		
353.15	0.06877	470.09	1.9325	373.15	0.05804	488.55	1.9660
363.15	0.07102	479.75	1.9595	383.15	0.05985	498.52	1.9924
373.15	0.07325	489.52	1.9860	393.15	0.06164	508.61	2.0184
383.15	0.0757	499.43	2.0122	403.15	0.06342	518.83	2.0440
393.15	0.07797	509.46	2.0381	413.15	0.06518	529.19	2.0694
403.15	0.07985	519.63	2.0636	423.15	0.06694	539.67	2.0945
413.15	0.08202	529.94	2.0889	433.15	0.06869	550.29	2.1193
423.15	0.08418	540.38	2.1139	443.15	0.07043	561.04	2.1438
	P = 0.7 MPa				*P = 0.9 MPa*		
299.97	0.02947	413.38	1.7163	308.8	0.02276	417.76	1.7137
303.15	0.03007	416.81	1.7277	323.15	0.02487	433.23	1.7627
313.15	0.03178	426.93	1.7606	333.15	0.02619	443.60	1.7943
323.15	0.03339	436.89	1.7919	343.15	0.02745	453.83	1.8246
333.15	0.03493	446.78	1.8220	353.15	0.02865	464.03	1.8539
343.15	0.03641	456.66	1.8512	363.15	0.02981	474.22	1.8823
353.15	0.03785	466.55	1.8796	373.15	0.03094	484.44	1.9101
363.15	0.03925	476.51	1.9074	383.15	0.03204	494.73	1.9373
373.15	0.04063	486.53	1.9347	393.15	0.03313	505.09	1.9640
383.15	0.04198	496.65	1.9614	403.15	0.03419	515.54	1.9902
393.15	0.04331	506.88	1.9878	413.15	0.03524	526.10	2.0161
403.15	0.04463	517.21	2.0137	423.15	0.03628	536.76	2.0416
413.15	0.04594	527.66	2.0393	433.15	0.03731	547.54	2.0668
423.15	0.04723	538.23	2.0646	443.15	0.03832	558.44	2.0917
433.15	0.04851	548.92	2.0896	453.15	0.03933	569.45	2.1162
443.15	0.04979	559.75	2.1143	463.15	0.04034	580.60	2.141
	P = 1 MPa				*P = 1.2 MPa*		
312.52	0.02038	419.54	1.7125	319.61	0.01676	422.49	1.7102
313.15	0.02047	420.25	1.7148	323.15	0.01724	426.84	1.7237
323.15	0.02185	431.24	1.7494	333.15	0.01844	438.21	1.7584
333.15	0.02311	441.89	1.7818	343.15	0.01953	449.18	1.7908
343.15	0.02429	452.34	1.8127	353.15	0.02055	459.92	1.8217
353.15	0.02542	462.70	1.8425	363.15	0.05151	470.55	1.8514
363.15	0.02650	473.03	1.8713	373.15	0.02244	481.13	1.8801
373.15	0.02754	483.36	1.8994	383.15	0.02333	491.70	1.9081
383.15	0.02856	493.74	1.9268	393.15	0.02420	502.31	1.9351
393.15	0.02956	504.17	1.9537	403.15	0.02504	512.97	1.9621
403.15	0.03053	514.69	1.9801	413.15	0.02587	523.70	1.9884
413.15	0.03150	525.30	2.0061	423.15	0.02669	534.51	2.0143
423.15	0.03244	536.02	2.0318	433.15	0.02750	545.43	2.0398
433.15	0.03338	546.84	2.0570	443.15	0.02829	556.44	2.0649
443.15	0.03431	557.77	2.0820	453.15	0.02907	567.57	2.0898
453.15	0.03523	568.83	2.1067	463.15	0.02985	578.80	2.1140

TABLE F-2 SI (CONTINUED)

Superheated R-134a Vapor

Temp. K T	Specific Volume m³/kg	Enthalpy kJ/kg	Entropy kJ/(kg.K)	Temp. K T	Specific Volume m³/kg	Enthalpy kJ/kg	Entropy kJ/(kg.K)
	P = 1.4 *MPa*				*P* = 1.6 *MPa*		
325.72	0.01414	424.78	1.7077	331.2	0.01215	426.54	1.7051
333.15	0.01503	434.08	1.7360	333.15	0.01239	429.32	1.7135
343.15	0.01608	445.72	1.7704	343.15	0.01345	441.89	1.7507
353.15	0.01704	456.94	1.8026	353.15	0.01438	453.72	1.7847
363.15	0.01793	467.93	1.8333	363.15	0.01522	465.15	1.8166
373.15	0.01878	478.79	1.8628	373.15	0.01601	476.33	1.8469
383.15	0.01958	489.59	1.8914	383.15	0.01676	487.39	1.8762
393.15	0.02036	500.38	1.9192	393.15	0.01748	498.39	1.9045
403.15	0.02112	511.19	1.9463	403.15	0.01817	509.37	1.9321
413.15	0.02186	522.05	1.9730	413.15	0.01884	520.38	1.9591
423.15	0.02258	532.98	1.9991	423.15	0.01949	531.43	1.9855
433.15	0.02329	543.99	2.0248	433.15	0.02013	542.54	2.0115
443.15	0.02399	555.1	2.0502	443.15	0.02076	553.73	2.0370
453.15	0.02468	566.3	2.0752	453.15	0.02138	565.02	2.0622
	P = 1.8 *MPa*				*P* = 2 *MPa*		
336.19	0.01057	427.85	1.7022	340.63	0.00930	428.75	1.6991
343.15	0.01134	437.56	1.7309	343.15	0.00958	432.53	1.7101
353.15	0.01227	450.20	1.7672	353.15	0.01055	446.30	1.7497
363.15	0.01310	432.16	1.8006	363.15	0.01137	458.95	1.7850
373.15	0.01385	473.74	1.8320	373.15	0.01211	471.00	1.8177
383.15	0.01456	485.09	1.8620	383.15	0.01279	482.69	1.8487
393.15	0.01523	496.32	1.8910	393.15	0.01342	494.19	1.8783
403.15	0.01587	507.50	1.9190	403.15	0.01406	505.57	1.9069
413.15	0.01649	518.66	1.9464	413.15	0.01461	516.90	1.9346
423.15	0.01709	529.84	1.9731	423.15	0.01517	528.22	1.9617
433.15	0.01768	541.07	1.9994	433.15	0.01571	539.57	1.9882
443.15	0.01825	552.36	2.0251	443.15	0.01624	550.96	2.0142
453.15	0.01881	563.72	2.0505	453.15	0.01676	562.42	2.0398
	P = 2.5 *MPa*				*P* = 3 *MPa*		
350.87	0.00694	429.41	1.6893	359.5	0.00528	427.67	1.6759
353.15	0.00722	433.80	1.7018	363.15	0.00575	436.19	1.6995
363.15	0.00816	449.50	1.7457	373.15	0.00665	453.73	1.7472
373.15	0.00891	463.28	1.7831	383.15	0.00734	468.5	1.7862
383.15	0.00956	476.13	1.8171	393.15	0.00792	482.04	1.8211
393.15	0.01015	488.46	1.8489	403.15	0.00845	494.91	1.8535
403.15	0.01069	500.47	1.8790	413.15	0.00893	507.39	1.884
413.15	0.01121	512.31	1.9080	423.15	0.00937	519.62	1.9133
423.15	0.01170	524.04	1.9361	433.15	0.00980	531.70	1.9415
433.15	0.01217	535.72	1.9634	443.15	0.01021	543.71	1.9689
443.15	0.01262	547.40	1.9900	453.15	0.01060	555.69	1.9956
453.15	0.01307	559.10	2.0161	463.15	0.01098	567.70	2.0220

TABLE F-2 SI (CONTINUED)

Superheated R-134a Vapor

Temp. K T	Specific Volume m³/kg	Enthalpy kJ/kg	Entropy kJ/(kg.K)	Temp. K T	Specific Volume m³/kg	Enthalpy kJ/kg	Entropy kJ/(kg.K)
	P = 3.5 *MPa*				*P* = 4 *MPa*		
367.02	0.00396	422.43	1.6562	373.63	0.00252	404.94	1.6036
373.15	0.00484	440.43	1.7039	383.15	0.00428	446.84	1.7148
383.15	0.00567	459.21	1.7535	393.15	0.00500	465.99	1.7642
393.15	0.00629	474.70	1.7935	403.15	0.00556	481.87	1.8040
403.15	0.00681	488.77	1.8288	413.15	0.00603	496.29	1.8394
413.15	0.00728	502.08	1.8614	423.15	0.00644	509.92	1.8720
423.15	0.00771	514.93	1.8922	433.15	0.00683	523.07	1.9027
433.15	0.00810	527.50	1.9215	443.15	0.00718	535.92	1.9320
443.15	0.00848	539.89	1.9498	453.15	0.00752	548.57	1.9603
453.15	0.00884	552.18	1.9772	463.15	0.00784	561.10	1.98775
	P = 5 *MPa*				*P* = 6 *MPa*		
363.15	0.001089	336.61	1.4163	363.15	0.001059	334.70	1.4081
373.15	0.001216	357.68	1.4735	373.15	0.001150	353.61	1.4595
383.15	0.001659	392.1	1.5644	383.15	0.001307	375.90	1.5184
393.15	0.002969	440.47	1.6892	393.15	0.001698	406.78	1.5979
403.15	0.003705	464.63	1.7499	403.15	0.002396	441.18	1.6843
413.15	0.004226	482.86	1.7946	413.15	0.002985	466.25	1.7458
423.15	0.004652	498.77	1.8327	423.15	0.003439	485.82	1.7926
433.15	0.005023	513.48	1.8670	433.15	0.003814	502.77	1.8322
443.15	0.005357	527.47	1.8990	443.15	0.004141	518.30	1.8676
453.15	0.005665	541	1.9292	453.15	0.004435	532.96	1.9004
	P = 7 *MPa*				*P* = 8 *MPa*		
363.15	0.001037	333.29	1.4013	363.15	0.001019	332.20	1.3955
373.15	0.001110	351.10	1.4497	373.15	0.001081	349.30	1.442
383.15	0.001215	370.68	1.5015	383.15	0.001163	367.57	1.4903
393.15	0.001393	393.45	1.5601	393.15	0.001282	387.56	1.5417
403.15	0.001720	420.73	1.6286	403.15	0.001465	409.98	1.5981
413.15	0.002169	448.28	1.6961	413.15	0.001736	434.40	1.6579
423.15	0.002599	471.55	1.7518	423.15	0.002061	458.21	1.7148
433.15	0.002968	491.16	1.7976	433.15	0.002384	479.59	1.7648
443.15	0.003287	508.52	1.8373	443.15	0.002680	498.61	1.8082
453.15	0.003569	524.51	1.8729	453.15	0.002946	515.93	1.8469
	P = 10 *MPa*				*P* = 20 *MPa*		
363.15	0.000991	330.62	1.3856	363.15	0.000912	327.89	1.3520
373.15	0.001040	346.85	1.4297	373.15	0.000939	342.49	1.3917
383.15	0.001100	363.73	1.4744	383.15	0.000969	357.33	1.4309
393.15	0.001175	381.44	1.5200	393.15	0.001002	372.38	1.4697
403.15	0.001272	400.16	1.5670	403.15	0.001037	387.65	1.5081
413.15	0.001400	419.98	1.6155	413.15	0.001076	403.13	1.5460
423.15	0.001564	440.63	1.6679	423.15	0.001118	418.83	1.5836
433.15	0.001758	461.34	1.7133	433.15	0.001164	434.72	1.6207
443.15	0.001965	481.30	1.7589	443.15	0.001214	450.80	1.6574
453.15	0.002172	500.12	1.8009	453.15	0.001268	467.03	1.6936

TABLE F-3 SI

Compressed Liquid R-134a

Temp. K T	Specific Volume m³/kg	Enthalpy kJ/kg	Entropy kJ/(kg.K)	Temp. K T	Specific Volume m³/kg	Enthalpy kJ/kg	Entropy kJ/(kg.K)
	P = 0.1 MPa				*P = 0.25 MPa*		
240	0.000717	157.24	0.8338	240	0.000717	157.30	0.8336
240.15	0.000718	157.43	0.8346	243.15	0.000722	161.18	0.8497
241.15	0.000719	158.65	0.8397	248.15	0.000729	167.43	0.8751
242.15	0.000721	159.89	0.8448	251.15	0.000734	171.23	0.8904
243.15	0.000722	161.12	0.8499	253.15	0.000737	173.78	0.9005
243.65	0.000723	161.40	0.8525	256.15	0.000742	177.63	0.9156
244.15	0.000724	162.37	0.8550	258.15	0.000746	180.22	0.9257
244.65	0.000724	162.99	0.8576	261.15	0.000751	184.12	0.9407
245.15	0.000725	163.61	0.8601	263.15	0.000754	186.74	0.9507
245.65	0.000726	164.24	0.8626	266.15	0.000760	190.68	0.9656
246.15	0.000727	164.86	0.8652	268.15	0.000764	193.33	0.9755
246.61	0.000727	165.44	0.8675	268.739	0.000765	194.11	0.9784
	P = 1 MPa				*P = 2 MPa*		
240.15	0.000716	157.58	0.8325	233.3	0.000715	157.95	0.8311
243.15	0.000721	161.45	0.8486	247.15	0.000725	166.78	0.8674
253.15	0.000736	174.03	0.8993	265.15	0.000755	189.85	0.9575
263.15	0.000753	186.95	0.9493	273.15	0.000769	200.4	0.9967
268.15	0.000761	193.52	0.9741	281.15	7.85E-05	211.2	1.0354
273.15	0.000771	200.2	0.9986	289.15	0.000802	222.1	1.0738
278.15	0.000781	206.9	1.023	297.15	0.000821	233.2	1.1118
283.15	0.000792	213.7	1.0472	303.15	8.37E-05	241.8	1.1403
293.15	0.000815	227.5	1.0953	315.15	0.000873	259.4	1.1972
303.15	0.000842	241.8	1.1431	325.15	0.000910	274.7	1.245
308.15	0.000857	249.1	1.167	343.15	0.000953	432.5	1.7101
312.67	0.000871	255.6	1.1879	373.15	0.001211	471.0	1.8177
	P = 3 MPa				*P = 4 MPa*		
243.15	0.000718	162.18	0.8457	243.15	0.000717	162.55	0.8442
263.15	0.000749	187.54	0.9459	258.15	0.000739	181.4	0.9195
273.15	0.000767	200.7	0.9948	273.15	0.000765	200.9	0.9930
283.15	0.000786	214.1	1.0430	283.15	0.000784	214.3	1.0409
293.15	0.000808	227.8	1.0905	293.15	0.009805	227.9	1.0882
303.15	0.000833	241.8	1.1376	303.15	0.000829	241.9	1.1350
313.15	0.000861	256.3	1.1845	318.15	0.000871	263.5	1.2048
323.15	0.000894	271.2	1.2316	333.15	0.000925	286.4	1.2750
333.15	0.000935	286.9	1.2793	343.15	0.000972	302.5	1.3228
343.15	0.000987	303.5	1.3285	353.15	0.001035	319.9	1.3726
353.15	0.001064	321.8	1.3811	363.15	0.001133	339.5	1.4273
359.34	0.001142	335.1	1.4183	373.51	0.001631	378.2	1.5320

TABLE F-3 SI (CONTINUED)

Compressed Liquid R-134a

Temp. K T	Specific Volume m³/kg	Enthalpy kJ/kg	Entropy kJ/(kg.K)	Temp. K T	Specific Volume m³/kg	Enthalpy kJ/kg	Entropy kJ/(kg.K)
	P = 5 MPa				*P = 6 MPa*		
233.3	0.000714	158.07	0.8307	248.15	0.000721	169.46	0.8665
247.15	0.000715	169.71	0.8559	265.15	0.000747	191.06	0.9507
265.15	0.000741	192.34	0.9443	273.15	0.000761	201.5	0.9894
273.15	0.000753	202.7	0.9826	281.15	0.000775	212.1	1.0276
281.15	0.000767	213.1	1.0203	289.15	0.000791	222.8	1.0652
289.15	0.000781	223.7	1.0575	298.15	0.000810	235.1	1.1071
297.15	0.000796	234.4	1.0941	303.15	0.000821	242.0	1.1302
303.15	0.000809	242.6	1.1212	315.15	0.000852	259.0	1.1851
315.15	0.000836	259.2	1.1755	325.15	0.000881	273.6	1.2307
325.15	0.000861	273.4	1.2192	343.15	0.000948	301.2	1.3132
343.15	0.000974	299.7	1.2982	363.15	0.001059	334.7	1.4081
373.15	0.001040	346.8	1.4297	373.15	0.001150	353.6	1.4595
	P = 7 MPa				*P = 10 MPa*		
248.15	0.000720	169.83	0.8651	248.15	0.000716	170.94	0.8609
265.15	0.000745	191.37	0.9490	265.15	0.000741	192.34	0.9443
273.15	0.000759	201.8	0.9877	273.15	0.000753	202.7	0.9826
281.15	0.000773	213.3	1.0257	281.15	0.000767	213.1	1.0203
289.15	0.000788	223.0	1.0632	289.15	0.000781	223.7	1.0575
298.15	0.000807	235.2	1.1049	298.15	0.000798	235.8	1.0986
303.15	0.000818	242.1	1.1278	303.15	0.000809	242.6	1.1213
315.15	0.000847	259.0	1.1825	315.15	0.000836	259.2	1.1750
325.15	0.000876	273.5	1.2277	325.15	0.000861	273.4	1.2192
343.15	0.000938	300.7	1.3090	343.15	0.000914	299.7	1.2982
363.15	0.001037	333.3	1.4013	363.15	0.000991	330.6	1.3856
373.15	0.001110	351.1	1.4497	373.15	0.001040	346.8	1.4297
	P = 15 MPa						
233.3	0.000714	158.07	0.8307				
247.15	0.000722	183.98	0.8983				
265.15	0.000734	194.05	0.9369				
273.15	0.000745	204.3	0.9748				
281.15	0.000758	214.6	1.0120				
289.15	0.000771	225.0	1.0486				
297.15	0.000785	235.6	1.0847				
303.15	0.000796	243.6	1.1114				
315.15	0.000819	259.9	1.1639				
325.15	0.000841	273.7	1.2071				
343.15	0.000885	299.2	1.2834				
373.15	0.000978	343.7	1.4078				

Adapted from Sonntag, R. E., Borgnakke, C., and Van Wylen, G. J., *Fundamentals of Thermodynamics*, John Wiley & Sons, New York, 1998. With permission.

AA-Series Tables

TABLE AA-1 U.S.

Critical Constants

Substance	Formula	Molec. Mass	Vol. ft^3/lb mol	Temp. °R	Pressure $lb_f/in.^2$
Ammonia	NH_3	17.031	1.161305	729.90	1646.20
Argon	Ar	39.948	1.199748	271.44	706.34
Bromine	Br_2	159.808	2.037490	1058.40	1493.91
Carbon dioxide	CO_2	44.01	1.504090	547.38	1070.40
Carbon monoxide	CO	28.01	1.492878	239.22	507.64
Chlorine	Cl_2	70.906	1.983028	750.42	1157.42
Fluorine	F_2	37.997	1.061993	259.74	757.11
Helium	He	4.003	0.919433	9.34	32.92
Hydrogen	H_2	2.016	1.042772	59.76	188.55
Krypton	Kr	83.8	1.460842	376.92	797.72
Neon	Ne	20.183	0.666349	79.92	400.31
Nitric oxide	NO	30.006	0.924239	324.00	939.86
Nitrogen	N_2	28.013	1.438416	227.16	491.69
Nitrogen dioxide	NO_2	46.006	2.687820	775.80	1464.90
Nitrous oxide	N_2O	44.013	1.560153	557.28	1050.09
Oxygen	O_2	31.999	1.175721	278.28	731.00
Sulfur dioxide	SO_2	64.063	1.957400	775.44	1142.92
Water	H_2O	18.015	0.914628	1165.14	3208.28
Xenon	Xe	131.3	1.896531	521.46	847.03
Benzene	C_6H_6	78.114	4.148662	1011.96	709.25
n-Butane	C_4H_{10}	58.124	4.084590	765.36	551.15
Chlorodifluoromethane (r-22)	$CHClF_2$	86.469	2.652581	664.74	720.85
Dichlorodifluoromethane (r-12)	CCl_2F_2	120.914	3.471101	693.00	600.47
Ethane	C_2H_6	30.07	2.375469	549.72	707.80
Ethyl alcohol	C_2H_5OH	46.069	2.676608	925.02	890.55
Ehylene	C_2H_4	28.054	2.088747	508.32	731.00
n-Heptane	C_7H_{16}	100.205	6.919776	972.54	397.41
n-Hexane	C_6H_{14}	86.178	5.926660	913.50	436.57
Methane	CH_4	16.043	1.588986	342.72	667.18
Methyl alcohol	CH_3OH	32.042	1.890124	922.68	1173.37
n-Octane	C_8H_{18}	114.232	7.880856	1023.84	361.15
n-Pentane	C_5H_{12}	72.151	4.869472	845.46	488.78
Propane	C_3H_8	44.094	3.251654	665.64	616.42
Tetrafluoroethane (r-134a)	CF_3CH_2F	102.03	3.216414	673.56	588.86

Adapted from Sonntag, R. E., Borgnakke, C., and Van Wylen, G. J., *Fundamentals of Thermodynamics*, John Wiley & Sons, New York, 1998. With permission.

TABLE AA-2 U.S.

Properties of Selected Ideal Gases at 77°F, 14.504 $lb_f/in.^2$ (or Saturation Pressure if it Is <14.504 $lb_f/in.^2$)

Gas	Formula	Molecular Mass	R Btu/(lb_m·°R)	v ft^3/lb_m	C_{vo} Btu/(lb_m·°R)	C_{po} Btu/(lb_m·°R)	k
Air	—	28.97	0.068549	13.70231	0.171253	0.239801	1.400
Ammonia	NH_3	17.031	0.116605	23.08069	0.392185	0.508742	1.297
Argon	Ar	39.948	0.049704	9.930564	0.074520	0.124200	1.667
Butane	C_4H_{10}	58.124	0.034155	6.654757	0.375705	0.409860	1.091
Carbon monoxide	CO	28.01	0.070889	14.17522	0.177701	0.248639	1.399
Carbon dioxide	CO_2	44.01	0.045118	9.024225	0.155966	0.201108	1.289
Ethane	C_2H_6	30.07	0.066041	13.10802	0.355880	0.421802	1.186
Ethanol	C_2H_5OH	46.069	0.043112	8.506638	0.297602	0.340833	1.145
Ethylene	C_2H_4	28.054	0.070794	14.07557	0.299035	0.369733	1.237
Helium	He	4.003	0.479388	99.18266	0.744244	1.240327	1.667
Hydrogen	H_2	2.016	0.985072	197.0234	0.240757	3.393761	1.409
Methane	CH_4	16.043	0.123794	24.71914	0.414636	0.538359	1.299
Methanol	CH_3OH	32.042	0.061981	12.22748	0.273717	0.335578	1.227
Neon	Ne	20.183	0.098405	19.67813	0.147607	0.246011	1.667
Nitric oxide	NO	30.006	0.066184	13.23802	0.171014	0.237174	1.387
Nitrogen	N_2	28.013	0.070889	14.17522	0.177940	0.248877	1.400
Nitrous oxide	N_2O	44.013	0.045118	9.024225	0.164804	0.209946	1.274
n-Octane	C_8H_{18}	114.23	0.017386	174.1087	0.391230	0.408665	1.044
Oxygen	O_2	31.999	0.062052	12.39783	0.158116	0.220216	1.393
Propane	C_3H_8	44.094	0.045046	8.859513	0.355880	0.401022	1.126
R-12	CCl_2F_2	120.914	0.016423	3.216466	0.130649	0.147129	1.126
R-134a	CF_3CH_2F	102.03	0.019464	3.813810	0.184150	0.203497	1.106
Steam	H_2O	18.015	0.110227	693.4199	0.336773	0.447120	1.327

TABLE AA-3 U.S.

Ideal-Gas Properties of Air, Standard Entropy at 14.504 $lb_f/in.^2$

T °R	v ft^3/lb_m	h Btu/lb_m	s° $Btu/(lb_m\ °R)$	P_r	v_r
360	9.19389	86.05868	1.543562	0.270	493.4700
380	9.70466	90.83535	1.556334	0.326	431.2700
400	10.21546	95.62306	1.568623	0.390	379.5200
420	10.72623	100.4108	1.58029	0.463	336.0700
440	11.237	105.2019	1.591439	0.545	299.2600
460	11.74777	109.9896	1.602071	0.636	267.8600
480	12.25854	114.7808	1.612289	0.738	240.8800
500	12.76931	119.5754	1.622057	0.852	217.5400
520	13.28008	124.37	1.631481	0.973	197.2440
536.67	13.7058	128.3804	1.639213	1.091	182.2880
540	13.79085	129.1795	1.640697	1.115	179.4910
560	14.30162	133.9662	1.649258	1.266	163.8850
580	14.81239	138.7677	1.65768	1.432	150.1040
600	15.32316	143.5761	1.665827	1.612	137.8800
620	15.83393	148.3845	1.673697	1.809	126.9930
660	16.85551	158.0186	1.688747	2.253	108.5270
700	17.87705	167.6735	1.702969	2.772	93.5670
720	18.38781	172.5264	1.709976	3.061	87.1370
760	19.40935	182.2023	1.722886	3.706	75.9780
800	20.43089	191.9227	1.735347	4.445	66.6780
850	21.70784	204.125	1.750121	5.515	57.0950
900	22.98474	216.4046	1.764342	6.766	49.2780
1000	25.53863	241.0942	1.790162	9.892	37.4520
1050	26.81557	253.5589	1.802347	11.812	32.9330
1100	28.09247	266.096	1.814014	14.003	29.1030
1150	29.36942	278.709	1.825233	16.492	25.8340
1200	30.64636	291.4049	1.836037	19.307	23.0260
1250	31.92327	304.1802	1.846462	22.480	20.6000
1300	33.20021	317.0314	1.85673	26.042	18.4939
1350	34.47715	329.9655	1.86631	30.028	16.6559
1400	35.75406	342.9789	1.875768	34.474	15.0451
1450	37.031	356.0683	1.88495	39.418	13.6278
1500	38.30794	369.2095	1.89389	44.902	12.3761
1550	39.58485	382.4784	1.902554	50.967	11.2669
1600	40.86179	395.7957	1.911011	57.658	10.2807
1650	42.13873	409.1819	1.919261	65.021	9.4013
1700	43.41567	422.6372	1.927304	73.107	8.6149
1750	44.69258	436.1408	1.93514	81.966	7.9098
1800	45.96952	449.7912	1.94301	91.651	7.2760
1850	47.24646	463.3966	1.950259	102.218	6.7050
1900	48.52337	477.1074	1.957577	113.726	6.1894
1950	49.80031	490.8768	1.964722	126.236	5.7228
2000	51.07725	504.7014	1.971729	139.809	5.2997
2050	52.35416	518.5813	1.978599	154.512	4.9153
2100	53.6311	532.5129	1.985295	170.412	4.5654
2150	54.90804	546.4964	1.991888	187.581	4.2463
2200	56.18495	560.5281	1.998343	206.090	3.9548
2250	57.46189	574.6641	2.004867	226.020	3.6880
2300	58.73883	588.7297	2.010873	247.440	3.4436

TABLE AA-3 U.S. (CONTINUED)

Ideal-Gas Properties of Air, Standard Entropy at 14.504 $lb_f/in.^2$

T °R	v ft^3/lb_m	h Btu/lb_m	s° $Btu/(lb_m\ °R)$	P_r	v_r
2350	60.01577	602.8996	2.006593	270.450	3.2192
2400	61.29268	617.1074	2.022955	295.110	3.0129
2450	62.56962	631.36	2.028823	321.520	2.8230
2500	63.84657	645.6507	2.034588	349.780	2.6479
2550	65.12347	659.9827	2.040283	379.970	2.4863
2600	66.40041	674.3493	2.045841	412.200	2.3368
2650	67.67736	688.7539	2.051329	446.560	2.1985
2700	68.95426	703.263	2.056955	483.160	2.0703
2750	70.2312	717.6735	2.062064	522.070	1.9515
2800	71.50815	732.1712	2.067276	563.430	1.8411
2850	72.78505	746.7035	2.07242	607.340	1.7385
2900	74.062	761.2703	2.077494	653.930	1.6430
2950	75.33894	775.8716	2.082465	703.300	1.5540
3000	76.61588	793.2344	2.087401	755.580	1.4710
3050	77.89279	805.1433	2.092233	810.890	1.3935
3100	79.16973	819.8136	2.096997	869.370	1.3210
3150	80.44667	834.593	2.101933	931.140	1.2533
3200	81.72358	849.2233	2.106351	996.330	1.1899
3250	83.00052	863.9972	2.110942	1065.100	1.1305
3300	84.27746	878.7711	2.11543	1137.570	1.0747
3350	85.55437	893.5796	2.119883	1213.890	1.0224
3400	86.83131	908.388	2.124266	1294.210	0.9733
3450	88.10825	923.2309	2.128616	1378.690	0.9271
3500	89.38516	938.1084	2.132896	1467.480	0.8836
3550	90.6621	952.9858	2.137107	1560.740	0.8427
3600	91.93904	967.9997	2.141526	1658.640	0.8041
3650	93.21595	982.8443	2.145392	1761.340	0.7677
3700	94.49289	997.7908	2.149465	1869.020	0.7334
3750	95.76983	1012.737	2.153504	1981.860	0.7010
3800	97.04677	1027.753	2.157473	2100.000	0.6704
3850	98.32368	1042.734	2.161374	2223.700	0.6414
3900	99.60062	1057.784	2.165274	2353.100	0.6140
3950	100.8776	1072.799	2.169106	2488.400	0.5881
4000	102.1545	1087.884	2.172868	2629.800	0.5635
4100	104.7084	1118.053	2.180324	2931.700	0.5181
4200	107.2622	1148.291	2.187608	3260.500	0.4772
4300	109.8161	1178.564	2.194753	3617.900	0.4403
4400	112.3699	1208.94	2.201726	4005.700	0.4069
4500	114.9238	1239.485	2.208802	4425.800	0.3767
4600	117.4777	1269.831	2.215257	4880.100	0.3492
4700	120.0315	1300.38	2.221816	5370.600	0.3242
4800	122.5854	1330.963	2.228271	5899.500	0.3014
4900	125.1392	1361.581	2.234588	6469.000	0.2806
5000	127.6931	1392.268	2.240801	7081.300	0.2616
5100	130.247	1422.989	2.246876	7738.700	0.2442
5200	132.8008	1453.78	2.252848	8443.700	0.2282
5300	135.3547	1484.605	2.258716	9198.900	0.2135
5400	137.9086	1515.622	2.264722	10006.600	0.1999

TABLE AA-4 U.S.

Ideal-Gas Properties of Nitrogen, Entropies at 14.504 $lb_f/in.^2$

		Diatomic Nitrogen (N_2) M = 28.013		Monatomic Nitrogen (N) M = 14.007	
T °R	v ft³/lb mol	$(\bar{h} \pm \overline{h^o_{536}})$ Btu/lb mol	$\overline{s^o}$ Btu/(lb mol°R)	$(\bar{h} \pm \overline{h^o_{536}})$ Btu/lb mol	$\overline{s^o}$ Btu/(lb mol°R)
0	0.0000	–3727.406	0.0000	–2664.214	0.0000
180	133.1737	–2479.779	38.1704	–1770.840	31.1916
360	266.3473	–1228.281	42.9887	–877.037	34.6329
536.4	396.8575	0.000	45.7650	0.000	36.6151
540	399.5210	23.216	45.8080	16.337	36.6459
720	532.6946	1277.292	47.8124	910.141	38.0742
900	665.8683	2541.257	49.3790	1803.944	39.1820
1080	799.0419	3823.708	50.6776	2697.318	40.0872
1260	932.2156	5131.955	51.7952	3591.122	40.8524
1440	1065.3892	6468.576	52.7888	4484.496	41.5152
1620	1198.5629	7834.432	53.6823	5378.299	42.1002
1800	1331.7365	9227.373	54.4977	6272.103	42.6232
1980	1464.9102	10644.819	55.2484	7165.477	43.0964
2160	1598.0838	12084.621	55.9442	8059.280	43.5282
2340	1731.2575	13543.770	56.5929	8953.084	43.9257
2520	1864.4311	15019.685	57.2005	9846.458	44.2935
2700	1997.6048	16511.078	57.7723	10740.261	44.6362
2880	2130.7784	18015.368	58.3116	11633.635	44.9565
3060	2263.9521	19531.266	58.8220	12527.439	45.2575
3240	2397.1257	21057.052	59.3064	13421.243	45.5412
3420	2530.2994	22591.866	59.7674	14314.616	45.8097
3600	2663.4730	24134.419	60.2071	15208.420	46.0645
3960	2929.8203	27240.591	61.0294	16996.457	46.5379
4320	3196.1676	30369.549	61.7856	18785.354	46.9702
4680	3462.5150	33517.853	62.4857	20575.971	47.3684
5040	3728.8623	36682.064	63.1370	22370.027	47.7377
5400	3995.2096	39860.033	63.7461	24169.243	48.0823
5760	4261.5569	43049.609	64.3179	25975.766	48.4062
6120	4527.9042	46249.504	64.8567	27792.608	48.7121
6480	4794.2515	49458.857	65.3662	29622.348	49.0026
6840	5060.5988	52676.378	65.8493	31467.564	49.2799
7200	5326.9461	55901.208	66.3089	33332.557	49.5455
7920	5859.6407	62371.934	67.1656	37130.901	50.0482
8640	6392.3353	68868.025	67.9505	41038.873	50.5204
9360	6925.0299	75387.332	68.6754	45061.205	50.9690
10080	7457.7245	81930.714	69.3487	49247.336	51.3984
10800	7990.4191	88498.172	69.9780	53563.733	51.8119

TABLE AA-5 U.S.

Ideal-Gas Properties of Oxygen, Entropies at 14.504 $lb_f/in.^2$

		Diatomic Oxygen (O_2) M = 31.999		Monatomic Oxygen (O) M = 16.00	
T °R	v ft³/lb mol	$(\bar{h} \pm \overline{h^o_{536}})$ Btu/lb mol	$\bar{s}^o$ Btu/(lb mol°R)	$(\bar{h} \pm \overline{h^o_{536}})$ Btu/lb mol	$\bar{s}^o$ Btu/(lb mol°R)
0	0	–3733	0	–2891.21	0
180	133.1737	–2483.65	41.39391	–1942.38	32.47038
360	266.3473	–1233.01	46.21262	–939.805	36.34112
536.4	396.8575	0	48.99876	0	38.46828
540	399.5210	23.21568	49.04190	17.62672	38.50053
720	532.6946	1301.368	51.08269	948.8334	39.99021
900	665.8683	2616.493	52.71162	1867.143	41.12879
1080	799.0419	3974.610	54.08665	2778.143	42.05121
1260	932.2156	5373.570	55.28447	3684.414	42.82746
1440	1065.389	6806.923	56.34852	4587.676	43.49766
1620	1198.563	8272.091	57.30654	5488.789	44.08737
1800	1331.737	9760.474	58.17785	6388.611	44.61403
1980	1464.910	11269.06	58.97655	7287.144	45.09005
2160	1598.084	12794.85	59.71400	8185.247	45.52403
2340	1731.257	14335.68	60.39887	9082.490	45.92290
2520	1864.431	15888.98	61.03850	9979.303	46.29216
2700	1997.605	17454.75	61.63848	10875.26	46.63562
2880	2130.778	19031.27	62.20359	11771.64	46.95687
3060	2263.952	20618.53	62.73837	12667.16	47.25853
3240	2397.126	22215.69	63.24568	13562.69	47.54299
3420	2530.299	23823.59	63.72862	14458.21	47.81193
3600	2663.473	25673.10	64.18936	15353.73	48.06702
3960	2929.820	28705.76	65.05350	17144.35	48.54113
4320	3196.168	32008.83	65.85173	18939.70	48.97416
4680	3462.515	35350.17	66.59454	20729.02	49.37279
5040	3728.862	38727.19	67.28982	22523.94	49.74252
5400	3995.210	42137.75	67.94354	24322.29	50.08718
5760	4261.557	45580.98	68.56072	26124.95	50.41034
6120	4527.904	49054.30	69.14565	27932.33	50.71463
6480	4794.251	52555.57	69.70144	29746.16	51.00268
6840	5060.599	56081.77	70.23096	31566.45	51.27592
7200	5326.946	59632.05	70.73684	33394.04	51.53650
7920	5859.641	66798.39	71.68554	37073.72	52.02350
8640	6392.335	74049.42	72.56162	40787.80	52.47229
9360	6925.030	81389.02	73.37752	44536.27	52.88884
10080	7457.725	88829.21	74.14326	48319.14	53.27840
10800	7990.419	96392.36	74.86792	52133.82	53.64407

TABLE AA-6

Beattie-Bridgeman Constants

Beattie-Bridgeman Equation of State: $$\mathbf{P} = \frac{\mathbf{R}_u T}{\bar{v}^2}\left(1-\frac{c}{\bar{v}T^3}\right)(\bar{v}+B)-\frac{A}{\bar{v}^2}$$

where $$A = A_0\left(1-\frac{a}{\bar{v}}\right) \text{ and } B = B_0\left(1-\frac{b}{\bar{v}}\right).$$

The constants in the Beattie-Bridgeman equation are as listed below for various gases, when T is in °R, P is in $lb_f/in.^2$, $\bar{v}$ is in ft^3/lb mol, and R_u = 10.73 $lb_f ft^3/(lb\ mol.°R)$.

Gas	A_o	a	B_o	b	c
Air	4,905.096	0.3093	0.7386	–0.01764	4,054,000
Carbon dioxide, CO_2	18,872.857	1.142	1.678	1.159	61,660,000
Hydrogen, H_2	744.510	–0.08105	0.3357	–0.6982	47,080
Nitrogen, N_2	5,068.324	0.4192	0.8083	–0.1107	3,924,000
Oxygen, O_2	5,620.956	0.4104	0.7407	0.06741	4,484,000

BB-Series Tables

Thermodynamic Properties of Water

TABLE BB-1 U.S.

Saturated Water

Temp. °R T	Pressure $lb_f/in.^2$ P	Specific Volume ft^3/lb_m Sat. Liquid v_f	Sat. Vapor v_g	Enthalpy Btu/lb_m Sat. Liquid h_f	Sat. Vapor h_g	Entropy $Btu/(lb_m \cdot °R)$ Sat. Liquid s_f	Sat. Vapor s_g
491.67	0.088663	0.016018	3301.822	0	1075.380	0	2.186921
500.67	0.126489	0.016018	2356.536	9.019722	1079.331	0.018176	2.155751
509.67	0.178051	0.016018	1703.931	18.05234	1083.287	0.036066	2.125824
518.67	0.247293	0.016034	1248.203	27.07636	1087.229	0.053621	2.097378
527.67	0.339249	0.016050	925.6754	36.08748	1091.163	0.070842	2.070101
536.67	0.459632	0.016066	694.5293	45.08571	1095.079	0.087728	2.044019
545.67	0.615840	0.016082	526.8833	54.07104	1098.983	0.104352	2.019036
554.67	0.816285	0.016114	403.9067	63.05207	1102.865	0.120665	1.995080
563.67	1.070975	0.016146	312.7178	72.02880	1106.726	0.136715	1.972127
572.67	1.391369	0.016178	244.4042	81.00553	1110.565	0.152527	1.950105
574.128	1.450400	0.016178	235.0417	82.46296	1111.184	0.155059	1.946618
581.67	1.791244	0.016210	192.7254	89.98656	1114.378	0.168076	1.928967
590.67	2.285540	0.016258	153.2658	98.96758	1118.162	0.183410	1.908665
599.67	2.892243	0.016290	122.8694	107.9572	1121.915	0.198505	1.889152
608.67	3.630351	0.016338	99.25650	116.9511	1125.634	0.213385	1.870378
617.67	4.523798	0.016386	80.76548	125.9494	1129.314	0.228050	1.852298
626.67	5.595643	0.016434	66.17404	134.9562	1132.960	0.242524	1.834910
635.67	6.873446	0.016483	54.57573	143.9716	1136.562	0.256807	1.818119
644.67	8.387663	0.016531	45.29202	152.9999	1140.118	0.270899	1.801949
653.67	10.17311	0.016595	37.81145	162.0368	1143.626	0.284800	1.786328
662.67	12.26313	0.016659	31.74543	171.0824	1147.082	0.298557	1.771233
670.986	14.50400	0.016707	27.13449	179.4658	970.7680	0.311097	1.757739
671.67	14.69255	0.016723	26.79651	180.1451	1150.487	0.312124	1.756664
680.67	17.52083	0.016771	22.73531	189.2207	1153.832	0.325523	1.742572
689.67	20.78423	0.016851	19.38402	198.3092	1157.117	0.338779	1.728910
698.67	24.52626	0.016915	16.60394	207.4192	1160.337	0.351892	1.715678
707.67	28.79044	0.016979	14.28581	216.5464	1163.492	0.364837	1.702852
716.67	33.66378	0.017059	12.34331	225.6908	1166.571	0.377663	1.690408
725.67	39.17530	0.017139	10.70803	234.8610	1169.580	0.390346	1.678346
734.67	45.39752	0.017219	9.325199	244.0527	1172.504	0.402909	1.666595
743.67	52.40295	0.017299	8.150759	253.2702	1175.345	0.415353	1.655154
752.67	60.24962	0.017380	7.149154	262.5135	1178.093	0.427677	1.644024
761.67	69.02454	0.017460	6.291550	271.7868	1180.749	0.439882	1.633180
770.67	78.77122	0.017556	5.554402	281.0903	1183.308	0.451992	1.622576
779.67	89.60571	0.017652	4.918487	290.4239	1185.758	0.463982	1.612234
788.67	101.6005	0.017748	4.367948	299.7918	1188.097	0.475877	1.602059
797.67	114.8282	0.017844	3.889651	309.1985	1190.320	0.487676	1.592218
806.67	129.3757	0.017956	3.472702	318.6395	1192.418	0.499403	1.582497
815.67	145.3591	0.018052	3.108293	328.1192	1194.387	0.511011	1.572967

TABLE BB-1 U.S. (CONTINUED)

Saturated Water

		Specific Volume ft³/lbm		Enthalpy Btu/lbm		Entropy Btu/(lbm·°R)	
Temp. °R T	Pressure lbf/in.² P	Sat. Liquid v_f	Sat. Vapor v_g	Sat. Liquid h_f	Sat. Vapor h_g	Sat. Liquid s_f	Sat. Vapor s_g
824.67	162.8364	0.018164	2.788574	337.6420	1196.222	0.522547	1.563581
833.67	181.9382	0.018277	2.507458	347.2077	1197.916	0.534012	1.554361
842.67	202.7369	0.018405	2.259339	356.8164	1199.460	0.545405	1.545261
851.67	225.3632	0.018517	2.021472	366.4767	1200.844	0.556702	1.536305
860.67	249.9039	0.018645	1.845434	376.1929	1202.069	0.567952	1.527443
869.67	276.4898	0.018789	1.672920	385.9607	1203.123	0.579130	1.518678
878.67	305.1932	0.018917	1.518346	395.7887	1203.995	0.590260	1.510008
887.67	336.1737	0.019061	1.380591	405.6768	1204.687	0.601342	1.501385
896.67	369.5184	0.019206	1.257253	415.6338	1205.182	0.612377	1.492835
905.67	405.3723	0.019366	1.146568	425.6638	1205.474	0.623364	1.484332
914.67	443.8369	0.019526	1.046936	435.7712	1205.552	0.634303	1.475853
923.67	485.0428	0.019686	0.957236	445.9603	1205.414	0.645242	1.467374
932.67	529.1349	0.019862	0.876185	456.2354	1205.044	0.656134	1.458942
941.67	576.2439	0.020039	0.802982	466.6094	1204.429	0.667025	1.450487
950.67	626.5003	0.020231	0.736508	477.0908	1203.565	0.677892	1.442008
959.67	680.0345	0.020439	0.675960	487.6798	1202.439	0.688760	1.433505
968.67	736.9918	0.020647	0.621018	498.3934	1201.029	0.699651	1.424979
977.67	797.5314	0.020855	0.570882	509.2359	1199.326	0.710567	1.416380
986.67	861.7987	0.021096	0.525230	520.2247	1197.314	0.721506	1.407686
995.67	929.9530	0.021336	0.483263	531.3682	1194.976	0.732469	1.398920
1004.67	1002.139	0.021592	0.444820	542.6837	1192.284	0.743503	1.390035
1013.67	1078.227	0.021881	0.409580	554.1841	1189.215	0.754586	1.381031
1022.67	1159.276	0.022169	0.377064	565.8908	1185.741	0.765764	1.371835
1031.67	1244.588	0.022489	0.347110	577.8168	1181.824	0.777037	1.362473
1040.67	1334.629	0.022826	0.319559	589.9921	1177.431	0.788406	1.352871
1049.67	1429.601	0.023178	0.293930	602.4426	1172.508	0.799943	1.343007
1058.67	1529.737	0.023578	0.270224	615.2026	1166.992	0.811622	1.332832
1067.67	1635.181	0.024011	0.248119	628.3066	1160.818	0.823517	1.322275
1076.67	1746.282	0.024476	0.227456	641.8018	1153.892	0.835650	1.311264
1085.67	1863.039	0.025004	0.208234	655.7527	1146.102	0.848046	1.299704
1094.67	1986.178	0.025581	0.189973	670.2367	1137.289	0.860801	1.287451
1103.67	2115.553	0.026237	0.172994	685.3570	1127.255	0.874009	1.274386
1112.67	2251.746	0.026990	0.156656	701.2726	1115.724	0.887766	1.260223
1121.67	2395.191	0.027871	0.141119	718.1986	1102.280	0.902264	1.244650
1130.67	2546.032	0.028945	0.126062	736.5088	1086.287	0.917813	1.227142
1139.67	2705.141	0.030306	0.111165	756.8656	1066.632	0.934986	1.206769
1148.67	2872.807	0.032212	0.095948	780.7175	1041.008	0.954978	1.181571
1157.67	3049.901	0.035448	0.078969	812.7079	1002.625	0.981752	1.145792
1165.05	3203.789	0.050537	0.050457	902.5139	902.5139	1.058016	1.058016

Adapted from Sonntag, R. E., Borgnakke, C., and Van Wylen, J., *Fundamentals of Thermodynamics*, John Wiley & Sons, New York, 1998. With permission.

TABLE BB-2 U.S.

Superheated Water Vapor

Temp. °R	Specific Volume ft³/lb$_m$	Enthalpy Btu/lb$_m$	Entropy Btu/(lb$_m$·°R)	Temp. °R	Specific Volume ft³/lb$_m$	Enthalpy Btu/lb$_m$	Entropy Btu/(lb$_m$·°R)
	P = 1.4504 lb$_f$/in.^{2} (574.13)				*P = 7.252 lb$_f$/in.^{2} (638.064)*		
574.13	235.0409	1111.184	1.946618	638.064	51.90377	1137.512	1.813772
581.67	238.1743	1114.593	1.952541	671.67	54.75481	1153.269	1.837848
671.67	275.4393	1155.393	2.017746	761.67	62.29993	1195.212	1.896436
851.67	349.5940	1237.963	2.126612	851.67	69.77361	1237.155	1.948481
1031.67	423.5973	1322.653	2.216777	1031.67	84.63767	1322.228	2.039075
1211.67	497.5594	1409.927	2.294736	1211.67	99.46041	1409.660	2.117164
1391.67	571.5056	1500.012	2.364025	1391.67	114.2666	1499.828	2.186515
1571.67	645.4434	1593.026	2.426865	1571.67	129.0647	1592.897	2.249403
1751.67	719.3767	1689.040	2.484666	1751.67	143.8583	1688.945	2.307251
1931.67	793.3073	1788.080	2.538478	1931.67	158.6493	1788.003	2.361039
2111.67	867.2365	1890.117	2.588970	2111.67	173.4386	1890.057	2.411555
2291.67	941.1647	1995.078	2.636668	2291.67	188.2270	1995.027	2.459253
2471.67	1015.092	2102.820	2.681929	2471.67	202.5863	2102.773	2.504490
2651.67	1089.020	2213.134	2.724993	2651.67	217.8027	2213.095	2.547578
2831.67	1162.947	2325.738	2.766074	2831.67	232.5900	2325.700	2.588660
2831.67	1162.947	2325.738	2.766074	2831.67	232.5900	2325.700	2.588660
	P = 14.504 lb$_f$/in.^{2} (670.986)				*P = 29.008 lb$_f$/in.^{2} (708.084)*		
670.986	27.13449	1150.234	1.757739	708.084	14.18762	1163.634	1.702279
761.67	31.01661	1193.621	1.818405	761.67	15.37151	1190.362	1.738679
851.67	34.79526	1236.136	1.871167	851.67	17.30489	1234.068	1.792921
941.67	38.53995	1278.724	1.918697	941.67	19.20238	1277.284	1.841144
1031.67	42.26766	1321.694	1.962286	1031.67	21.08225	1320.624	1.885115
1121.67	45.98768	1364.996	2.002723	1121.67	22.95219	1364.566	1.925814
1211.67	49.69793	1409.325	2.040556	1211.67	24.81669	1408.654	1.963719
1391.67	57.11170	1499.600	2.109989	1391.67	28.53431	1499.144	2.033343
1571.67	64.51746	1592.733	2.172901	1571.67	32.24375	1592.406	2.096327
1751.67	71.91858	1688.825	2.230773	1751.67	35.94866	1688.580	2.164247
1931.67	79.31697	1787.913	2.284609	1931.67	39.65080	1787.723	2.208130
2111.67	86.71392	1889.980	2.335125	2111.67	43.35160	1889.829	2.258646
2291.67	94.10991	1994.962	2.382822	2291.67	47.05127	1994.833	2.306368
2471.67	101.5054	2102.717	2.428060	2471.67	50.75047	2102.601	2.351629
2651.67	108.7403	2213.039	2.471147	2651.67	54.44951	2212.936	2.394717
2831.67	116.2955	2325.648	2.512229	2831.67	58.14822	2325.549	2.435798
	P = 43.512 lb$_f$/in.^{2} (732.06)				*P = 58.016 lb$_f$/in.^{2} (750.204)*		
732.06	9.704025	1171.661	1.669963	750.204	7.407684	1177.349	1.647034
761.67	10.15349	1186.988	1.690504	761.67	7.541915	1183.492	1.655178
851.67	11.47353	1231.953	1.746322	851.67	8.557136	1229.790	1.712668
941.67	12.75609	1275.826	1.795285	941.67	9.532632	1274.352	1.762396
1031.67	14.02040	1319.545	1.839639	1031.67	10.48923	1318.457	1.807132
1121.67	15.27476	1363.706	1.880434	1121.67	11.43525	1362.846	1.848190
1211.67	16.52273	1407.979	1.918625	1211.67	12.37583	1407.304	1.886500
1391.67	19.00840	1498.684	1.988392	1391.67	14.24545	1498.224	1.956435
1571.67	21.48590	1592.080	2.051495	1571.67	16.10690	1591.753	2.019609

TABLE BB-2 U.S. (CONTINUED)

Superheated Water Vapor

Temp. °R	Specific Volume ft³/lb$_m$	Enthalpy Btu/lb$_m$	Entropy Btu/(lb$_m$·°R)	Temp. °R	Specific Volume ft³/lb$_m$	Enthalpy Btu/lb$_m$	Entropy Btu/(lb$_m$·°R)
	P = 43.512 lb$_f$/in.^{2} (732.06)				*P* = 58.016 lb$_f$/in.^{2} (750.204)		
1751.67	23.95860	1688.339	2.109463	1751.67	17.96371	1688.094	2.077697
1931.67	26.42874	1787.534	2.163347	1931.67	19.81779	1787.349	2.131556
2111.67	28.89743	1889.679	2.213887	2111.67	21.67043	1889.524	2.182120
2291.67	31.36517	1994.704	2.261608	2291.67	23.52195	1994.575	2.229865
2471.67	33.83226	2102.485	2.306893	2471.67	25.37315	2102.373	2.275127
2651.67	36.29903	2212.837	2.349981	2651.67	27.22387	2212.725	2.318214
2831.67	38.76564	2325.450	2.391062	2831.67	29.07443	2325.351	2.359320
	P = 72.52 lb$_f$/in.^{2} (765.018)				*P* = 87.024 lb$_f$/in.^{2} (777.60)		
765.018	6.004988	1181.708	1.629216	777.60	5.056402	1185.203	1.614598
851.67	6.806369	1227.581	1.686061	851.67	5.638656	1225.324	1.663920
941.67	7.598298	1272.856	1.736601	941.67	6.308369	1271.342	1.715296
1031.67	8.370366	1317.361	1.781743	1031.67	6.957739	1316.256	1.760844
1121.67	9.132182	1361.836	1.823063	1121.67	7.596376	1360.981	1.802403
1211.67	9.887591	1406.625	1.861493	1211.67	8.228767	1405.946	1.840976
1301.67	10.63916	1451.840	1.897631	1301.67	8.856352	1451.410	1.877090
1391.67	11.38768	1497.764	1.931595	1391.67	9.482496	1497.304	1.911245
1571.67	12.87943	1591.422	1.994865	1571.67	10.72790	1591.095	1.974611
1751.67	14.36670	1687.853	2.052928	1751.67	11.96865	1687.612	2.032746
1931.67	15.85125	1787.160	2.106884	1931.67	13.20684	1786.971	2.086725
2111.67	17.33420	1889.374	2.157471	2111.67	14.44327	1889.223	2.137313
2291.67	18.81618	1994.446	2.205216	2291.67	15.67890	1994.317	2.185058
2471.67	20.29769	2102.257	2.250502	2471.67	16.91405	2102.141	2.230367
2651.67	21.77871	2212.618	2.294831	2651.67	18.14871	2212.514	2.273455
2831.67	23.25974	2325.252	2.334695	2831.67	19.38322	2325.154	2.314560
	P = 116.032 lb$_f$/in.^{2} (798.444)				*P* = 145.04 lb$_f$/in.^{2} (815.508)		
798.444	3.851208	1190.504	1.591359	815.508	3.114540	1194.352	1.573135
851.67	4.177494	1220.650	1.627926	851.67	3.299067	1215.754	1.598811
941.67	4.695517	1268.251	1.681093	941.67	3.727068	1265.078	1.653912
1031.67	5.191594	1314.020	1.727501	1031.67	4.131683	1311.750	1.701252
1121.67	5.676619	1359.269	1.769562	1121.67	4.524604	1357.537	1.743814
1211.67	6.155077	1404.579	1.808422	1211.67	4.910959	1403.207	1.782985
1301.67	6.629850	1450.120	1.844846	1301.67	5.292347	1449.260	1.819528
1391.67	7.100940	1496.380	1.879048	1391.67	5.672134	1495.451	1.853946
1571.67	8.038473	1590.437	1.942581	1571.67	6.424660	1589.780	1.917670
1751.67	8.971201	1687.122	2.000812	1751.67	7.174462	1686.636	1.975996
1931.67	9.901206	1786.597	2.054863	1931.67	7.918018	1786.223	2.030095
2111.67	10.82977	1888.918	2.105498	2111.67	8.661734	1888.613	2.080778
2291.67	11.75737	1994.055	2.153267	2291.67	9.404488	1993.797	2.128571
2471.67	12.68433	2101.913	2.198553	2471.67	10.14660	2101.685	2.173880
2651.67	13.61114	2212.304	2.241664	2651.67	6.083156	2212.093	2.216992
2831.67	14.53762	2324.960	2.282770	2831.67	11.63035	2324.754	2.258097

TABLE BB-2 U.S. (CONTINUED)

Superheated Water Vapor

Temp. °R	Specific Volume ft³/lb$_m$	Enthalpy Btu/lb$_m$	Entropy Btu/(lb$_m$·°R)	Temp. °R	Specific Volume ft³/lb$_m$	Enthalpy Btu/lb$_m$	Entropy Btu/(lb$_m$·°R)
	P = 174.048 lb$_f$/in.² (830.052)				*P = 203.056 lb$_f$/in.² (842.796)*		
830.052	2.616220	1197.250	1.558063	842.796	2.255975	1199.477	1.545142
851.67	2.711847	1210.612	1.573947	851.67	2.290894	1205.203	1.551901
941.67	3.081062	1261.819	1.631150	941.67	2.618943	1258.470	1.611422
1031.67	3.424969	1309.450	1.679469	1031.67	2.919761	1307.107	1.660767
1121.67	3.756541	1355.791	1.722557	1121.67	3.207765	1354.029	1.704380
1211.67	4.081386	1401.823	1.762038	1211.67	3.488720	1400.430	1.744172
1301.67	4.401746	1447.971	1.798748	1301.67	3.765832	1446.681	1.781074
1391.67	4.719383	1494.522	1.833333	1391.67	4.038939	1493.589	1.815850
1571.67	5.348891	1589.122	1.897249	1571.67	4.580507	1588.460	1.879956
1751.67	5.973753	1686.151	1.955694	1751.67	5.117270	1685.660	1.938473
1931.67	6.595732	1785.845	2.009864	1931.67	5.651311	1785.471	1.992715
2111.67	7.216750	1888.308	2.060571	2111.67	6.183909	1888.007	2.043446
2291.67	7.835845	1993.539	2.108388	2291.67	6.715386	1993.285	2.091287
2471.67	8.454781	2101.458	2.153697	2471.67	7.246383	2101.230	2.136620
2651.67	9.073556	2211.883	2.196833	2651.67	7.777059	2211.672	2.179755
2831.67	9.692011	2324.556	2.237938	2831.67	8.307576	2324.358	2.220861
	P = 290.08 lb$_f$/in.² (874.026)				*P = 362.6 lb$_f$/in.² (894.852)*		
874.026	1.595873	1203.565	1.514474	894.852	1.281120	1205.096	1.494554
941.670	1.785046	1247.826	1.563294	941.670	1.393566	1238.195	1.530620
1031.67	2.009778	1299.863	1616103	1031.67	1.584180	1293.548	1.586821
1121.67	2.219614	1348.642	1.661460	1121.67	1.758136	1344.033	1.633754
1211.67	2.421922	1396.208	1.702255	1211.67	1.923762	1392.631	1.675432
1301.67	2.619424	1443.448	1.739849	1301.67	2.084583	1440.563	1.713600
1391.67	2.814042	1490.769	1.775007	1391.67	2.242200	1488.400	1.749140
1481.67	3.006579	1538.254	1.808063	1481.67	2.397895	1536.534	1.782507
1571.67	3.197193	1586.465	1.839663	1571.67	2.551667	1584.793	1.814273
1751.67	3.575698	1684.190	1.898514	1751.67	2.856330	1682.961	1.873388
1931.67	3.951320	1784.340	1.952947	1931.87	3.158109	1783.394	1.927964
2111.67	4.325501	1887.091	2.003798	2111.67	3.458286	1886.334	1.978910
2291.67	4.698560	1992.512	2.051686	2291.67	3.757502	1991.871	2.026846
2471.67	5.071139	2100.542	2.097067	2471.67	4.056078	2099.974	2.072251
2651.67	5.443557	2211.044	2.140203	2651.67	4.354493	2210.520	2.115434
2831.67	5.815495	2323.761	2.181332	2831.67	4.652588	2323.266	2.156564
	P = 435.12 lb$_f$/in.² (912.69)				*P = 725.2 lb$_f$/in.² (966.852)*		
912.69	1.068080	1205.556	1.477716	966.852	0.631750	1201.338	1.426698
941.67	1.130550	1227.744	1501648	1031.67	0.725936	1257.314	1.482827
1031.67	1.299701	1286.957	1.561789	1121.67	0.831975	1319162	1.540365
1121.67	1.450110	1339.308	1.610466	1211.67	0.926001	1373.870	1.587322
1211.67	1.591548	1388.994	1.653076	1301.67	1.013939	1425.679	1.628571
1301.67	1.727862	1437.652	1.691817	1391.67	1.098354	1476.242	1.666141
1391.67	1.861131	1486.010	1.727740	1481.67	1.180206	1526.315	1.700989
1481.67	1.992639	1534.384	1.761489	1571.67	1.260456	1576.289	1.733735
1571.67	2.121264	1583.112	1.793351	1661.67	1.339425	1626.387	1.764832

TABLE BB-2 U.S. (CONTINUED)

Superheated Water Vapor

Temp. °R	Specific Volume ft³/lb$_m$	Enthalpy Btu/lb$_m$	Entropy Btu/(lb$_m$.°R)	Temp. °R	Specific Volume ft³/lb$_m$	Enthalpy Btu/lb$_m$	Entropy Btu/(lb$_m$.°R)
	P = 435.12 lb$_f$/in.² (912.69)				*P = 725.2 lb$_f$/in.² (966.852)*		
1751.67	2.376751	1681.727	1.852752	1751.67	1.417433	1676.744	1.794258
1931.67	2.629195	1782.448	1.907471	1931.67	1.571526	1778.652	1.849623
2111.67	2.880036	1885.573	1.958512	2111.67	1.723857	1882.542	1.901046
2291.67	3.130077	1991.230	2.006521	2291.67	1.875227	1988.677	1.949269
2471.67	3.379478	2099.411	2.051949	2471.67	2.025957	2097.158	1.994817
2651.67	3.628397	2209.999	2.095132	2651.67	2.176366	2207.927	2.038072
2831.67	3.877317	2322.776	2.136285	2831.67	2.326775	2320.811	2.079273
	P = 1015.28 lb$_f$/in.² (1006.254)				*P = 1160.32 lb$_f$/in.² (1022.778)*		
1006.254	0.438413	1191.768	1.388459	1022.778	0.376743	1185.694	1.371716
1031.67	0.472050	1220.285	1.416452	1031.67	0.388597	1197.319	1.383037
1121.67	0.564474	1296.647	1.487580	1121.67	0.479739	1284.300	1.464125
1211.67	0.639599	1357.717	1.540007	1211.67	0.549738	1349.209	1.519848
1301.67	0.707355	1413.164	1.584169	1301.67	0.611407	1406.694	1.565635
1391.67	0.771107	1466.152	1.623531	1391.67	0.668752	1460.984	1.605976
1481.67	0.832135	1517.992	1.659645	1481.67	0.723373	1513.753	1.642734
1571.67	0.891402	1569.320	1.693274	1571.67	0.776072	1565.782	1.676818
1661.67	0.949387	1620.368	1.724945	1661.67	0.827490	1617.359	1708704
1751.67	1.006411	1671.697	1.754944	1751.67	0.877947	1669.152	1.731919
1931.67	1.118217	1774.839	1.810977	1931.67	0.976617	1772.921	1.795476
2111.67	1.228420	1879.511	1.862783	2111.67	1.073526	1878.002	1.847473
2291.67	1.337503	1986.144	1.911245	2291.67	1.169474	1984.885	1.896054
2471.67	1.445945	2094.927	1.956936	2471.67	1.264781	2093.818	1.941793
2651.67	1.126065	2205.877	2.000263	2651.67	1.359928	2204.862	1.985168
2831.67	1.662188	2318.864	2.041464	2831.67	1.454434	2317.896	2.026416
	P = 1305.36 lb$_f$/in.² (1037.79)				*P = 1450.4 lb$_f$/in.² (1051.578)*		
1037.79	0.328049	1178.888	1.355952	1051.573	0.288805	1172.250	1.340881
1121.67	0.413264	1271.080	1.441698	1121.67	0.359124	1256.824	1.419748
1211.67	0.479419	1340.387	1.501218	1211.67	0.423035	1331.230	1.483687
1301.67	0536603	1400.073	1.548743	1301.67	0.476536	1393.298	1.533128
1391.67	0.588982	1455.731	1.590117	1391.67	0.525230	1450.391	1.575547
1481.67	0.638638	1509.458	1.627520	1481.67	0570882	1505.116	1.613667
1571.67	0.686371	1562.213	1.662081	1571.67	0.614611	1558.606	1.648705
1661.67	0.732663	1614.487	1.694444	1661.67	0.656898	1611.456	1.681403
1751.67	0.777994	1666.589	1.724969	1751.67	0.698064	1664.014	1.712215
1841.67	0.822524	1718.820	1.754084	1841.67	0.738590	1716.671	1.741425
1931.67	0.866414	1771.004	1.781719	1931.67	0.778315	1769.082	1.769299
2111.67	0.953071	1876.489	1.833907	2111.67	0.856803	1874.984	1.821725
2291.67	1.038767	1983.629	1.882607	2291.67	0.934170	1982.378	1.870522
2471.67	1.123823	2092.717	1.928418	2471.67	1.011056	2091.621	1.916404
2651.67	1.208398	2203.847	1.971816	2651.67	1.087462	2202.841	1.959826
2831.67	1.292973	2316.933	2.013065	2831.67	1.163708	2315.975	2.001123

TABLE BB-2 U.S. (CONTINUED)

Superheated Water Vapor

Temp. °R	Specific Volume ft^3/lb_m	Enthalpy Btu/lb_m	Entropy $Btu/(lb_m \cdot °R)$	Temp. °R	Specific Volume ft^3/lb_m	Enthalpy Btu/lb_m	Entropy $Btu/(lb_m \cdot °R)$
	P = 1813 $lb_f/in.^2$ (1081.872)				*P = 2175.6 $lb_f/in.^2$ (1107.702)*		
1081.872	0.216243	1149.507	1.304648	1107.702	0.165626	1122.302	1.268200
1121.67	0.258370	1215.018	1.364216	1121.67	0.183726	1157.521	1.299799
1211.67	0.320360	1306.656	1.443011	1211.67	0.250682	1279.201	1.404653
1301.67	0.478938	1375.649	1.497994	1301.67	0.295532	1356.892	1.466585
1391.67	0.400450	1436.672	1.543351	1391.67	0.333174	1422.403	1.515286
1481.67	0.448504	1494.028	1.583286	1481.67	0.367293	1482.626	1.557227
1571.67	0.483744	1549.453	1.619614	1571.67	0.399008	1540.102	1.594893
1661.67	0.518983	1603.791	1.653244	1661.67	0.429282	1596.001	1.629478
1751.67	0.554383	1657.518	1.684723	1751.67	0.458275	1650.944	1.661699
1841.67	0.587380	1711.082	1.714436	1841.67	0.486467	1705.493	1.691984
1931.67	0.618455	1764.258	1.742739	1931.67	0.514178	1759.418	1.720646
2111.67	0.672756	1871.218	1.795667	2111.67	0.567998	1867.465	1.774123
2291.67	0.736828	1979.270	1.844774	2291.67	0.620698	1976.183	1.823517
2471.67	0.800900	2088.904	1.890824	2471.67	0.672756	2086.213	1.869733
2651.67	0.869777	2200.339	1.934341	2651.67	0.724494	2197.867	1.913347
2831.67	0.929044	2313.589	1.975662	2831.67	0.776072	2311.224	1.954691
	P = 2538.2 $lb_f/in.^2$ (1130.22)				*P = 2900.8 $lb_f/in.^2$ (1150.128)*		
1130.22	0.112126	1087.177	1.228093	1150.128	0.093385	1035.995	1.176770
1211.67	0.192216	1247.980	1.366485	1211.67	0.159219	1211.545	1.326526
1301.67	0.241872	1336.918	1.437422	1301.67	0.203429	1315.581	1.409573
1391.67	0.277111	1407.567	1.489968	1391.67	0.236586	1392.158	1.466514
1481.67	0.307546	1470.915	1.534083	1481.67	0.265258	1458.912	1.513017
1571.67	0.336378	1530.571	1.573182	1571.67	0.291207	1520.872	1.553645
1661.67	0.352396	1588.099	1.608770	1661.67	0.315394	1580.094	1.590284
1751.67	0.384432	1644.302	1.646508	1751.67	0.338460	1637.604	1.623985
1841.67	0.400450	1699.680	1.672542	1841.67	0.360565	1694.001	1.655393
1931.67	0.432486	1754.559	1.701658	1931.67	0.382029	1749.688	1.684915
2021.67	0.462279	1809.103	1.729244	2021.67	0.403013	1804.804	1.712764
2111.67	0.480540	1863.725	1.755685	2111.67	0.423676	1859.993	1.739515
2291.67	0.528594	1973.126	1.805388	2291.67	0.464041	1970.087	1.789553
2471.67	0.560630	2083.551	1.851772	2471.67	0.503766	2080.916	1.836104
2651.67	0.608684	2195.425	1.895457	2651.67	0.543170	2193.005	1.879861
2831.67	0.664747	2308.885	1.936873	2831.67	0.582414	2306.564	1.9213
	P = 3626 $lb_f/in.^2$				*P = 4351.2 $lb_f/in.^2$*		
1166.67	0.016018	794.4621	0.963003	1166.67	0.028656	770.1716	0.938736
1211.67	0.096108	1109.262	1.228098	1211.67	0.044690	924.7751	1.068310
1256.67	0.112126	1206.463	1.307013	1256.67	0.084959	1123.884	1.230128
1301.67	0.144162	1268.135	1.355283	1301.67	0.107881	1212.955	1.299871
1391.67	0.176198	1359.575	1.423330	1391.67	0.139020	1324.596	1.383013
1481.67	0.192216	1434.050	1.475208	1481.67	0.162871	1408.143	1.441244
1571.67	0.224252	1501.005	1.519108	1571.67	0.183342	1480.606	1.488726
1661.67	0.240270	1563.817	1.557968	1661.67	0.201763	1547.252	1.529975

TABLE BB-2 U.S. (CONTINUED)

Superheated Water Vapor

Temp. °R	Specific Volume ft³/lb$_m$	Enthalpy Btu/lb$_m$	Entropy Btu/(lb$_m$·°R)	Temp. °R	Specific Volume ft³/lb$_m$	Enthalpy Btu/lb$_m$	Entropy Btu/(lb$_m$·°R)
	P = 3626 lb$_f$/in.^{2}				*P = 4351.2 lb$_f$/in.^{2}*		
1751.67	0.256288	1624.049	1.593269	1751.67	0.218822	1610.338	1.566972
1931.67	0.302948	1739.921	1.656277	1931.67	0.250249	1730.131	1.632082
2021.67	0.320200	1796.636	1.684819	2021.67	0.265098	1788.037	1.661412
2111.67	0.337099	1852.564	1.712024	2111.67	0.279482	1845.187	1.689047
2291.67	0.370048	1964.077	1.762707	2291.67	0.307482	1958.148	1.740398
2471.67	0.402356	2075.718	1.809616	2471.67	0.334824	2070.615	1.787642
2651.67	0.434328	2188.233	1.853540	2651.67	0.361831	2183.551	1.831733
2831.67	0.466140	2301.981	1.895051	2831.67	0.388693	2297.471	1.873316
	P = 5076.4 lb$_f$/in.^{2}				*P = 5801.6 lb$_f$/in.^{2}*		
1166.67	0.027231	757.6781	0.924835	1166.67	0.026286	749.2259	0.914517
1211.67	0.033638	854.4746	1.006114	1211.67	0.030562	830.1024	0.982469
1256.67	0.054910	1020.376	1.140418	1256.67	0.040558	945.0115	1.075475
1301.67	0.079481	1148.901	1.241091	1301.67	0.059154	1080.299	1.181308
1391.67	0.110957	1287.327	1.344249	1391.67	0.090069	1248.170	1.306463
1481.67	0.133670	1381.337	1.409788	1481.67	0.111870	1353.840	1.380147
1571.67	0.152603	1459.789	1.46121i	1571.67	0.129650	1438.676	1.435774
1661.67	0.169390	1530.477	1.504968	1661.67	0.145187	1513.568	1.482134
1751.67	0.184736	1596.525	1.543685	1751.67	0.159251	1582.660	1.522643
1931.67	0.212687	1720.342	1.611016	1931.67	0.184575	1710.566	1.592195
2021.67	0.225694	1779.439	1.641110	2021.67	0.196317	1771.270	1.622958
2111.67	0.238396	1837.852	1.669198	2111.67	0.207641	1830.569	1.651619
2291.67	0.262855	1952.288	1.721195	2291.67	0.229442	1946.501	1.704309
2471.67	0.286642	2065.589	1.768797	2471.67	0.250570	2060.645	1.752269
2651.67	0.310108	2178.946	1.813079	2651.67	0.271345	2174.415	1.796694
2831.67	0.333415	2293.030	1.854686	2831.67	0.291992	2288.658	1.838373
	P = 7252 lb$_f$/in.^{2}				*P = 8702.4 lb$_f$/in.^{2}*		
1166.67	0.024972	737.9663	0.898968	1166.67	0.024075	730.6533	0.887074
1211.67	0.027721	805.9194	0.956100	1211.67	0.026157	792.4930	0.939070
1256.67	0.032148	885.6266	1.020660	1256.67	0.029105	860.5494	0.994196
1301.67	0.039821	981.8986	1.095897	1301.67	0.033398	936.7785	1.053764
1391.67	0.062342	1169.412	1.235430	1391.67	0.047349	1103.983	1.177988
1481.67	0.081980	1298.148	1.325236	1481.67	0.063383	1245.117	1.276392
1571.67	0.097902	1396.204	1.389534	1571.67	0.077447	1354.768	1.348309
1661.67	0.111581	1479.716	1.441244	1661.67	0.089621	1446.487	1.405107
1751.67	0.123771	1554.982	1.485359	1751.67	0.100465	1527.747	1.452756
1931.67	0.145379	1691.142	1.559425	1931.67	0.119478	1672.010	1.531241
2021.67	0.155262	1754.504	1.591669	2021.67	0.128080	1738.167	1.564679
2111.67	0.164713	1816.158	1.621334	2111.67	0.136281	1801.997	1.595610
2291.67	0.182781	1935.109	1.675408	2291.67	0.151851	1923.961	1.651046
2471.67	0.200177	2050.955	1.724061	2471.67	0.166731	2041.522	1.700440
2651.67	0.217220	2165.572	1.768845	2651.67	0.181276	2156.990	1.745534
2831.67	0.234119	2280.107	1.810619	2831.67	0.195660	2271.818	1.787451

TABLE BB-3 U.S.

Compressed Liquid Water

Temp. °R	Specific Volume ft³/lbm	Enthalpy Btu/lbm	Entropy Btu/(lbm·°R)	Temp. °R	Specific Volume ft³/lbm	Enthalpy Btu/lbm	Entropy Btu/(lbm·°R)
	P = 725.2 lbf/in.² (966.852)				*P = 1450.4 lbf/in.² (1051.578)*		
491.67	0.015986	2.158198	2.39E-05	491.67	0.015938	4.320696	7.17E-05
527.67	0.016018	38.10811	0.070579	527.67	0.01597	40.12013	0.070340
563.67	0.016114	73.92474	0.136262	563.67	0.016066	75.82069	0.135784
599.67	0.016258	109.7500	0.197860	599.67	0.016226	111.5513	0.197239
635.67	0.016450	145.6698	0.256019	635.67	0.016418	147.3809	0.255255
671.67	0.016675	181.7315	0.311216	671.67	0.016643	183.3523	0.310309
707.67	0.016947	217.9995	0.363810	707.67	0.016899	219.5215	0.362759
743.67	0.017251	254.5685	0.414207	743.67	0.017203	255.9744	0.412988
779.67	0.017604	291.5288	0.462740	779.67	0.019142	292.8056	0.461355
815.67	0.018004	328.9920	0.509721	815.67	0.017940	330.1055	0.508121
851.67	0.018469	367.0872	0.555412	851.67	0.018389	367.9986	0.553597
887.67	0.019013	405.9993	0.600172	887.67	0.018917	406.6527	0.598022
923.67	0.019638	446.0291	0.644358	923.67	0.019526	446.3128	0.641827
941.67	0.020010	466.6000	0.666400	959.67	0.020263	487.3917	0.685440
959.67	0.020423	487.6583	0.688565	995.67	0.021176	530.5686	0.729603
966.852	0.020599	496.2180	0.697454	1051.578	0.023258	605.1253	0.802403
	P = 2175.6 lbf/in.² (1107.702)				*P = 2900.8 lbf/in.² (1150.128)*		
491.67	0.015906	6.465997	9.55E-05	491.67	0.015858	8.598400	9.55E-05
527.67	0.015938	42.11926	0.070077	527.67	0.015906	44.11409	0.069791
563.67	0.016032	77.70804	0.135306	563.67	0.016002	79.59539	0.134852
599.67	0.016194	113.3484	0.196594	599.67	0.016146	115.1412	0.195973
635.67	0.016370	149.0920	0.254490	635.67	0.016338	150.8073	0.253726
671.67	0.016595	184.9774	0.309401	671.67	0.016563	186.6025	0.308517
743.67	0.017155	257.3931	0.411794	743.67	0.017107	258.8247	0.410624
779.67	0.017492	294.0954	0.459993	779.67	0.017444	295.4023	0.458656
815.67	0.017876	331.2448	0.506568	851.67	0.018245	369.9333	0.550086
851.67	0018309	368.9487	0.551806	887.67	0.018725	408.1102	0.593986
923.67	0.019398	446.6826	0.639390	923.67	0.019302	447.1340	0.637074
959.67	0.020103	487.2756	0.682502	995.67	0.020775	529.0682	0.722461
995.67	0.020952	529.7001	0.725900	1031.67	0.021784	573.2080	0.766003
1031.67	0.022057	574.9019	0.770493	1103.67	0.025116	675.4086	0.861613
1103.67	0.026125	684.3810	0.872862	1139.67	0.029201	747.7298	0.926006
1107.702	0.026558	692.3647	0.880075	1150.128	0.032597	785.1113	0.958656
	P = 4351.2 lbf/in.²				*P = 7252 lbf/in.²*		
491.67	0.015794	12.82021	2.39E-05	491.67	0.015650	21.07898	–0.00033
527.67	0.015842	48.07365	0.069218	527.67	0.015698	55.88960	0.067999
563.67	0.015938	83.34859	0.133897	563.67	0.015810	90.79910	0.131986
599.67	0.016082	118.7267	0.194731	599.67	0.015954	125.8677	0.192295
635.67	0.016274	154.2338	0.252245	635.67	0.016130	161.0824	0.249331
671.67	0.016483	189.8656	0.306774	671.67	0.016338	196.4176	0.303406
707.67	0.016723	225.6693	0.358675	743.67	0.016851	267.5521	0.404008
779.67	0.017331	298.0506	0.455240	779.67	0.017139	303.4848	0.451180

TABLE BB-3 U.S. (CONTINUED)

Compressed Liquid Water

Temp. °R	Specific Volume ft^3/lb_m	Enthalpy Btu/lb_m	Entropy $Btu/(lb_m \cdot °R)$	Temp. °R	Specific Volume ft^3/lb_m	Enthalpy Btu/lb_m	Entropy $Btu/(lb_m \cdot °R)$
	$P = 4351.2\ lb_f/in.^2$				$P = 7252\ lb_f/in.^2$		
815.67	0.017700	334.7830	0.502150	815.67	0.017476	339.7400	0.496632
851.67	0.018100	371.9840	0.546766	851.67	0.017860	376.3778	0.540604
887.67	0.018565	409.7525	0.590188	923.67	0.018741	451.0721	0.624773
959.67	0.019702	487.6540	0.674549	959.67	0.019270	489.3478	0.665425
1031.67	0.021304	570.8478	0.758097	1031.67	0.020599	568.7627	0.745199
1067.67	0.022425	615.9163	0.801041	1067.67	0.021448	610.5595	0.785015
1139.67	0.026061	720.2708	0.895481	1139.67	0.023771	700.8384	0.866772
1175.67	0.029938	789.9479	0.955622	1175.67	0.025437	750.8725	0.910003

CC-Series Tables

Thermodynamic Properties of Ammonia

TABLE CC-1 U.S.

Saturated Ammonia

Temp. °R T	Pressure $lb_f/in.^2$ P	Specific Volume ft^3/lb_m Sat. Liquid v_f	Sat. Vapor v_g	Enthalpy Btu/lb_m Sat. Liquid h_f	Sat. Vapor h_g	Entropy $Btu/(lb_m{\cdot}°R)$ Sat. Liquid s_f	Sat. Vapor s_g
399.67	5.547	0.02277	44.7625	–20.89	589.30	–0.0510	1.4758
409.67	7.663	0.02299	33.0932	–10.48	593.26	–0.0252	1.4485
419.67	10.404	0.02322	24.8696	0.00	597.08	0.0000	1.4227
429.67	13.898	0.02345	18.9724	10.54	600.77	0.0248	1.3985
431.67	14.696	0.02350	18.0068	12.65	601.49	0.0297	1.3938
439.67	18.289	0.02369	14.6747	21.15	604.31	0.0492	1.3755
449.67	23.737	0.02394	11.4953	31.84	607.69	0.0731	1.3538
459.67	30.415	0.02420	9.1103	42.60	610.92	0.0967	1.3331
469.67	38.508	0.02446	7.2979	53.43	613.97	0.1200	1.3134
479.67	48.218	0.02474	5.9039	64.34	616.84	0.1429	1.2947
489.67	59.756	0.02502	4.8195	75.33	619.52	0.1654	1.2768
499.67	73.346	0.02532	3.9671	86.41	622.00	0.1877	1.2596
509.67	89.226	0.02564	3.2903	97.58	624.26	0.2097	1.2431
519.67	107.641	0.02597	2.7481	108.84	626.30	0.2314	1.2271
529.67	128.849	0.02631	2.3098	120.21	628.09	0.2529	1.2117
539.67	153.116	0.02668	1.9526	131.68	629.62	0.2741	1.1968
549.67	180.721	0.02706	1.6594	143.26	630.86	0.2951	1.1822
559.67	211.949	0.02747	1.4168	154.97	631.80	0.3159	1.1679
569.67	247.098	0.02790	1.2149	166.80	632.40	0.3366	1.1539
579.67	286.473	0.02836	1.0456	178.79	632.63	0.3571	1.1400
589.67	330.392	0.02885	0.9028	190.93	632.47	0.3774	1.1262
599.67	379.181	0.02938	0.7818	203.26	631.87	0.3977	1.1125
609.67	433.181	0.02995	0.6785	215.80	630.80	0.4180	1.0987
619.67	492.742	0.03057	0.5899	228.58	629.19	0.4382	1.0847
629.67	558.231	0.03124	0.5135	241.65	627.00	04586	1.0705
639.67	630.029	0.03199	0.4472	255.06	624.14	0.4790	1.0560
649.67	708.538	0.03281	0.3895	268.88	620.51	0.4997	1.0410
659.67	794.183	0.03375	0.3388	283.20	616.00	0.5208	1.0253
669.67	887.424	0.03482	0.2941	298.14	610.42	0.5424	1.0087
679.67	988.761	0.03608	0.2542	313.88	603.51	0.5647	0.9909
689.67	1098.766	0.03759	0.2183	330.67	594.89	0.5882	0.9713
699.67	1218.113	0.03950	0.1855	348.95	583.87	0.6132	0.9490
709.67	1347.668	0.04206	0.1547	369.52	569.17	0.6410	0.9224
719.67	1488.694	0.04599	0.1241	394.41	547.66	0.6743	0.8872
129.77	1643.742	0.06816	0.0682	466.83	466.83	0.7718	0.7718

TABLE CC-2 U.S.

Superheated Ammonia Vapor

Temp. °R	Specific Volume ft^3/lb_m	Enthalpy Btu/lb_m	Entropy $Btu/(lb_m{\cdot}°R)$	Temp. °R	Specific Volume ft^3/lb_m	Enthalpy Btu/lb_m	Entropy $Btu/(lb_m{\cdot}°R)$
	$P = 10\ lb_f/in.^2$				*$P = 20\ lb_f/in.^2$*		
418.27	25.8065	596.58	1.4260	443	13.4963	605.47	1.3680
419.67	25.8962	597.27	1.4277	459.67	14.0774	614.54	1.3881
439.67	27.2401	607.60	1.4581	479.67	14.7635	625.30	1.4111
459.67	28.5674	617.88	1.4746	499.67	15.4385	635.94	1.4328
479.67	29.8814	628.12	1.4964	519.67	16.1051	646.51	1.4535
499.67	32.1852	638.34	1.5173	539.67	16.7651	657.02	1.4734
519.67	32.4809	648.56	1.5374	559.67	17.4200	667.51	1.4925
539.67	33.7703	658.80	1.5567	579.67	18.0709	678.01	1.5109
559.67	35.0549	669.07	1.5754	599.67	18.7187	688.53	1.5287
579.67	36.3356	679.38	1.5935	619.67	18.7187	699.09	1.5461
	$P = 30\ lb_f/in.^2$				*$P = 40\ lb_f/in.^2$*		
459.07	9.2285	610.74	1.3342	465.57	7.0414	614.45	1.3103
459.67	9.2423	611.06	1.3349	479.67	7.1964	619.39	1.3206
479.67	9.7206	622.39	1.3591	499.67	7.5596	630.96	1.3443
499.67	10.1872	633.49	1.3817	519.67	7.9132	642.26	1.3665
519.67	10.6447	644.41	1.4032	539.67	8.2596	653.37	1.3874
539.67	11.0954	655.21	1.4236	559.67	8.6004	664.33	1.4074
559.67	11.5407	665.93	1.4431	579.67	8.9370	675.21	1.4265
579.67	11.9820	676.62	1.4618	599.67	9.2702	686.04	1.4449
599.67	12.4200	687.29	1.4799	619.67	9.6008	696.86	1.4626
619.67	12.8554	697.98	1.4975	639.67	9.9294	707.69	1.4798
	$P = 50\ lb_f/in.^2$				*$P = 60\ lb_f/in.^2$*		
481.27	5.7049	617.30	1.2917	489.87	4.8009	619.57	1.2764
519.67	5.9814	628.37	1.3142	519.67	5.1787	625.69	1.3126
539.67	6.2731	640.07	1.3372	539.67	5.4217	637.82	1.3348
559.67	6.5573	651.49	1.3588	559.67	5.6586	649.57	1.3557
579.67	6.8356	662.70	1.3792	579.67	5.8909	661.05	1.3755
599.67	7.1096	673.79	1.3986	599.67	6.1197	672.34	1.3944
619.67	7.3800	687.78	1.4173	619.67	6.3456	683.50	1.4126
639.67	7.6478	695.73	1.4352	639.67	6.5694	694.59	1.4302
659.67	7.9135	706.67	1.4526	659.67	6.7915	705.64	1.4472
679.67	8.1775	717.61	1.4695	679.67	7.0121	716.68	1.4637
	$P = 70\ lb_f/in.^2$				*$P = 80\ lb_f/in.^2$*		
497.27	4.1473	621.44	1.2635	504.07	3.6520	623.02	1.2523
519.67	4.3961	635.52	1.2912	519.67	3.8083	633.16	1.2721
539.67	4.6099	647.62	1.3140	539.67	4.0005	645.63	1.2956
559.67	4.8174	659.37	1.3354	559.67	4.1861	657.66	1.3175
579.67	5.0201	670.88	1.3556	579.67	4.3667	669.39	1.3381
599.67	5.2191	682.21	1.3749	599.67	4.5435	680.90	1.3577

TABLE CC-2 U.S. (CONTINUED)

Superheated Ammonia Vapor

Temp. °R	Specific Volume ft³/lb$_m$	Enthalpy Btu/lb$_m$	Entropy Btu/(lb$_m$·°R)	Temp. °R	Specific Volume ft³/lb$_m$	Enthalpy Btu/lb$_m$	Entropy Btu/(lb$_m$·°R)
	$P = 70\ lb_f/in.^2$				$P = 80\ lb_f/in.^2$		
619.67	5.4153	693.44	1.3933	619.67	4.7174	692.27	1.3763
639.67	5.6093	704.60	1.4110	639.67	4.8890	703.55	1.3942
659.67	5.8014	715.73	1.4281	659.67	5.5088	714.79	1.4115
679.67	5.9921	726.87	1.4448	679.67	5.2270	726.00	1.4283
	$P = 90\ lb_f/in.^2$				$P = 100\ lb_f/in.^2$		
510.07	3.2632	624.36	1.2423	515.7	2.9497	625.52	1.2334
539.67	3.5260	643.59	1.2790	539.67	3.1459	641.51	1.2637
559.67	3.6947	655.92	1.3014	559.67	3.3013	654.16	1.2867
579.67	3.8583	667.88	1.3224	579.67	3.4513	666.36	1.3082
599.67	4.0179	679.58	1.3423	599.67	3.5972	678.24	1.3283
619.67	4.1745	691.10	1.3612	619.67	3.7400	689.91	1.3475
639.67	4.3287	702.50	1.3793	639.67	3.8804	701.44	1.3658
659.67	4.4811	713.83	1.3967	659.67	4.0188	712.87	1.3834
679.67	4.6319	725.13	1.4136	679.67	4.1558	724.25	1.4004
699.67	4.7816	736.43	1.4300	699.67	4.2915	735.63	1.4169
	$P = 125\ lb_f/in.^2$				$P = 150\ lb_f/in.^2$		
528	2.3787	627.80	1.2143	538.47	1.9923	629.45	1.1986
539.67	2.4597	636.11	1.2299	559.67	2.1170	644.81	1.2265
559.67	2.5917	649.59	1.2544	579.67	2.2275	658.37	1.2504
579.67	2.7177	662.44	1.2770	599.67	2.3331	671.31	1.2723
599.67	2.8392	674.83	1.2980	619.67	2.4351	683.80	1.2928
619.67	2.9574	686.90	1.3178	639.67	2.5343	695.99	1.3122
639.67	30730	698.74	1.3366	659.67	2.6313	707.96	1.3306
659.67	3.1865	710.44	1.3546	679.67	2.7267	719.79	1.3483
679.67	3.2985	722.04	1.3720	699.67	2.8207	731.54	1.3653
699.67	3.4091	733.59	1.3887	719.67	2.9136	743.24	1.3818
	$P = 175\ lb_f/in.^2$				$P = 250\ lb_f/in.^2$		
547.67	1.5010	631.49	1.1731	570.41	1.2006	632.43	1.1528
559.67	1.7762	639.77	1.2015	599.67	1.3150	655.95	1.1930
579.67	1.8762	654.13	1.2267	619.67	1.3863	670.53	1.2170
599.67	1.9708	667.67	1.2497	639.67	1.4539	684.34	1.2389
619.67	2.0614	680.62	1.2710	659.67	1.5188	697.59	1.2593
639.67	2.1491	693.17	1.2909	679.67	1.5815	710.45	1.2785
659.67	2.2345	705.44	1.3098	699.67	1;6426	723.05	1.2968
679.67	2.3181	717.51	1.3278	719.67	1.7024	735.46	1.3142
699.67	2.4002	729.46	1.3451	739.67	1.7612	747.76	1.3311
719.67	2.4813	741.33	1.3619	759.67	1.8191	759.98	1.3474

TABLE CC-2 U.S. (CONTINUED)

Superheated Ammonia Vapor

Temp. °R	Specific Volume ft³/lb$_m$	Enthalpy Btu/lb$_m$	Entropy Btu/(lb$_m$·°R)	Temp. °R	Specific Volume ft³/lb$_m$	Enthalpy Btu/lb$_m$	Entropy Btu/(lb$_m$·°R)
	P = 300 lb$_f$/in.^{2}				*P = 350 lb$_f$/in.^{2}*		
582.87	0.9973	632.63	1.1356	593.87	0.8503	632.28	1.1205
599.67	1.0568	647.32	1.1605	599.67	0.0870	637.87	1.1299
619.67	1.1217	663.27	1.1866	619.67	0.9309	655.48	1.1588
639.67	1.1821	678.07	1.2101	639.67	0.9868	671.46	1.1842
659.67	1.2394	692.08	1.2317	659.67	1.0391	686.34	1.2071
679.67	1.2943	705.55	1.2518	679.67	1.0886	700.47	1.2282
699.67	1.3474	718.63	1.2708	699.67	1.1362	714.08	1.2479
719.67	1.3991	731.44	1.2888	719.67	1.1822	727.32	1.2666
739.67	1.4497	744.07	1.3062	739.67	1.2270	740.31	1.2844
759.67	1.4994	756.58	1.3228	759.67	1.2708	753.12	1.3015
	P = 400 lb$_f$/in.^{2}				*P = 600 lb$_f$/in.^{2}*		
603.57	0.7388	631.50	1.1070	635.57	0.4731	625.39	1.0620
679.67	0.0934	695.21	1.2067	679.67	0.5680	671.78	1.1327
699.67	0.9773	709.40	1.2273	699.67	0.6035	689.03	1.1577
719.67	1.0192	723.10	1.2466	719.67	0.6366	705.06	1.1803
739.67	1.0597	736.47	1.2650	739.67	0.6678	720.26	1.2011
759.67	1.0992	749.60	1.2825	759.67	0.6976	734.88	1.2206
779.67	1.1379	762.58	1.2993	779.67	0.7264	749.09	1.2391
799.67	1.1758	775.45	1.3156	799.67	0.7542	763.02	1.2567
819.67	1.2131	788.27	1.3315	819.67	0.7814	776.75	1.2737
839.67	1.2499	801.06	1.3469	839.67	0.8079	790.34	1.2901
	P = 800 lb$_f$/in.^{2}						
660.27	0.3358	615.67	1.0242				
679.67	0.3769	642.62	1.0645				
699.67	0.4115	665.08	1.0971				
719.67	0.4419	684.62	1.1246				
739.67	0.4694	702.36	1.1489				
759.67	0.4951	718.93	1.1710				
779.67	0.5193	734.69	1.1915				
799.67	0.5425	749.89	1.2108				
819.67	0.5648	767.68	1.2290				
839.67	0.5864	779.19	1.2465				

TABLE CC-3 U.S.

Compressed Liquid Ammonia

Temperature °R	Specific Volume ft^3/lb_m	Enthalpy Btu/lb_m	Entropy Btu/$(lb_m{\cdot}°R)$	Temperature °R	Specific Volume ft^3/lb_m	Enthalpy Btu/lb_m	Entropy Btu/$(lb_m{\cdot}°R)$
	$P = 60\ lb_f/in.^2$				*$P = 80\ lb_f/in.^2$*		
464.67	0.02432	48.11	0.10827	464.67	0.02432	48.15	0.10817
469.67	0.02446	53.52	0.11989	469.67	0.02445	53.56	0.11976
474.67	0.02456	58.96	0.13137	474.67	0.02459	59.00	0.13127
476.67	0.02465	61.14	0.13595	479.67	0.02473	64.45	0.14270
479.67	0.02473	64.41	0.14280	484.67	0.02487	69.92	0.15405
481.67	0.02479	66.60	0.14735	489.67	0.02502	75.42	0.16532
484.67	0.02488	69.88	0.15415	494.67	0.02517	80.93	0.01765
486.67	0.02494	72.08	0.15867	488.67	0.02529	85.36	0.18544
489.67	0.02502	75.38	0.16543	491.67	0.02538	88.69	0.19210
489.87	0.02503	75.63	0.16594	504.07	0.02546	91.36	0.19742
	$P = 100\ lb_f/in.^2$				*$P = 200\ lb_f/in.^2$*		
464.67	0.02432	48.19	0.10807	469.67	0.02443	53.82	0.11916
469.67	0.02445	53.61	0.11966	479.67	0.02470	64.70	0.14207
474.67	0.02459	59.04	0.13116	489.67	0.02500	75.67	0.16467
479.67	0.02473	54.49	0.14259	499.67	0.02530	86.90	0.18698
484.67	0.02487	69.96	0.15394	509.67	0.02561	97.81	0.20900
489.67	0.02502	75.46	0.16521	519.67	0.02594	109.03	0.23080
494.67	0.02516	80.97	0.17642	529.67	0.02629	120.35	0.25240
504.67	0.05470	92.06	0.19862	539.67	0.02666	131.79	0.27380
514.67	0.02580	103.25	0.22060	549.67	0.02706	143.32	0.29500
515.67	0.02583	104.43	0.22280	556	0.02732	150.69	0.30830
	$P = 300\ lb_f/in.^2$				*$P = 400\ lb_f/in.^2$*		
469.67	0.02442	54.04	0.11866	469.67	0.02440	54.26	0.11816
479.67	0.02469	64.91	0.14155	479.67	0.02467	65.12	0.14103
489.67	0.02498	75.85	0.16413	489.67	0.02496	76.05	0.16359
499.67	0.02527	86.87	0.18641	499.67	0.02525	87.06	0.18595
509.67	0.02559	97.98	0.20840	509.67	0.02556	98.15	0.20780
519.67	0.02592	109.18	0.23020	519.67	0.02589	109.34	0.22960
529.67	0.02626	120.48	0.25170	529.67	0.02624	120.62	0.25110
539.67	0.02663	131.89	0.27310	539.67	0.02660	132.01	0.27240
549.67	0.02702	143.41	0.29420	549.67	0.02699	143.51	0.29350
582.87	0.02852	182.69	0.36360	603.57	0.02960	208.30	0.40580
	$P = 500\ lb_f/in.^2$				*$P = 600\ lb_f/in.^2$*		
469.67	0.02438	54.38	0.11767	469.67	0.02436	54.70	0.11717
479.67	0.02465	65.33	0.14052	479.67	0.02463	65.54	0.14001
489.67	0.02494	76.25	0.16305	489.67	0.02492	76.45	0.16252
499.67	0.02523	87.24	0.18528	499.67	0.02521	87.43	0.18473
509.67	0.02554	98.33	0.20720	509.67	0.02552	98.50	0.20670
519.67	0.02587	109.50	0.22890	519.67	0.02584	109.65	0.22830
529.67	0.02621	120.76	0.25040	529.67	0.02618	120.90	0.24980
539.67	0.02657	138.13	0.27170	539.67	0.02654	132.25	0.27100
549.67	0.02695	143.60	0.29270	549.67	0.02692	143.70	0.29200
620.77	0.03064	230.10	0.44060	635.57	0.03167	249.60	0.47070

TABLE CC-3 U.S. (CONTINUED)

Compressed Liquid Ammonia

Temperature °R	Specific Volume ft^3/lb_m	Enthalpy Btu/lb_m	Entropy Btu/ $(lb_m \cdot °R)$	Temperature °R	Specific Volume ft^3/lb_m	Enthalpy Btu/lb_m	Entropy Btu/ $(lb_m \cdot °R)$
	P = *700* $lb_f/in.^2$				*P* = *800* $lb_f/in.^2$		
469.67	0.02435	54.92	0.11668	469.67	0.02433	55.14	0.11619
479.67	0.02462	65.75	0.13950	479.67	0.02460	65.96	0.13899
489.67	0.02490	76.65	0.16198	489.67	0.02488	76.85	0.16145
499.67	0.02519	87.62	0.18417	499.67	0.02517	87.81	0.18362
509.67	0.02550	98.68	0.20610	509.67	0.02548	98.85	0.20550
519.67	0.05820	109.81	0.22770	519.67	0.05790	109.98	0.22710
529.67	0.02616	121.04	0.24910	529.67	0.02613	121.19	0.24850
539.67	0.02651	132.37	0.27030	539.67	0.02648	132.50	0.26960
549.67	0.02689	143.80	0.29130	549.67	0.02686	143.91	0.29060
648.57	0.03272	267.50	0.49760	660.37	0.03382	284.20	0.52230

Adapted from Sonntag, R. E., Borgnakke, C., and Van Wylen, G. J., *Fundamentals of Thermodynamics*, John Wiley & Sons, New York, 1998. With permission.

DD-Series Tables

Thermodynamic Properties of Carbon Dioxide

TABLE DD-1 U.S.

Saturated Carbon Dioxide

		Specific Volume ft^3/lb_m		Enthalpy Btu/lb_m		Entropy $Btu/(lb_m\ °R)$	
Temp. °R T	Pressure $lb_f/in.^2$ P	Sat. Liquid v_f	Sat. Vapor v_g	Sat. Liquid h_f	Sat. Vapor h_g	Sat. Liquid s_f	Sat. Vapor s_g
389.772	75.029	0.013567	1.16579	0.0000	151.2760	0.000000	0.388125
396	87.024	0.013727	1.01138	3.0094	151.8134	0.007595	0.383348
405	106.836	0.013952	0.82989	7.5322	152.5055	0.018749	0.376708
414	129.796	0.014176	0.68669	12.0679	153.0902	0.029689	0.370331
423	156.208	0.014432	0.57232	16.5605	153.5631	0.040246	0.364144
432	186.376	0.014705	0.48006	21.0231	153.9114	0050492	0.358102
441	220.606	0.014993	0.40477	25.5029	154.1177	0.060547	0.352178
450	259.332	0.015313	0.34279	30.0299	154.1650	0.070460	0.346327
459	302.698	0.015650	0.29153	34.6387	154.0360	0.080324	0.340451
468	351.142	0.016034	0.24860	39.3334	153.7093	0.090140	0.334551
477	404.952	0.016450	0.21224	44.1356	153.1504	0.099981	0.328509
486	464.708	0.016915	0.18148	49.0797	152.3164	0.107481	0.322275
495	530.701	0.017460	0.15497	54.2215	151.1470	0.119901	0.315730
504	603.366	0.018084	0.13202	59.6385	149.5606	0.130290	0.308708
513	683.283	0.018853	0.11184	65.4381	147.4196	0.141158	0.300970
522	770.743	0.019846	0.09377	71.7751	144.4918	0.152790	0.292085
531	866.904	0.021208	0.07701	78.9634	140.2743	0.165735	0.281217
540	972.638	0.023562	0.06005	88.1594	133.2537	0.182096	0.265597
547.578	1070.830	0.034519	0.03452	110.6227	110.6227	0.222413	0.222413

TABLE DD-2 U.S.

Superheated Carbon Dioxide Vapor

Temp. °R	Specific Volume ft^3/lb_m	Enthalpy Btu/lb_m	Entropy $Btu/(lb_m \cdot °R)$	Temp. °R	Specific Volume ft^3/lb_m	Enthalpy Btu/lb_m	Entropy $Btu/(lb_m \cdot °R)$
	$P = 145.04\ lb_f/in.^2$				$P = 290.08\ lb_f/in.^2$		
419.4	0.615732	153.3869	0.366628	456.48	0.304823	154.0919	0.342099
540	0.861608	180.5449	0.423641	540	0.406056	176.0135	0.386309
720	1.188215	220.8284	0.487962	720	0.583055	218.5928	0.454404
900	1.501848	263.6355	0.540962	900	0.745478	262.2555	0.508479
1080	1.810034	309.0695	0.586940	1080	0.903095	308.1366	0.554911
1260	2.117580	356.7906	0.627783	1260	1.058469	356.1328	0.596016
1440	2.423523	406.4206	0.664589	1440	1.212723	405.9520	0.632942
1620	2.727865	457.6326	0.698099	1620	1.366175	457.2973	0.666547
1800	3.032207	510.1474	0.728838	1800	1.519147	509.9152	0.697334
	$P = 725.2\ lb_f/in.^2$				$P = 1450.4\ lb_f/in.^2$		
517.5	0.102675	146.1040	0.296790	720	0.099312	199.1346	0.361302
540	0.124780	157.7720	0.318883	900	0.141759	251.3871	0.426244
720	0.220087	211.6324	0.405871	1080	0.178120	301.0472	0.476545
900	0.292168	258.1369	0.463528	1260	0.212239	351.2317	0.519514
1080	0.358963	305.4023	0.511369	1440	0.245075	402.4911	0.557538
1260	0.423356	354.2240	0.553167	1620	0.277272	454.8640	0.591788
1440	0.486787	404.5977	0.590523	1800	0.308987	508.2557	0.623053
1620	0.549257	456.3386	0.624367				
1800	0.611407	509.2531	0.655345				
	$P = 2900.8\ lb_f/in.^2$						
720	0.041967	173.2707	0.301758				
900	0.068237	238.7346	0.383443				
1080	0.088740	293.1796	0.438664				
1260	0.107321	345.9437	0.483854				
1440	0.124780	398.8497	0.523096				
1620	0.141759	452.3747	0.558111				
1800	0.158258	506.6263	0.589878				

TABLE DD-3 U.S.

Compressed Liquid Carbon Dioxide

Temp. °R T	Specific Volume ft^3/lb_m	Enthalpy Btu/lb_m	Entropy $Btu/(lb_m{\cdot}°R)$
	$P = 464.708\ lb_f/in.^2$		
396	0.0136521	3.30050	0.005897
414	0.0141071	12.23982	0.041081
432	0.0146228	21.07898	0.048868
450	0.0152283	29.98692	0.069074
468	0.0154926	33.94218	0.077076
486	0.0169150	49.07967	0.109869
	$P = 725.2\ lb_f/in.^2$		
396	0.0136080	3.50385	0.004753
414	0.0140496	12.38600	0.026703
432	0.0145486	21.14347	0.047387
450	0.0151242	29.93963	0.067355
468	0.0158140	38.98085	0.087035
486	0.0166819	48.51647	0.107027
504	0.0178876	59.05811	0.128332
	$P = 1450.4\ lb_f/in.^2$		
396	0.0134923	4.12293	0.001744
414	0.0139045	12.87180	0.023335
432	0.0143621	21.44441	0.043613
450	0.0148769	29.98692	0.062984
468	0.0154733	38.63691	0.081829
486	0.0161803	47.54485	0.100506
504	0.0170586	56.90421	0.119399
522	0.0182238	67.05462	0.139176
540	0.0199713	78.70545	0.161125
	$P = 603.366\ lb_f/in.^2$		
396	0.0136281	3.40884	0.005288
414	0.0140766	12.31721	0.027300
432	0.0145844	21.11337	0.048080
450	0.0151723	29.96112	0.068167
468	0.0156111	35.85533	0.080825
486	0.0162102	42.55778	0.094965
504	0.0180872	59.63850	0.130290
	$P = 972.638\ lb_f/in.^2$		
396	0.0135688	3.71494	0.003726
414	0.0139997	12.55366	0.025557
432	0.0144851	21.24665	0.046097
450	0.0150393	29.95683	0.065874
468	0.0156976	38.86477	0.085268
486	0.0165146	48.19403	0.104806
504	0.0176038	58.34014	0.125275
522	0.0178601	61.95147	0.131079
540	0.0235545	88.15940	0.182096

Adapted from Reynolds, W. C., *Thermodynamic Properties in SI*, Stanford University Press, Stanford, CA, 1979. With permission.

EE-Series Tables

Thermodynamic Properties of R-12

TABLE EE-1 U.S.

Saturated R-12

		Specific Volume ft^3/lb_m		Enthalpy Btu/lb_m		Entropy $Btu/(lb_m.°R)$	
Temp. °R T	Pressure $lb_f/in.^2$ P	Sat. Liquid v_f	Sat. Vapor v_g	Sat. Liquid h_f	Sat. Vapor h_g	Sat. Liquid s_f	Sat. Vapor s_g
329.67	0.4061	0.009739	70.71851	–18.607	62.9661	–0.0498	0.1976
347.67	0.8992	0.009883	34.25209	–14.927	64.9265	–0.039	0.1907
365.67	1.784	0.010043	18.057	–11.234	66.9127	–0.0286	0.1851
383.67	3.2778	0.010203	10.21804	–7.5193	68.9119	–0.0187	0.1805
401.67	5.6709	0.01038	6.1365	–3.7747	70.9153	–0.0092	0.1768
410.67	7.3099	0.010476	4.84833	–1.8916	71.917	–0.0045	0.1752
419.67	9.3114	0.010556	3.87491	0	72.9101	0	0.1737
428.67	11.7045	0.010668	3.12992	1.9002	73.9032	0.0045	0.1724
437.67	14.5037	0.010764	2.55279	3.8091	74.8921	0.0089	0.1713
438.03	14.6922	0.010764	2.53132	3.8908	74.9308	0.0091	0.1712
446.67	17.9411	0.010876	2.10108	5.7308	75.8723	0.0132	0.1702
455.67	21.8861	0.010972	1.74356	7.6741	76.8439	0.0175	0.1693
464.67	26.4837	0.011100	1.45796	9.6001	77.8026	0.0216	0.1684
473.67	31.7631	0.011213	1.22778	11.5133	78.7570	0.0258	0.1676
482.67	37.8546	0.011341	1.04053	13.5210	79.6943	0.0299	0.1670
491.67	44.6713	0.011469	0.88724	15.4986	80.6229	0.0339	0.1664
500.67	52.5033	0.011597	0.76069	17.4934	81.5300	0.0379	0.1658
509.67	61.3506	0.011741	0.65530	19.5055	82.4329	0.0418	0.1653
518.67	71.2131	0.011901	0.56720	21.5390	83.3099	0.0457	0.1649
527.67	82.2359	0.012046	0.49303	23.5897	84.1697	0.0496	0.1644
536.67	94.4190	0.012222	0.43008	25.6662	85.0081	0.0535	0.1640
545.67	107.9074	0.012398	0.37658	27.7685	85.8206	0.0573	0.1637
554.67	122.8462	0.012590	0.33061	29.9009	86.6074	0.0611	0.1633
563.67	139.2354	0.012782	0.29105	32.0677	87.3597	0.0649	0.1630
572.67	157.2199	0.012991	0.25677	34.2689	88.0777	0.0687	0.1627
581.67	176.7999	0.013231	0.22698	36.5174	88.7570	0.0725	0.1623
590.67	198.1203	0.013471	0.20087	38.8089	89.3890	0.0764	0.1620
599.67	221.1812	0.013743	0.17796	41.1605	89.9651	0.0802	0.1616
608.67	246.2726	0.014048	0.15778	43.5767	90.4810	0.0848	0.1611
617.67	273.3944	0.014368	0.13984	46.0659	90.9195	0.0880	0.1606
626.67	302.6919	0.014737	0.12366	48.6454	91.2677	0.0921	0.1601
635.67	334.1649	0.015153	0.10924	51.3281	91.4999	0.0962	0.1594
644.67	368.1035	0.015634	0.09611	54.1398	91.5902	0.1004	0.1585
653.67	404.3627	0.016210	0.08425	57.1106	91.4870	0.1048	0.1574
662.67	443.2326	0.016915	0.07304	60.2877	91.1172	0.1095	0.1560
671.67	485.0032	0.017828	0.06263	63.8904	90.3348	0.1145	0.1540
680.67	529.3845	0.019174	0.05190	67.7210	88.8086	0.1201	0.1511
689.67	576.9565	0.021849	0.03940	72.8929	85.1199	0.1274	0.1451
693.27	597.0877	0.028704	0.02867	78.8602	78.8602	0.1359	0.1359

TABLE EE-2 U.S.

Superheated R-12 Vapor

Temp. °R T	Specific Volume ft³/lb$_m$	Enthalpy Btu/lb$_m$	Entropy Btu/(lb$_m$.°R)	Temp. °R T	Specific Volume ft³/lb$_m$	Enthalpy Btu/lb$_m$	Entropy Btu/(lb$_m$.°R)
	$P = 3.63\ lb_f/in.^2$				$P = 7.25\ lb_f/in.^2$		
387.072	9.31126	69.26	0.1798	410.346	4.88789	71.88	0.1752
437.67	10.6005	75.75	0.1955	455.67	5.47591	77.89	0.1891
455.67	6.24718	78.13	0.2009	473.67	5.70609	80.35	0.1944
473.67	11.5026	80.57	0.2061	491.67	5.93483	82.85	0.1996
491.67	11.9515	83.05	0.2112	509.67	6.16244	85.39	0.2046
509.67	12.3993	85.57	0.2163	527.67	6.38894	87.97	0.2096
527.67	12.8461	88.13	0.2212	545.67	6.61479	90.59	0.2145
545.67	13.292	90.73	0.2261	563.67	6.83985	93.25	0.2193
563.67	13.7373	93.33	0.2308	581.67	7.06442	95.94	0.2240
581.67	14.1821	96.07	0.2355	599.67	7.28851	98.67	0.2286
599.67	14.6263	98.78	0.2401	617.67	7.51212	101.44	0.2332
617.67	15.0702	101.54	0.2447	635.67	7.73541	104.24	0.2376
635.67	15.5137	104.33	0.2491	653.67	7.95838	107.07	0.2420
653.67	15.9568	107.16	0.2535	671.67	8.18103	109.93	0.2463
671.67	16.3998	110.02	0.2578	689.67	8.40352	112.83	0.2506
689.67	16.8426	112.91	0.2621	707.67	8.62585	115.76	0.2548
	$P = 14.5\ lb_f/in.^2$				$P = 29.01\ lb_f/in.^2$		
437.76	2.56272	74.87	0.1713	469.386	1.33814	78.28	0.1680
455.67	2.68622	77.38	0.1769	491.67	1.41935	81.60	0.1750
473.67	2.80667	79.90	0.1823	509.67	1.48247	84.27	0.1803
491.67	2.92565	82.45	0.1876	527.67	1.54446	86.96	0.1855
509.67	3.04326	85.03	0.1927	545.67	1.60548	89.68	0.1906
527.67	3.16003	87.64	0.1978	563.67	1.66571	92.42	0.1955
545.67	3.27584	90.29	0.2027	581.67	1.7253	95.19	0.2003
563.67	3.39085	92.97	0.2076	599.67	1.78441	97.98	0.2050
581.67	3.50538	95.69	0.2123	617.67	1.84303	100.8	0.2097
599.67	3.61943	98.44	0.2170	635.67	1.90118	103.65	0.2142
617.67	3.73299	101.23	0.2216	653.67	1.95900	106.52	0.2187
635.67	3.84608	104.04	0.2260	671.67	2.01667	109.43	0.2231
653.67	3.95901	106.89	0.2305	689.67	2.07401	112.36	0.2274
671.67	4.07162	109.77	0.2348	707.67	2.13119	115.31	0.2142
689.67	4.18406	112.67	0.2391	725.67	2.18822	118.29	0.2357
707.67	4.29619	115.61	0.2433	743.67	2.24508	121.30	0.2398
	$P = 43.51\ lb_f/in.^2$				$P = 58.01\ lb_f/in.^2$		
490.392	0.91142	80.46	0.1665	506.61	0.69214	82.10	0.1665
509.67	0.96076	83.48	0.1725	509.67	0.69887	82.63	0.1665
527.67	1.00481	86.26	0.1778	527.67	0.73427	85.52	0.1721
545.67	1.04790	89.05	0.1830	545.67	0.76838	88.38	0.1774
563.67	1.09002	91.85	0.1881	563.67	0.80170	91.25	0.1826
581.67	1.13151	94.66	0.1930	581.67	0.83406	94.13	0.1876

TABLE EE-2 U.S. (CONTINUED)

Superheated R-12 Vapor

Temp. °R T	Specific Volume ft^3/lb_m	Enthalpy Btu/lb_m	Entropy $Btu/(lb_m \cdot °R)$	Temp. °R T	Specific Volume ft^3/lb_m	Enthalpy Btu/lb_m	Entropy $Btu/(lb_m \cdot °R)$
599.67	1.17236	97.50	0.1978	599.67	0.86593	97.01	0.1925
	$P = 43.51\ lb_f/in.^2$				$P = 58.01\ lb_f/in.^2$		
617.67	1.21272	100.36	0.2025	617.67	0.89717	99.91	0.1973
635.67	1.25261	103.25	0.2071	635.67	0.92808	102.83	0.2019
653.67	1.29217	106.15	0.2116	653.67	0.95868	105.77	0.2065
671.67	1.33158	109.08	0.2184	671.67	0.98879	108.68	0.2109
689.67	1.37066	112.03	0.2204	689.67	1.01874	111.70	0.2153
707.67	1.40942	115.00	0.2246	707.67	1.04854	114.70	0.2196
725.67	1.44819	118.00	0.2288	725.67	1.07801	117.72	0.2238
743.67	1.48663	121.03	0.2329	743.67	1.10732	120.76	0.2280
761.67	1.52491	124.07	0.2370	761.67	1.13648	123.81	0.2320
	$P = 72.52\ lb_f/in.^2$				$P = 108.78\ lb_f/in.^2$		
519.75	0.55775	83.42	0.1648	546.408	0.37402	85.86	0.1637
545.67	0.60003	87.69	0.1728	563.67	0.39516	88.95	01692
581.67	0.65530	93.57	0.1832	581.67	0.41551	92.08	0.1747
599.67	0.68189	96.51	0.1882	599.67	0.43537	95.17	0.1799
617.67	0.70768	99.45	0.1931	617.67	0.45443	98.25	0.1850
635.67	0.73314	102.41	0.1978	635.67	0.47285	101.31	0.1899
653.67	0.75829	105.38	0.2024	653.67	0.49079	104.37	0.1946
671.67	0.78312	108.36	0.2069	671.67	0.50841	107.43	0.1992
689.67	0.80747	111.36	0.2113	689.67	0.52571	110.50	0.2037
707.67	0.83181	114.38	0.2156	707.67	0.54269	113.58	0.2082
725.67	0.85584	117.42	0.2199	725.67	0.55951	116.67	0.2125
743.67	0.87971	120.48	0.2240	743.67	0.57601	119.77	0.2167
761.67	0.90342	123.55	0.2281	761.67	0.59251	122.89	0.2208
779.67	0.92712	126.65	0.2321	779.67	0.60884	126.02	0.2249
797.67	0.95051	129.76	0.2361	797.67	0.62502	129.17	0.2289
815.67	0.97389	132.90	0.2399	815.67	0.64104	132.33	0.2328
	$P = 145.04\ lb_f/in.^2$				$P = 217.56\ lb_f/in.^2$		
566.892	0.27935	87.60	0.1629	598.266	0.18132	89.88	0.1617
581.67	0.29425	90.42	0.1678	635.67	0.20903	97.48	0.174
599.67	0.31091	93.71	0.1734	653.67	0.22057	100.93	0.1793
617.67	0.32677	96.94	0.1787	671.67	0.23162	104.32	0.1844
635.67	0.34182	100.13	0.1838	689.67	0.24219	107.66	0.1894
653.67	0.35640	103.30	0.1887	707.67	0.25228	110.96	0.1941
671.67	0.37050	106.45	0.1935	725.67	0.26205	114.25	0.1987
689.67	0.38427	109.60	0.1981	743.67	0.27167	117.52	0.2031
707.67	0.39773	112.75	0.2026	761.67	0.28096	120.78	0.2075
725.67	0.41102	115.89	0.2070	779.67	0.29009	124.04	0.2117
743.67	0.42400	119.04	0.2113	797.67	0.29906	127.30	0.2158

TABLE EE-2 U.S. (CONTINUED)

Superheated R-12 Vapor

Temp. °R T	Specific Volume ft^3/lb_m	Enthalpy Btu/lb_m	Entropy $Btu/(lb_m \cdot °R)$
	$P = 145.04\ lb_f/in.^2$		
761.67	0.43697	122.21	0.2155
779.67	0.44963	125.38	0.2196
797.67	0.46212	128.56	0.2236
815.67	0.47461	131.76	0.2276
	$P = 290.07\ lb_f/in.^2$		
622.854	0.13023	91.13	0.1603
635.67	0.13936	94.16	0.1651
653.67	0.15073	98.12	0.1713
671.67	0.16066	101.87	0.1769
689.67	0.16995	105.48	0.1822
707.67	0.17876	109.00	0.1873
725.67	0.18709	112.46	0.1921
743.67	0.19494	115.88	0.1968
761.67	0.20263	119.26	0.2013
779.67	0.21016	122.63	0.2056
797.67	0.21736	125.98	0.2099
815.67	0.22441	129.33	0.2140
833.67	0.23146	132.67	0.2181
851.67	0.23835	136.01	0.2221
869.67	0.24508	139.35	0.2259
887.67	0.25180	142.70	0.2297
	$P = 217.56\ lb_f/in.^2$		
815.67	0.30787	130.57	0.2199
833.67	0.31668	133.84	0.2238
851.67	0.32533	137.11	0.2277
869.67	0.33382	140.40	0.2315
887.67	0.34230	143.70	0.2353
	$P = 580.15\ lb_f/in.^2$		
690.516	0.03828	84.65	0.1444
707.67	0.05991	96.81	0.1619
725.67	0.06936	16.63	0.1700
743.67	0.07657	107.45	0.1766
761.67	0.08281	111.83	0.1824
779.67	0.08842	115.95	0.1877
797.67	0.09371	119.90	0.1927
815.67	0.09851	123.74	0.1975
833.67	0.10300	127.49	0.2020
851.67	0.10748	131.19	0.2064
869.67	0.11165	134.83	0.2107
887.67	0.11581	138.44	0.2148

TABLE EE-3

Compressed Liquid R-12

Temp. °R T	Pressure $lb_f/in.^2$ P	Specific Volume ft^3/lb_m	Enthalpy Btu/lb_m	Entropy $Btu/(lb_m \cdot °R)$
329.67	0.41	0.00974	–18.61	–0.0498
383.67	3.28	0.0102	–7.52	–0.0187
438.03	14.69	0.01076	3.89	0.0091
464.67	26.48	0.0111	9.6	0.0216
491.67	44.67	0.01147	15.5	0.0339
518.67	71.21	0.0119	21.54	0.0457
545.67	107.91	0.0124	27.77	0.0573
572.67	157.22	0.01299	34.27	0.0687
599.67	221.18	0.01374	41.16	0.0802
626.67	302.69	0.01474	48.65	0.0921
653.67	404.36	0.01621	57.11	0.1048
671.67	485	0.01783	63.89	0.1145
693.27	596.97	0.0287	78.86	0.1359

Adapted from Sonntag, R. E., Borgnakke, C., and Van Wylen, G. J., *Fundamentals of Thermodynamics*, John Wiley & Sons, New York, 1998. With permission.

FF-Series Tables

Thermodynamic Properties of R-134a

TABLE FF-1 U.S.

Saturated R-134a

Temp. °R T	Pressure $lb_f/in.^2$ P	Specific Volume ft^3/lb_m Sat. Liquid v_f	Sat. Vapor v_g	Enthalpy Btu/lb_m Sat. Liquid h_f	Sat. Vapor h_g	Entropy $Btu/(lb_m \cdot °R)$ Sat. Liquid s_f	Sat. Vapor s_g
365.67	1.2	0.010812	31.5993	51.36	152.46	0.158713	0.435177
374.67	1.7	0.010876	22.9030	52.96	153.80	0.163012	0.432168
383.67	2.4	0.010956	16.8618	54.83	155.14	0.167933	0.429397
392.67	3.2	0.011036	12.6026	56.91	156.49	0.173307	0.426913
401.67	4.3	0.011133	9.55586	59.17	157.85	0.178967	0.424668
410.67	5.7	0.011229	7.34473	61.56	159.21	0.184867	0.422638
419.67	7.5	0.011341	5.71779	64.05	160.57	0.190862	0.425624
428.67	9.7	0.011453	4.50458	66.63	161.93	0.196929	0.419246
437.67	12.3	0.011565	3.58835	69.27	163.28	0.202995	0.417813
444.33	14.7	0.011661	3.04662	71.28	164.30	0.207557	0.416858
446.67	15.5	0.011693	2.88805	71.96	164.64	0.209086	0.416571
455.67	19.4	0.011821	2.34648	74.69	165.98	0.215129	0.415473
464.67	23.9	0.011949	1.92328	77.47	167.32	0.221124	0.414493
473.67	29.3	0.012094	1.58915	80.27	168.65	0.227071	0.413657
482.67	35.5	0.012238	1.32261	83.11	169.96	0.232994	0.412917
491.67	42.6	0.012382	1.10829	85.98	171.26	0.238846	0.412296
500.67	50.9	0.012542	0.93433	88.89	172.54	0.244650	0.411747
509.67	60.3	0.012718	0.79209	91.82	173.79	0.250430	0.411245
518.67	71.0	0.012894	0.67484	94.79	175.01	0.256162	0.410815
527.67	83.1	0.013087	0.57761	97.80	176.20	0.261847	0.410409
536.67	96.6	0.013279	0.49624	100.85	177.35	0.267531	0.410051
545.67	111.8	0.013503	0.42784	103.95	178.45	0.273168	0.409693
554.67	128.7	0.013727	0.37002	107.09	179.50	0.278805	0.409358
563.67	147.5	0.013984	0.32068	110.29	180.49	0.284442	0.408976
572.67	168.3	0.014256	0.27855	113.55	181.41	0.290102	0.408570
581.67	191.2	0.014544	0.24219	116.87	182.25	0.295954	0.408140
590.67	216.3	0.014865	0.21080	120.26	183.00	0.301400	0.407615
599.67	243.9	0.015233	0.18357	123.73	183.63	0.307084	0.406994
608.67	274.1	0.015634	0.15970	127.30	184.13	0.312864	0.406229
617.67	307.0	0.016098	0.13872	130.97	184.48	0.318692	0.405322
626.67	342.9	0.016627	0.11997	134.78	184.63	0.324639	0.404199
635.67	382.0	0.017267	0.10332	138.77	184.52	0.330778	0.402742
644.67	424.4	0.018068	0.08810	143.01	184.05	0.337179	0.400831
653.67	470.6	0.019142	0.07384	147.63	183.02	0.344034	0.398180
662.67	520.9	0.020775	0.05975	152.98	180.91	0.351892	0.394048
671.67	576.3	0.024940	0.04229	161.11	175.07	0.363715	0.384494
673.83	589.4	0.031539	0.03156	168.09	168.09	0.373985	0.373985

TABLE FF-2 U.S.

Superheated R-134a Vapor

Temp. °R T	Specific Volume ft^3/lb_m	Enthalpy Btu/lb_m	Entropy $Btu/(lb_m \cdot °R)$	Temp. °R T	Specific Volume ft^3/lb_m	Enthalpy Btu/lb_m	Entropy $Btu/(lb_m \cdot °R)$
	$P = 7.3\ lb_f/in.^2$				$P = 14.5\ lb_f/in.^2$		
418.734	5.908880	160.39	0.4211	443.898	3.084586	164.22	0.4169
455.67	6.488411	167.16	0.4366	455.67	3.181655	166.47	0.4219
473.67	6.763120	170.52	0.4438	473.67	3.326138	169.93	0.4294
491.67	7.035266	173.94	0.4509	491.67	3.468217	173.43	0.4366
509.67	7.305489	177.43	0.4579	59.67	3.608375	176.99	0.4437
527.67	7.574432	180.98	0.4647	527.67	3.746931	180.59	0.4507
545.67	7.842092	184.60	0.4715	545.67	3.884365	184.25	0.4575
563.67	8.108792	188.29	0.4781	563.67	4.020678	187.97	0.4642
581.67	8.374851	192.05	0.4847	581.67	4.156351	191.76	0.4708
599.67	8.640269	195.87	0.4911	599.67	4.291382	195.61	0.4767
617.67	8.905207	199.77	0.4975	617.67	4.425934	199.53	0.4838
635.67	9.169825	203.73	0.5039	635.67	4.560004	203.51	0.4901
653.67	9433961	207.76	0.5101	653.67	4.693594	207.56	0.4964
671.67	9.697938	211.87	0.5163	671.67	4.827024	211.67	0.5026
689.67	9.961594	216.03	0.5225	689.67	4.960294	215.85	0.5159
707.67	10.22509	220.27	0.5285	707.67	5.093244	220.10	0.5148
	$P = 21.8\ lb_f/in.^2$				$P = 29\ lb_f/in.^2$		
460.818	2.104605	166.71	0.4149	473.544	1.602120	168.59	0.4137
473.67	2.178768	169.32	0.4205	491.67	1.682050	172.36	0.4215
491.67	2.278080	172.91	0.4279	509.67	1.757815	176.05	0.4289
509.67	2.375149	176.53	0.4352	527.67	1.831819	179.77	0.4360
527.67	2.470616	180.18	0.4422	545.67	1.904380	183.52	0.4430
545.67	2.564642	183.89	0.4491	563.67	1.975820	187.32	0.4499
563.67	2.657707	187.65	0.4559	581.67	2.046460	191.17	0.4566
581.67	2.749970	191.46	0.4626	599.67	2.116458	195.07	0.4632
599.67	2.841593	195.34	0.4691	617.67	2.185816	199.03	0.4697
617.67	2.932575	199.28	0.4756	635.67	2.254694	203.05	0.4761
635.67	3.023237	203.28	0.4820	653.67	2.323251	207.13	0.4825
653.67	3.113419	207.34	0.4883	671.67	2.391487	206.97	0.4887
671.67	3.203440	211.47	0.4945	689.67	2.459564	215.48	0.4949
689.67	3.293141	215.67	0.5007	707.67	2.527160	219.75	0.5010
707.67	3.382521	219.92	0.5068	725.67	2.594756	224.08	0.5070
725.67	3.471902	224.25	0.5128	743.67	2.662192	228.47	0.5130
	$P = 58\ lb_f/in.^2$				$P = 72.5\ lb_f/in.^2$		
507.852	0.822685	173.50	0.4114	520.128	0.660903	175.17	0.4108
509.67	0.827810	173.97	0.4123	545.67	0.712160	181.09	0.4219
527.67	0.870739	177.97	0.4200	563.67	0.745798	185.18	0.4292
545.67	0.911905	181.95	0.4274	581.67	0.778154	189.25	0.4364
563.67	0.951469	185.92	0.4346	599.67	0.809710	193.34	0.4433
581.67	0.831494	189.92	0.4416	617.67	0.840465	197.46	0.4501
599.67	1.027875	193.94	0.4484	635.67	0.870578	201.61	0.4567
617.67	1.064877	198.00	0.4550	653.67	0.900212	205.8	0.4632
635.67	1.101558	202.10	0.4616	671.67	0.929685	210.04	0.4696

TABLE FF-2 U.S. (CONTINUED)

Superheated R-134a Vapor

Temp. °R T	Specific Volume ft³/lb$_m$	Enthalpy Btu/lb$_m$	Entropy Btu/(lb$_m$.°R)	Temp. °R T	Specific Volume ft³/lb$_m$	Enthalpy Btu/lb$_m$	Entropy Btu/(lb$_m$.°R)
	P = 58 lb$_f$/in.^{2}				*P = 72.5 lb$_f$/in.^{2}*		
653.67	1.137598	206.25	0.4680	689.67	0.958677	214.32	0.4759
671.67	1.173319	210.45	0.4743	707.67	0.987350	218.66	0.4821
689.67	1.208879	214.71	0.4806	725.67	1.015862	223.06	0.4882
707.67	1.248924	219.03	0.4868	743.67	1.044053	227.51	0.4943
725.67	1.279037	223.40	0.4929	761.67	1.072245	232.01	0.5003
743.67	1.313796	227.83	0.4989	779.67	1.100276	236.58	0.5062
761.67	1.348395	232.32	0.5049	797.67	1.128148	241.20	0.5120
	P = 101.5 lb$_f$/in.^{2}				*P = 130.5 lb$_f$/in.^{2}*		
539.946	0.472051	177.72	0.4099	555.84	0.364570	179.60	0.4093
545.67	0.481661	179.19	0.4127	581.67	0.398368	186.25	0.4210
563.67	0.509052	183.55	0.4205	599.67	0.419511	190.71	0.4286
581.67	0.534841	187.83	0.4280	617.67	0.439694	195.11	0.4358
599.67	0.559509	192.08	0.4352	635.67	0.458916	199.50	0.4428
617.67	0.583215	196.33	0.4422	653.67	0.477497	203.88	0.4496
635.67	0.606281	200.58	0.4489	671.67	0.495597	208.27	0.4562
653.67	0.628707	204.86	0.4556	689.67	0.513217	212.69	0.4627
671.67	0.650811	209.17	0.4621	707.67	0.530676	217.15	0.4691
689.67	0.672436	213.52	0.4685	725.67	0.547655	221.64	0.4754
707.67	0.693740	217.92	0.4748	743.67	0.564474	226.18	0.4815
725.67	0.714883	222.36	0.4810	761.67	0.581133	230.76	0.4876
743.67	0.735867	226.85	0.4871	779.67	0.597632	235.40	0.4936
761.67	0.756530	231.40	0.4931	797.67	0.61381	240.08	0.4996
779.67	0.777033	235.9g	0.4991	815.67	0.629988	244.82	0.5054
797.67	0.797536	240.65	0.5050	833.67	0.6462	249.61	0.5114
	P = 145 lb$_f$/in.^{2}				*P = 174 lb$_f$/in.^{2}*		
562.536	0.326447	180.37	0.4090	575.298	0.268462	181.64	0.4085
563.67	0.327889	180.67	0.4096	581.67	0.276150	183.51	0.4117
581.67	0.349993	185.40	0.4178	599.67	0.295372	188.40	0.4200
599.67	0.370176	189.98	0.4256	617.67	0.312832	193.11	0.4277
617.67	0.389077	194.47	0.4330	635.67	0.329170	197.73	0.4351
635.67	0.407178	198.92	0.4401	653.67	0.825087	202.30	0.4422
653.67	0.424477	203.37	0.4470	671.67	0.359444	206.85	0.4491
671.67	0.441136	207.81	0.4537	689.67	0.373700	211.39	0.4557
689.67	0.457474	212.27	0.4602	707.67	0.387636	215.95	0.4622
707.67	0.473492	216.75	0.4666	725.67	0.401091	220.54	0.4686
725.67	0.489030	221.28	0.4729	743.67	0.414386	225.15	0.4749
743.67	0.504567	225.84	0.4791	761.67	0.427520	229.80	0.4811
761.67	0.519624	230.45	0.4853	779.67	0.440495	234.49	0.4872
779.67	0.534681	235.10	0.4913	797.67	0.453149	239.22	0.4932
797.67	0.549578	239.80	0.4973	815.67	0.465643	244.01	0.4991
815.67	0.564314	244.55	0.5032	833.67	0.478300	248.80	0.5050

TABLE FF-2 U.S. (CONTINUED)

Superheated R-134a Vapor

Temp. °R T	Specific Volume ft³/lb$_m$	Enthalpy Btu/lb$_m$	Entropy Btu/(lb$_m$.°R)	Temp. °R T	Specific Volume ft³/lb$_m$	Enthalpy Btu/lb$_m$	Entropy Btu/(lb$_m$.°R)
	P = 203.1 lb$_f$/in.²				*P = 232.1 lb$_f$/in.²*		
586.296	0.226495	182.62	0.4099	596.16	0.194619	183.38	0.4073
599.67	0.240751	186.62	0.4127	599.67	0.198463	184.57	0.4093
617.67	0.257569	191.62	0.4205	617.67	0.215442	189.98	0.4181
635.67	0.272947	196.45	0.4280	635.67	0.230339	195.06	0.4263
653.67	0.287203	201.17	0.4352	653.67	0.243794	199.98	0.4339
671.67	0.300818	205.84	0.4422	671.67	0.256448	204.78	0.4411
689.67	0.313632	210.48	0.4489	689.67	0.268462	209.54	0.4481
707.67	0.326127	215.12	0.4556	707.67	0.279995	214.27	0.4549
725.67	0.338300	219.77	0.4621	725.67	0.291047	218.99	0.4615
743.67	0.350154	224.44	0.4685	743.67	0.301779	223.72	0.4679
761.67	0.361686	229.14	0.4748	761.67	0.312191	228.47	0.4742
779.67	0.373059	233.87	0.4810	779.67	0.322442	233.25	0.4804
797.67	0.384272	238.65	0.4871	797.67	0.332534	238.06	0.4865
815.67	0.395324	243.46	0.4931	815.67	0.342465	242.91	0.4925
	P = 261.1 lb$_f$/in.²				*P = 290.1 lb$_f$/in.²*		
605.142	0.16931	183.94	0.4066	613.134	0.148967	184.33	0.4058
617.67	0.181644	188.12	0.4134	617.67	0.153452	185.95	0.4085
635.67	0.196541	193.55	0.4221	635.67	0.168990	191.87	0.4179
653.67	0.209836	185.79	0.4301	653.67	0.182125	197.31	0.4263
671.67	0.221849	203.67	0.4376	671.67	0.193978	202.49	0.4342
689.67	0.233222	208.55	0.4447	689.67	0.204870	207.52	0.4416
707.67	0.243954	213.38	0.4517	707.67	0.214962	212.46	0.4486
725.67	0.254206	218.18	0.4583	725.67	0.225213	217.35	0.4555
743.67	0.264137	222.98	0.4649	743.67	0.234023	222.23	0.4621
761.67	0.273748	227.79	0.4713	761.67	0.242993	227.09	0.4685
779.67	0.283198	232.62	0.4775	779.67	0.251643	231.97	0.4749
797.67	0.292329	237.47	0.4837	797.67	0.260132	236.87	0.4811
815.67	0.301299	242.35	0.4898	815.67	0.268462	241.80	0.4872
	P = 362.6 lb$_f$/in.²				*P = 435.1 lb$_f$/in.²*		
631.566	0.111165	184.61	0.4035	647.1	0.084575	183.86	0.4003
635.67	0.11565	186.50	0.4065	653.67	0.092104	187.53	0.4059
653.67	0.130707	193.25	0.417	671.67	0.106520	195.07	0.4173
671.67	0.142720	199.17	0.4259	689.67	0.117572	201.42	0.4266
689.67	0.153132	204.70	0.4340	707.67	0.126863	207.24	0.4350
707.67	0.162583	210.00	0.4416	725.67	0.135352	212.77	0.4427
725.67	0.171232	215.16	0.4488	743.67	0.143041	218.14	0.4500
743.67	0.179562	220.25	0.4557	761.67	0.150089	223.40	0.4570
761.67	0.187411	225.30	0.4624	779.67	0.156976	228.59	0.4637
779.67	0.194939	230.32	0.4690	797.67	0.163544	233.75	0.4703
797.67	0.202147	235.34	0.4753	815.67	0.169791	238.90	0.4766
815.67	0.209355	240.37	0.4815	833.67	0.175910	244.10	0.4829

TABLE FF-2 U.S. (CONTINUED)

Superheated R-134a Vapor

Temp. °R T	Specific Volume ft³/lb$_m$	Enthalpy Btu/lb$_m$	Entropy Btu/(lb$_m$.°R)	Temp. °R T	Specific Volume ft³/lb$_m$	Enthalpy Btu/lb$_m$	Entropy Btu/(lb$_m$.°R)
	P = 507.6 lb$_f$/in.^{2}				*P* = 580.1 lb$_f$/in.^{2}		
660.636	0.063431	181.61	0.3953	672.534	0.040365	174.09	0.3830
671.67	0.077527	189.35	0.4070	689.67	0.068557	192.11	0.4096
689.67	0.090822	197.42	0.4188	707.67	0.080090	200.34	0.4214
707.67	0.100753	204.08	0.4284	725.67	0.089060	207.17	0.4309
725.67	0.109083	210.13	0.4368	743.67	0.096589	213.36	0.4393
743.67	0.116611	215.85	0.4446	761.67	0.103156	219.22	0.4471
761.67	0.123499	221.38	0.4519	779.67	0.109403	224.88	0.4545
779.67	0.129746	226.78	0.4589	797.67	0.115009	230.40	0.4615
797.67	0.135833	232.11	0.4657	815.67	0.120455	235.84	0.4682
815.67	0.141599	237.39	0.4722	833.67	0.125550	241.20	0.4747
	P = 725.2 lb$_f$/in.^{2}				*P* = 870.2 lb$_f$/in.^{2}		
653.67	0.0174436	144.72	0.3383	653.67	0.0169631	143.89	0.3363
671.67	0.0194779	153.77	0.3519	671.67	0.0184207	152.02	0.3486
689.67	0.0265739	168.57	0.3737	689.67	0.0209355	161.61	0.3627
707.67	0.0475574	189.37	0.4035	707.67	0.0271986	174.88	0.3817
725.67	0.0593467	199.75	0.4180	725.67	0.0383791	189.67	0.4023
743.67	0.0676921	207.59	0.4286	743.67	0.0478137	200.45	0.4170
761.67	0.0745157	214.43	0.4377	761.67	0.0550859	208.86	0.4282
779.67	0.0804584	220.76	0.4459	779.67	0.0610927	216.15	0.4376
797.67	0.0858084	226.77	0.4536	797.67	0.0663305	222.83	0.4461
815.67	0.090742	232.59	0.4608	815.67	0.0710398	229.13	0.4539
	P = 1015.3 lb$_f$/in.^{2}				*P* = 1160.3 lb$_f$/in.^{2}		
653.67	0.0166107	143.29	0.3347	653.67	0.0163223	142.82	0.3333
671.67	0.0177800	150.94	0.3463	671.67	0.0173155	150.17	0.3444
689.67	0.0194619	159.36	0.3586	689.67	0.0186289	158.03	0.3560
707.67	0.0223131	169.15	0.3726	707.67	0.0205351	166.62	0.3682
725.67	0.0275510	180.88	0.3890	725.67	0.0234664	176.26	0.3817
743.67	0.0347430	192.72	0.4051	743.67	0.0278073	186.76	0.3960
761.67	0.0416308	202.73	0.4184	761.67	0.0330131	196.99	0.4096
779.67	0.0475414	211.16	0.4293	779.67	0.0381869	206.19	0.4215
797.67	0.0526512	218.62	0.4388	797.67	0.0429282	214.36	0.4319
815.67	0.0571682	225.50	0.4473	815.67	0.0471890	221.81	0.4411
	P = 1450.4 lb$_f$/in.^{2}				*P* = 2900.7 lb$_f$/in.^{2}		
653.67	0.0158738	142.14	0.3309	653.67	0.0146084	140.97	0.3229
671.67	0.0166587	149.12	0.3415	671.67	0.0150409	147.24	0.3324
689.67	0.0176198	156.37	0.3522	689.67	0.0155214	153.62	0.3418
707.67	0.0188212	163.99	0.3630	707.67	0.0160500	160.09	0.3510
725.67	0.0203749	172.04	0.3743	725.67	0.0166107	166.66	0.3602
743.67	0.0224252	180.56	0.3859	743.67	0.0172354	173.31	0.3693
761.67	0.0250522	189.44	0.3984	761.67	0.0179081	180.06	0.3782
779.67	0.0281596	198.34	0.4092	779.67	0.0186450	186.89	0.3871
797.67	0.0314754	206.92	0.4201	797.67	0.0194459	193.81	0.3959
815.67	0.0347911	215.01	0.4301	815.67	0.0203108	200.79	0.4045

TABLE FF-3 U.S.

Compressed Liquid R-134a

Temp. °R T	Specific Volume ft^3/lb_m	Enthalpy Btu/lb_m	Entropy $Btu/(lb_m \cdot °R)$	Temp. °R T	Specific Volume ft^3/lb_m	Enthalpy Btu/lb_m	Entropy $Btu/(lb_m \cdot °R)$
	$P = 14.5\ lb_f/in.^2$				$P = 36.3\ lb_f/in.^2$		
432	0.011491	67.60	0.1991	432	0.011487	67.63	0.1991
432.27	0.011495	67.68	0.1993	437.67	0.011562	69.29	0.2029
434.07	0.011517	68.21	0.2006	446.67	0.011684	71.98	0.2090
435.87	0.011541	68.74	0.2018	452.07	0.011759	73.62	0.2127
437.67	0.011565	69.27	0.2030	455.67	0.011812	74.71	0.2151
438.57	0.011576	69.39	0.2036	461.07	0.011890	76.37	0.2187
439.47	0.011589	69.81	0.2042	464.67	0.011945	77.48	0.2211
440.37	0.01160	70.07	0.2048	470.07	0.012028	79.16	0.2247
441.27	0.011613	70.34	0.2054	473.67	0.012084	80.28	0.2271
442.17	0.011626	70.61	0.206	479.07	0.012172	81.98	0.2306
443.07	0.011637	70.88	0.2066	482.67	0.012231	83.12	0.2330
443.898	0.011648	71.13	0.2072	483.73	0.012249	83.45	0.2337
	$P = 145\ lb_f/in.^2$				$P = 290.1\ lb_f/in.^2$		
432.27	0.011471	67.75	0.1988	419.94	0.011448	67.91	0.1985
437.67	0.011544	69.41	0.2027	444.87	0.011616	71.70	0.2072
455.67	0.011791	74.82	0.2148	477.27	0.012086	81.62	0.2287
473.67	0.012060	80.37	0.2267	491.67	0.012319	86.16	0.2381
482.67	0.012196	83.20	0.2327	506.07	0.001257	90.80	0.2473
491.67	0.012356	86.07	0.2385	520.47	0.012851	95.49	0.2565
500.67	0.012517	88.95	0.2443	534.87	0.013156	100.26	0.2655
509.67	0.012686	91.87	0.2501	545.67	0.001341	103.95	0.2724
527.67	0.013056	97.81	0.2616	567.27	0.013987	111.52	0.2859
545.67	0.013481	103.95	0.2730	585.27	0.014578	118.10	0.2974
554.67	0.013719	107.09	0.2787	617.67	0.015347	185.94	0.4085
562.806	0.013945	109.89	0.2837	671.67	0.019399	202.49	0.4342
	$P = 435.1\ lb_f/in.^2$				$P = 580.1\ lb_f/in.^2$		
437.67	0.011499	69.72	0.2020	437.67	0.011477	69.88	0.2016
473.67	0.011999	80.63	0.2259	464.67	0.011839	77.99	0.2196
491.67	0.012284	86.28	0.2376	491.67	0.012249	86.37	0.2372
509.67	0.012597	92.05	0.2491	509.67	0.012555	92.13	0.2486
527.67	0.012946	97.94	0.2605	527.67	0.157057	97.98	0.2599
545.67	0.013340	103.95	0.2717	545.67	0.013276	104.00	0.2711
563.67	0.013792	110.19	0.2829	572.67	0.013950	113.28	0.2878
581.67	0.014325	116.59	0.2942	599.67	0.014813	123.13	0.3045
599.67	0.014974	123.34	0.3056	617.67	0.015565	130.05	0.3159
617.67	0.015816	130.48	0.3173	635.67	0.016571	137.53	0.3278
635.67	0.017040	138.35	0.3299	653.67	0.018147	145.96	0.3409
646.812	0.018289	144.07	0.3388	672.318	0.02613	162.6	0.3659

TABLE FF-3 U.S. (CONTINUED)

Compressed Liquid R-134a

Temp. °R T	Specific Volume ft³/lb$_m$	Enthalpy Btu/lb$_m$	Entropy Btu/(lb$_m$.°R)	Temp. °R T	Specific Volume ft³/lb$_m$	Enthalpy Btu/lb$_m$	Entropy Btu/(lb$_m$.°R)
	P = 725.2 lb$_f$/in.²				*P* = 870.2 lb$_f$/in.²		
419.94	0.011442	67.96	0.1984	446.67	0.011549	72.85	0.2070
444.87	0.011445	72.96	0.2044	477.27	0.011967	82.14	0.2271
477.27	0.011861	82.69	0.2255	491.67	0.012185	86.63	0.2363
491.67	0.012065	87.14	0.2347	506.07	0.012417	91.19	0.2454
506.07	0.012281	91.62	0.2437	520.47	0.012667	95.79	0.2544
520.47	0.012512	96.17	0.2526	536.67	0.012973	101.07	0.2644
534.87	0.012757	100.77	0.2613	545.67	0.013157	104.04	0.2699
545.67	0.012954	104.30	0.2678	567.27	0.013646	111.35	0.2831
567.27	0.013383	111.44	0.2808	585.27	0.014115	117.63	0.2939
585.27	0.013784	117.54	0.2912	617.67	0.015177	129.49	0.3137
617.67	0.015597	128.85	0.3101	653.67	0.016965	143.89	0.3363
671.67	0.016656	149.1	0.3415	671.67	0.018413	152.02	0.3486
	P = 1015.3 lb$_f$/in.²				*P* = 1450.4 lb$_f$/in.²		
446.67	0.011527	73.01	0.2066	446.67	0.011467	73.49	0.2056
477.27	0.011940	82.27	0.2267	477.27	0.011861	82.69	0.2255
491.67	0.012153	86.76	0.2359	491.67	0.012065	87.14	0.2347
506.07	0.012380	91.70	0.2450	506.07	0.012281	91.62	0.2437
520.47	0.012625	95.87	0.2539	520.47	0.012512	96.17	0.2526
536.67	0.012925	101.12	0.2639	536.67	0.012789	101.38	0.2624
545.67	0.013103	104.08	0.2694	545.67	0.012954	104.30	0.2678
567.27	0.013574	111.35	0.2824	567.27	0.013383	111.44	0.2806
585.27	0.014024	117.58	0.2932	585.27	0.013784	117.54	0.2912
617.67	0.015019	129.28	0.3126	617.67	0.014636	128.85	0.3101
653.67	0.016608	143.29	0.3347	653.67	0.015867	142.13	0.3309
671.67	0.017772	150.94	0.3463	671.67	0.016656	149.10	0.3415
	P = 2175.6 lb$_f$/in.²						
419.94	0.011442	67.96	0.1984				
444.87	0.011565	79.10	0.2146				
477.27	0.011752	83.43	0.2238				
491.67	0.011935	87.83	0.2328				
506.07	0.012134	92.26	0.2417				
520.47	0.012344	96.73	0.2505				
534.87	0.012566	101.29	0.2591				
545.67	0.012742	104.73	0.2655				
567.27	0.013120	111.74	0.2780				
585.27	0.013465	117.67	0.2883				
617.67	0.014168	128.63	0.3065				
671.67	0.015664	147.76	0.3362				

Adapted from Sonntag, R. E., Borgnakke, C., and Van Wylen, G. J., *Fundamentals of Thermodynamics*, John Wiley & Sons, New York, 1998. With permission.

Answers to Problems

Chapter 1

1.6	698 kPa
1.7	3.3129 bars
1.8	114.33 psia
1.9	74.96 kPa
1.10	63 kPa
1.11	0.4996 bar
1.12	10.33 psia
1.13	232.4 m
1.14	2110 kPa
1.15	281.2 psia
1.16	199 kPa
1.17	99.35 kPa
1.18	15.125 psia
1.19	(a) 9.18 m, (b) 0.675 m
1.20	0.91 bar
1.21	(a) 99 kPa, (b) 4.00 kPa
1.22	0.341 bar
1.23	1821 kPa
1.24	412.3 kJ
1.25	60 J
1.26	–1.5 kJ
1.27	–20 J
1.28	–10.9 Btu
1.29	244.5 kJ
1.30	(a) 172.33 kPa, (b) 0.22 kJ, (c) 3.22 kJ
1.31	+0.0625 kJ
1.32	113.4 kJ

1.33	67 W, 783.9 kJ
1.34	1.257 kW
1.35	1.10×10^{-3} J
1.36	2.68 μJ
1.40	37°C
1.41	68°F, 528°R
1.42	295 K
1.43	12 K
1.44	8.33 K
1.45	–18°F, –18°R
1.46	–12.22 K
1.49	50°C, 75°C, 2°A
1.50	419.67°R

Chapter 2

2.1	6.504 m^3
2.8	28.8 ft^3
2.9	(a) 0.173, (b) 0.406, (c) 0.517, (d) 0.570, (e) 0.677
2.10	(a) 0.85, (b) 0.80, (c) 0.67, (d) 0.13, (e) 0.89
2.12	4.19 m^3
2.13	0.0783 m^3
2.14	0.0130 m^3/kg
2.15	0.328 m^3
2.16	67.12 ft^3
2.17	0.93 ft^3
2.18	0.34 ft^3/lb_m
2.19	3.92 ft^3
2.20	4.007 MPa, 0.4946 m^3
2.21	0.25, 0.2208 m^3
2.22	0.1667, 0.300 ft^3
2.23	0.713 MPa, 0.998
2.24	28.3 psia, 0.9992
2.25	0.080 MPa
2.26	0.000942, 179.9°C, 0.0013

2.27	0.31
2.29	(a) 66.65 kJ, (b) 100 kJ, (c) 81.1 kJ
2.30	(a)1.81 Btu, (b) 2.72 Btu, (c) 2.21 Btu
2.31	–114.94 kJ/kg
2.32	–46.1 kJ/kg
2.33	–49.9 Btu/lb_m
2.34	–20 Btu/lb_m
2.35	511 kJ
2.36	12 J
2.37	390 Btu
2.38	0.010 Btu
2.39	(a) 101 kPa, (b) –125.8 kJ
2.40	0.0353 m^3/kg, 153°C
2.41	(a) 17.22 m^3/kg, (b) 17.038 m^3/kg
2.42	(a) 90 psia, (b) 95.9 psia, (c) 94.384 psia
2.43	(a) 0.060827 m^3/kg, (b) 0.05839 m^3/kg
2.44	(a) 100 psia, (b) 103.8 psia, (c) 96.1 psia
2.45	(a) 151.11 psia, (b) 162.8 psia, (c) 160 psia
2.47	(a) 19,486 kPa, (b) 17,539 kPa, (c) 18,553 kPa
2.48	(a) 22,270 kPa, (b) 53,675 kPa, (c) 13,638 kPa
2.49	(a) 11,810 kPa, (b) 3856 kPa, (c) 2382 kPa
2.50	(a) 108.2°C, (b) 54°C

Chapter 3

3.2	1041 kJ/kg, 99.62°C
3.3	478.5 kJ/kg, 13.98%
3.4	196.6 Btu lb_m, 13.85%
3.5	10.38 kJ/kg
3.6	488°C
3.7	529°C
3.8	901°F
3.9	912°F
3.10	–15 kJ/s, 6185 kJ/s, 0.24%
3.11	–16.96 Btu/s, 9528 Btu/s, 0.18%

3.12	72,900 kJ
3.13	50,900 Btu
3.14	1 kJ
3.15	1000 Btu
3.16	660 kJ/kg
3.17	364 kJ/kg
3.18	244.8 Btu/lb_m
3.19	162.7 Btu/lb_m
3.20	49.6 kJ/kg
3.21	49 kJ/kg
3.22	422 kJ/kg
3.23	19.56 Btu/lb_m
3.24	16.7 Btu/lb_m
3.25	134.7 Btu/lb_m
3.26	13.8°C
3.27	131.8°F
3.28	287°C, 0.122 m^3/kg
3.29	544°F, 1.94 ft^3/lb_m
3.30	451.9°C, 0.119 m^2
3.31	571°F, 0.201 ft^2
3.32	17.3 kg/s
3.33	33.3 lb_m/s
3.34	33.98 kg/s
3.35	935 kg/s
3.36	–37,360 kJ
3.37	15,753 kJ, 931.5 kJ
3.38	3797 kJ, 1.9 kg
3.39	–41 Btu
3.40	a = 20, b = 50, c = –70, d = 90
3.41	a = 20, b = 100, c= –180, d = 120
3.42	2.105 kJ/s
3.43	6 kJ/s
3.44	2.105 Btu/s
3.45	5.5 Btu/s
3.46	13.98%
3.47	13.85%
3.48	0.667

3.49	600 W
3.50	1.0
3.51	0.4725 Btu/s
3.52	4.0
3.53	6 kW
3.54	2.67
3.55	5.4 Btu/s

Chapter 4

4.3	–14.8 kJ/K
4.4	–8.1 kJ/°R
4.5	0.04675 kJ/K, 2.383 kJ/K
4.6	0.867 Btu/°R
4.7	–0.0188 kJ/K, 0.0054 kJ/kg, 0.0134 kJ/K
4.8	30°C, 11.26 kW
4.9	12 Btu/s, 5.77 Btu/s
4.10	8.77
4.12	4.86, 3.88, –0.98
4.13	4.89, 3.50
4.14	7.66
4.15	8
4.16	359 kJ/kg
4.17	$22.88
4.18	0.279 kW, 11°C
4.19	1.74 bars
4.20	81.7 $lb_f/in.^2$
4.21	3.45 bars
4.22	84.8 $lb_f/in.^2$
4.23	607 K
4.24	873°F
4.25	1085 kJ/kg, 4.9 kJ/kg
4.26	317 kJ/kg
4.27	136 Btu/lb_m
4.28	0.35 kJ/kg

4.29	0.18 Btu/lb_m
4.30	84%
4.31	84%
4.32	$517.60
4.33	0.613 kJ/(kg.K)
4.34	1.8653 Btu/(lb_m.°R), 1.9213 Btu/(lb_m.°R)
4.35	37.3%, 0.77 kJ/(kg.K), 0.002 kJ/(kg.K), 29.8%
4.36	35.3%, 0.173 Btu/(lb_m.°R), 0.0008 Btu/(lb_m.°R), 27.5%
4.42	–1115 kJ/kg, 4.44 kJ/(kg.K)
4.43	–482.1 Btu/lb_m, 1.07 Btu/(lb_m.°R)
4.44	–2291 kJ/kg, 8.02 kJ/(kg.K)
4.45	–982.1 Btu/lb_m, 1.896 Btu/(lb_m.°R)
4.46	8.8 kJ/kg
4.47	10.1 Btu/lb_m
4.48	24.3 kJ/kg
4.49	9.22 Btu/lb_m
4.53	–199 kJ/kg, –136 kJ/kg
4.54	–87 Btu/lb_m, –59 Btu/lb_m.

Chapter 5

5.1	1458 kJ, 1.86 kJ, –1456 kJ
5.2	1253 Btu, 1.34 Btu, –1251.7 Btu
5.3	853.56 kJ, 640.8 kJ, –212.8 kJ
5.4	1105 Btu, 831 Btu, –274 Btu
5.5	397 kJ/kg
5.6	328 kJ/kg
5.7	141 Btu/lb_m
5.8	–236.4 kJ/kg, 236.4 kJ/kg
5.9	–104.4 Btu/lb_m, 104.4 Btu/lb_m
5.10	2335 kJ/kg, 95 kJ/kg
5.11	962 Btu/lb_m, 37 Btu/lb_m
5.12	499 kJ/kg, 426 kJ/kg, 73 kJ/kg
5.13	172 Btu/lb_m, 141 Btu/lb_m, 31 Btu/lb_m
5.14	402 kW, 511 kJ/kg

5.15	434.6 Btu/lb_m, 253 Btu/s, 1140 Btu/s, 887 Btu/s
5.17	0.0839 kJ/(kg.K), 24.6 kJ/kg, 20.1 kJ/kg
5.18	–0.597 kW, –12.46 kW, 0.597 kW, 0 kW, –12.46 kW, 0.597 kW
5.19	196 kJ, 132 kJ, –64 kJ, 5.5 kJ
5.20	190.1 Btu, 127.4 Btu, 4.9 Btu
5.21	16°C, 32.8 W, 61.4 W
5.22	50°F, 0.072 Btu/s, 0.178 Btu/s
5.23	195°C, 0.1 MPa, 835 kJ
5.24	132 kJ, 136.02 kJ, 132 kJ, 86.75 kJ
5.25	0.24, 0.10
5.26	0.195, 0.074
5.27	0.26, 0.23
5.28	0.18, 0.16
5.29	1368.3 kJ/kg
5.30	391 Btu/lb_m
5.31	0.276, 0.095
5.32	48%
5.33	116.7°C, 39%
5.34	0.8
5.35	0.5
5.36	0.89
5.37	0.87
5.39	0.676, 0.99, 0.844, 0.65, –5.0%
5.40	0.676, 0.99, 0.86, 0.71, –6.0%

Chapter 6

6.1	120 MW, –152 MW, 32°C, 44%
6.2	146.7 lb_m/s, –127.6 × 10^3 Btu/s, 4911 lb_m/s, 43.9%
6.3	850.56 kJ, 635.2 kJ, –215.3 kJ
6.6	0.93 kg/s
6.7	29%
6.8	44%
6.9	12.8 MPa or lower, 45.6%
6.10	2191 psia or lower, 45.4%

6.11	22.5%, 44.7%
6.12	21.6%, 44.9%
6.14	37.5%, 40.6%, 41.4%
6.15	45.3%, 49.3%, 50.3%
6.16	34.5%
6.19	67%, –21.2%
6.20	21.43
6.21	1807 kJ/kg, 163 kJ/kg
6.22	734 Btu/lb_m, 48 Btu/lb_m
6.23	0.687 kJ/(kg.K), 204.7 kJ/kg
6.24	0.1629 Btu/(lb_m.°R), 87.5 Btu/lb_m
6.25	1.113 kJ/(kg.K), 331.7 kJ/kg
6.26	0.2652 Btu/(lb_m.°R), 142.4 Btu/lb_m
6.27	855 kJ/kg, 1186.5 kJ/kg
6.29	0.976 kg/s
6.30	17.42 MW, 34.22 MW, 20.63 MW
6.31	15,935 Btu/s, 31,130 Btu/s, 18,873 Btu/s
6.32	3.23 kg/s, 475.9 kW
6.34	12.2, 4631.4 kJ/kg, 58.2%
6.35	12.2, 1991.1 Btu/lb_m, 58.2%
6.37	493.21 kg of steam/min, 3945.7 kg of air/min, 61.9%, 75.8%
6.38	1295 lb_m/min of steam, 9525 lb_m/min of air, 61.6%, 76.1%
6.41	24.65°C

Chapter 7

7.1	1.476 MW/m^2
7.2	27°C
7.3	752°F
7.4	209 Btu/(ft.°F.h)
7.5	16°C
7.6	8°F
7.7	17.45 kW/m^2
7.8	254°C
7.9	1110°F

7.10	3066 ft^2
7.11	523 K
7.12	0.15
7.13	0.78
7.14	8680 Btu/h
7.15	2.491 kW
7.16	18 W/(m^2.°C)
7.17	0.3375 m^2
7.18	150°C
7.19	72°F
7.20	3.17 Btu/(ft^2.°F.h)
7.21	8976 Btu/h
7.22	40 in.
7.23	20°C
7.24	1.1
7.25	0.7
7.26	85.6°F

Index

A

B

C

D

E

F

G

H

I

K

L

M

N

O

P

Q

R

S

T

U

V

W